Studies in Log

Volume 57

Proof-theoretic Semantics

Studies in Logic Series Editor
Dov Gabbay   dov.gabbay@kcl.ac.uk

Proof-theoretic Semantics

Nissim Francez

ISBN 978-1-84890-183-4

College Publications
Scientific Director: Dov Gabbay
Managing Director: Jane Spurr

http://www.collegepublications.co.uk

Original cover design by Orchid Creative   www.orchidcreative.co.uk
Printed by Lightning Source, Milton Keynes, UK

To Tikva, for a life-long trans-verificational love,
in need of no proof

Preface

Ever since having encountered proof-theoretic semantics (PTS), I became fascinated by it. When delving deeper into it, two perplexities emerged.

- In view of the extensive activity in PTS as an area of research, to which several ph.d theses by young researchers were dedicated, in addition to numerous publications advancing it appearing frequently, why isn't there a book on this topic describing it in detail? The only reference resource on PTS I could find is a Staford Encyclopedia of Philosophy (SEP) entry, which, while fairly detailed, cannot replace a book.

- Coming from formal semantics for natural language (NL), with a computational background in semantics of programming languages, I came to wonder why this powerful theory of meaning is not extended beyond logic, to cover also NL semantics?

The present book is the culmination of a research programme on PTS trying to face those two perplexities, extending over a number of years.

The book attempts to present as a coherent story both the main line of research pertaining to the PTS for logic, as well as the extension of PTS to NL – an ongoing endeavour. The book is structured in two parts:

- Part I is dedicated to a presentation of PTS for logic (and some related formal systems). I rely here on the extensive literature on the topic, summarizing it but occasionally presenting some of the issues involved in accordance with my own work. A notable such example is the issue of the way PTS defines meaning. The traditional view is that the rules of a meaning-conferring proof-system define meanings *implicitly*, while my own preference is for a view emphasising an *explicit* definition of meaning, a kind of a proof-theoretic semantic value.

- Part II is dedicated to a PTS for some fragments of NL, establishing, I hope, the viability of this approach as an alternative to the model-theoretic semantics in which, so it seems, most linguistic semanticists are entrenched. This part presents mostly my own work with co-authors.

There is an attempt in this book to downplay the role of philosophical arguments and disputes, such as realism vs. anti-realism, underlying many of the discussions of PTS, and concentrate instead on the technical details involved in developing PTS.

Similarly, I tried not to enter into the issues of rivalry between logics, e.g., Intuitionistic Logic vs. Classical Logic, and rather view PTS *in a variety of forms* as a universal *definitional tool*.

The book starts from basics, not even assuming prior acquaintance with natural-deduction, the main technical tool employed. However, prior acquaintance with standard model-theoretic semantics, both for logic and for NL, would be helpful in order to appreciate the contribution of PTS as an alternative theory of meaning.

The audience I have in mind are researchers, in particular graduate students, in any of the following disciplines (where interdisciplinarity would be an advantage, as in many other similar situations):

- Logic

- Linguistics

- Philosophy (of language)

- Computer Science

Acknowledgements: I have interacted with many colleagues during the period of writing this book, all of whom contributed to enhancing my understanding of the topics at issue, and to all of whom I would like to express my thanks.

Cooperation: I thank my foremost coauthor, Roy Dyckhoff, who gave my wife and me a wonderful time during my St Andrews sabbatical, where a major part of the PTS for NL was conceived and developed. He was invaluable in helping me dwell into the required general background in proof theory.

I thank Gilad Ben-Avi, who worked with me at the Technion in advancing the PTS for NL, in particular in the study of determiners.

Bartosz Wieckowski, during his visits, was also very cooperative.

Encouragement: I thank Peter Schroeder-Heister for a lot of encouragement and "moral support", as well as technical help during the period of our acquaintance.

Extensive Communication: I thank Stephen Read for a really extensive exchange of ideas in various forms. He always lent his ear to me, and never failed to reply to a message.

Communication: I thank the following colleagues with whom I communicated over the years on topics related to PTS.

Nicholas Asher, Arnon Avron, Hanoch Ben-Yami, Carlo Celluci, Andreas Fjellstad, Itamar Francez, Ole Hjortland, Lloyd Humberstone, Michael Kaminski, Steven Kuhn, Niels Kürbis, Ed Mares, Peter Milne, Enrico Moriconi, Larry Moss, Julien Murzi, Alberto Naibo, Sara Negri, Rick Nouwen, Francesco Paoli,

Mattia Petrolo, Ivo Pezlar, Frank Pfenning, Dag Prawitz, Ruy de Queiroz, Hartley Slater, Florian Steinberger, Anna Szabolcsi, Laura Tesconi, Luca Tranchini, Heinrich Wansing,

Technical Assistance: I thank Jane Spurr for invaluable assistance in complying with the formatting requirements of College Publications.

Financial Support: I thank ISF, the Israeli Science Foundation, (grant number 2006938), and EPSRC (grant number EP/D064015/1), the former partially supporting my NL-PTS research at the Technion, the latter supporting my work at St Andrews.

Coffee: I thank the owner and stuff of the *Pinguin* coffee house in Nahariya, where large parts of this book were written, for good service and pleasant music. Only they can estimate the amount of coffee needed for completing such a task ...

Finally, and most importantly, I thank my beloved wife, Tikva, to whom this book is dedicated, for everything.

Contents

I   Proof-Theoretic Semantics in Logic   49

2   A variety of logics   51

Chapter 1

Introduction I

1.1 On meaning

What are meanings?

Meanings, by which I mean *literal meanings* (also called *standing meanings*) are *abstract, theoretical entities*, not directly observable, or otherwise accessible, but playing a crucial explanatory role in the scientific branch studying meanings, *semantics*, as a part of general linguistics, the study of (natural *and* formalized language). As such, they play a similar role to, say, various kinds of elementary particles in Physics, used to explain observable natural phenomena. I would like to stress on the onset that the meanings I am referring to here are meanings *of* linguistic carriers (formulas in logic, declarative sentences and their components, words, in natural languages), to which I refer as *meaning bearers* (see more in Section 1.1.1), and *not* meanings of *users* of those linguistic carriers. Such meanings depend only on language, *not* on actual uses of language in *circumstances of utterance* that depend on *intentions*, *plans* and many other context-dependent and user-dependent features, the study of which is usually referred to as *pragmatics* (however, see Chapter 12 for a semantic theory of a context-dependent linguistic meaning, giving rise to *contextual meaning variation*). For emphasis, this distinction is repeated again (cf . p. 240), pertaining especially to natural language.

Some linguistic observable phenomena, explainable in terms of meanings, are the following.

- *Entailment* among (affirmative, declarative) sentences is, probably, the most agreed upon ingredient of the use of natural language explained by alluding to meanings, allowing drawing consequences and reasoning in general.

- *communication* is conceived as the transfer of meanings from one member of a linguistic community to another. Here is where language is seen as a *public* enterprise, transcending subjectivity and individual linguistic capabilities. Thus, the grasping of meanings is a major criterion for language understanding. A generalization of communication is participation in *dialogues*, giving rise to a variety of acts related to meanings: asserting, denying, asking, answering, obeying, and many more.

- *Translation* from one language to another (in particular for non-literary uses of language) uses the preservation of meaning as its main *correctness criterion.*

- With the advent of computers playing a major role in life in general, meanings of formalized languages are tools for many applications, like *knowledge representation, question answering, data mining* in text (a blooming area of application in the internet era), *machine translation*, and many more. *Computational Linguistics* is the name of the scientific discipline carrying such tasks.

- Meanings are also exploited to specify the *normative* role of language; its *correct* use, its *justification* of what consists *rational* linguistic behavior, and more.

In what follows, I will not be directly concerned with all of those observable phenomena, but they are always in the background.

A major goal I see in the study of meanings is the ability to incorporate them into *grammars*, viewed as the formal vehicle for *defining* (specific) languages, whether natural or artificial. Thus, a central attribute of a grammar is its ability to bridge between form and content, its *syntax-semantics interface*.

This book presents, in Dummett's terms ([35], p. 22), both a general *theory of meaning* and a *meaning theory* for various logics and (fragments of) natural language (NL), mostly of English. A meaning theory gives

> a complete specification of the meanings of ... all expressions of one particular language

while a theory of meaning enquires

> into the general principles upon-which a meaning theory is to be constructed.

In other words, a theory of meaning specifies *how* (or in terms of what) is a meaning of a certain meaning-bearing expression defined, while a meaning theory specifies *what* the meaning of that expression is in terms of the theory of meaning chosen.

My main concern is with presenting a theory of meaning; the theory of meaning unfolded in this book, *proof-theoretic semantics (PTS)*[1] will be illustrated and exemplified with various meaning theories, both for logics and for NL.

---

[1] According to Schroeder-Heister [191], the origin of the term is in [185].

The following is a brief description of the main theories of meanings that have been proposed in the literature.

Model-Theoretic Semantics (MTS): This is the traditional, most common, theory of meaning, identifying as its central concept *truth*. As it is the main "rival" of PTS, I present it in some more detail.

> **Logic:** The MTS for logic originates from Tarski's analysis of truth [212] (later migrating to natural language semantics).
>
> > **Sentential meanings:** MTS regards the meaning of formulas in logic, or affirmative sentences in NL, as based on their *truth-conditions* (or a generalization thereof for multiple truth values). It comes in two (not unrelated) versions.
> >
> > > **Relative truth conditions:** Truth-conditions are formulated in terms of *arbitrary models* (of a suitable kind), and an *interpretation* providing meanings for the basic terms. The truth-conditions for compound expressions are obtained *compositionally* by a recursive definition.
> > >
> > > In classical propositional logic, MTS bases the meaning of connectives on *truth-tables*. An interpretation $\mathbf{I}$ assigns truth-values $[\![p]\!]^{\mathbf{I}}$ (**t** for truth and **f** for falsity) to atomic propositions p, yielding the ground clause for the satisfaction relation (of a formula in a model), where here models $\mathcal{M}$ can be identified with assignments.
> > >
> > > $$\mathcal{M} \vDash p \text{ iff } [\![p]\!]^{\mathbf{I}} = \mathbf{t}$$
> > >
> > > Satisfaction is then extended recursively, e.g. for conjunction
> > >
> > > $$\mathcal{M} \vDash \varphi \wedge \psi \text{ iff } \mathcal{M} \vDash \varphi \text{ and } \mathcal{M} \vDash \psi$$
> > >
> > > For classical first-order logic, a model $\mathcal{M}$ is a pair $\langle E, \mathbf{I} \rangle$, where E is a *domain* (or *carrier*) of the model, an *arbitrary* non-empty set, and the interpretation $\mathbf{I}$ assigns an arbitrary element $[\![c]\!]^{\mathbf{I}} \in E$ to every constant a (its *denotation*, or *reference* in $\mathcal{M}$), an arbitrary subset $[\![P]\!]^{\mathbf{I}} \subseteq E$ to every unary predicate symbol P (its *extension* in $\mathcal{M}$), and similarly subset of the Cartesian product $[\![R]\!]^{\mathbf{I}} \subseteq E^n$ to every n-ary relational symbol R. To cope with (free) variables, an assignment σ assigns a value $\sigma[\![x]\!]$ to every free variable x. This naturally induces a value $t^{\sigma,\mathbf{I}} \in E$ for every term t.
> > >
> > > – The meaning of predication is given by
> > >
> > > $$\mathcal{M}, \sigma \vDash P(t) \text{ iff } t^{\sigma,\mathbf{I}} \in [\![P]\!]^{\mathbf{I}}$$
> > >
> > > – The meaning of (first-order) quantifiers as ranging over all the entities in a model is captured by a recursive clause
> > >
> > > $$\mathcal{M}, \sigma \vDash \forall x.\varphi \text{ iff } \mathcal{M}, \sigma' \vDash \varphi$$
> > >
> > > where σ' differs from σ at most by the value assigned to x.

Absolute truth conditions: This is a less frequent approach, essentially similar to the relative truth condition, but *without* varying models and/or interpretation. It uses a *fixed* model, known as the *intended model*. For example, the meaning of *Arithmetical statements* is defined over Peano's interpretation.

Sub-sentential meanings: There is not much interest in sub-sentential meanings in logic, except in the meaning of the logical operators themselves, neither in the relative nor in the absolute approaches. They are seen as arising indirectly from interpretations and assignments as discussed above.

Natural language: The approaches here are again divided into relative and absolute specification of truth conditions. This division reflects a fundamental distinction between *lexical semantics* and *compositional semantics*. Here sub-sentential meanings are of significant interest.

Sentential meanings: MTS regards the meaning of (affirmative) sentences as based on their truth-conditions. It also comes in the above mentioned two versions.

Relative truth-conditions: This is the dominant approach, originating from Montague's seminal work [223]. For simple extensional fragments, models and interpretations look very much like those for logic, using a domain over which quantifiers range (within determiner-phrases – to be explained below). Viewed in a highly simplified way, proper names are assigned an element of the domain, this assignment is referred to as *reference*. Words of open categories, like nouns, (intransitive) verbs and adjectives are *all* conceived as predicates, assigned arbitrary subsets of the domain, referred to as their *extensions*. Those abstract over the "real" lexical semantics for words of those categories. The means for composing meanings is via *functional composition*. For example, a predication like Rachel smiles obtains the following truth-condition in a model $\mathcal{M}$ and interpretation $\mathbf{I}$: $\text{Rachel}^{\mathbf{I}} \in \text{smile}^{\mathbf{I}}$. Words of closed categories such as determiners receive a fixed interpretation in every model, known as *generalized quantifiers*, (binary) relations among subsets of the domain. For example, a sentence such as every girl smiles obtains the following truth-condition in a model $\mathcal{M}$: $\text{girl}^{\mathbf{I}} \subseteq \text{smile}^{\mathbf{I}}$.

For more complicated fragments, more committing ontologies are modelled, like *possible worlds*, *times and locations*, *events* and many more.

Absolute truth-conditions: This approach, referred also as neo-Davidsonian ([23]) uses the meta-language (reflecting the "real world") to specify truth-conditions in what is known as *T*-sentences (also due to Tarski). For example,

'Rachel smiles' *is true* iff Rachel smiles

Those truth-conditions are obtained by having 'Rachel' refer to

Rachel (the actual person), and having the extension of 'smile' the (actual) set of smiling entities.

Sub-sentential meanings: In contrast to logic, the semantics of NL regards sub-sentential meanings as highly significant, as sentences may have non-trivial parts. In the relative truth-conditions approach, such meanings are usually expressed in *higher-order logics*, by variants of *typed* λ-terms. The typical types used by Montague are e (for entities) and t (for truth-values), and (τ_1, τ_2) for the type of functions from τ_1 to τ_2. Thus, predicates have type (e, t) and generalized quantifiers - $((e, t), t)$. A noun like girl will have meaning of type (e, t), and a determiner-phrase like every girl - $((e, t), t)$. The actual sub-sentential meaning of girl would be $\lambda x^e . girl^{(e,t)}(x)$, and that of every girl - $\lambda P^{(e,t)} \forall x^e . (girl(x) \rightarrow P(x))^t$.

There is a brand of MTS, called *Situation Semantics*, whereby instead of models, one relativizes truth to situations, partial (possibly inconsistent) "parts" of models (see Barwise and Perry [8]), also including entities.

I will assume that the reader is sufficiently familiar with MTS both for logic and for NL.

Proof-Theoretic Semantics: PTS attempts to provide a revisionary alternative theory of meaning, the central concept of which is *proof* (and more generally *derivation*) as described throughout the book.

Categorial Semantics: This theory is based on a mathematical (algebraic) discipline of *Category Theory*, and as far as I know was never attempted for a meaning theory of natural language. It defines connectives and quantifiers by mapping them to categories. See Lambek and Scott [100].

Game-Theoretic Semantics (GTS): GTS (also known as *Dialogical Semantics*) specifies meanings by means of abstract *games* between two players, a *proponent* and an *opponent*, taking alternating moves on an abstract board. Connectives and quantifiers are defined by the *rules* for the next step in a game. See, for example, Lorenzen and Lorenz [107], Hintika [75] and Marion [111].

I shall be concerned only with PTS, with occasional comparison and contrast to MTS, and say nothing further on the other approaches.

1.1.1 On meaning bearers and their inter-relationship

Recall that meaning bearers are linguistic expressions of various kinds.

1. **Syntactically related meaning bearers:** There is a natural (partial) ordering amongst syntactically related meaning bearers, that of being a sub-expression, or sub-phrase. This relationship naturally induces a relationship among the meaning of a phrase and the meaning of one of its sub-phrases. I will refer

to one as the *primary* meaning bearer and to the other as the *secondary* bearer of meaning, with the understanding that the meaning of a secondary meaning bearer is defined in terms of the meaning of the primary one.

As it turns out, the two theories of meaning, MTS and PTS, differ, in addition to the identification of their underlying central concept, also on the following two related questions[2].

(a) Which are the primary carriers of meanings, and which are the secondary, derived carriers of meaning?

(b) What principle determines the direction of flow of meaning from primary carriers to secondary ones?

In MTS, the primary carriers of meaning are logical operators in logic and words in NL. The principle governing the flow of meaning from operators to formulas in logic, and from words to (affirmative) sentences in NL, is Frege's principle of *compositionality*, whereby (in the case of NL) the meaning of a sentence (its truth-conditions) is obtained compositionally from the meanings of the words in the sentence and the way they are composed (specified by syntax). On the other hand, in PTS (according to *my* conception of it) the primary carriers of meaning are formulas in logic and sentences in NL (where only affirmative sentences are considered here). The principle governing the flow of meaning from formulas to operators in logic, and from sentences to words in NL, is Frege's principle of *context*, whereby *compound* complete (affirmative) sentences obtain their meaning from a meaning-conferring rule-system, *relative to some* primitive *sentences that obtain their meanings externally to the rule-system* (to be discussed more below), and the meaning of words is their *contributions* to sentential meanings, obtained by abstraction from the latter.

Thus, PTS can be viewed as a theory of meaning for extending primitive sentential meanings to compound ones, *even when compounding is not governed by an explicit operator*; an example of the latter kind are the rules for *relative clauses* (in Chapter 9). Thus,

MTS: words $\rightsquigarrow$ phrases, sentences

PTS: sentences $\rightsquigarrow$ phrases, words

Those issues are elaborated in detail in Chapter 5 (for logic) and Chapters 8 and 9 (for NL).

2. **Syntactically unrelated meaning bearers:** When two meaning bearers are syntactically mutually independent, there still may arise a dependence among their meanings, with the natural question as to which, if any, dependencies are allowed/(un)desirable/(un)attainable. This issue is closely related to the issue, potentially important for logic but mostly negligible for natural language, namely extending a given language with a "new" meaning bearing expression. The primary example is extending a logic with a new operator. The way I see it is that a logical operator can be added to an *arbitrary* logic of a suitable kind.

---

[2]This issue is not explicitly discussed in detail in the literature, and emerged from my own work with my coauthors.

In the MTS literature, this issue is not frequently discussed. On the other hand, in the PTS literature, this is a frequently discussed issue, and several relationships among the meaning definition of syntactically independent meaning bearing expressions have been discussed. As we shall see as the theory unfolds, these approaches manifest themselves in the allowed/attainable (etc.) dependencies amongst the rules of a meaning-conferring proof-system.

Holism: *Holism* is the approach according to which *all* meaning bearers of a given language can only be defined simultaneously, their meanings being mutually dependent. This approach is considered unfavourable by most adherents of PTS.

Molecularism: *Molecularism* is the approach that does allow *some*, preferably small, dependence among the meanings of syntactically unrelated expressions. an important restriction that must be imposed on the allowed dependence is its *well-roundedness*, namely the absence of any *circularity* in the dependence. The presence of such circularity, under typical understanding of definitions, prohibits the attempted defined meaning to be well-defined.

Atomism: *Atomism* is an approach that prohibits *any* dependence of the definition of the meaning of one expression on the meaning of some other, syntactically unrelated, expression. We shall see that this approach imposes very strict constraints on the form of the rules in the meaning-conferring proof-system. Clearly, if attained, this approach best serves the goal of arbitrary extendability mentioned above. An important desideratum of this approach is that when a language is extended with more than one operator, the order of extension should not matter.

Another point about PTS as a theory of meaning relates to the mode of definition of meanings. The standard way PTS is conceived in the literature is as providing an *implicit definition* of meanings by means of meaning-conferring rules. The approach I am proposing in this book is rather to conceive of PTS as providing an *explicit definition* of meanings by meaning-conferring rules. Thus, if ξ is some meaning bearing expression, PTS should provide some *proof-theoretic semantic value* of the form $[[\xi]] =^{df.} \cdots$ as the meaning of ξ. I refer to this semantic value as a *reified* meaning. This issue is further elaborated in Section 1.6.3 and developed as the book progresses.

1.2   Criticisms of MTS as a theory of meaning

There is a vast literature about various critical arguments against MTS as a theory of meaning. I present here briefly only some of the main ones. Some involve philosophical considerations, and others - not. My personal position is closely related to the latter sort of criticism. I will say more about this preference and its origin, as well as a consequence of it, in Section 1.6.3.

1.2.1   The manifestation argument

The most famous criticism is Dummett's *manifestation argument*, associating meaning of an affirmative sentence with the understanding of that sentence, an understanding manifesting itself as involving the ability (at least in principle) to verify the sentence as a condition for its assertability. See, amongst others, Chapter 13 of [33], [35],
[34]. Trans-verificational truth is rejected since it is not reflecting a cognitive process
of understanding (this is where the philosophical position of anti-realism emerges);
this is followed by a rejection of *bivalence*, according to which every (affirmative)
sentence has a truth value (either true or false), independently of any ability to verify
what that value is. The argument is not mathematical in nature, and has been open
to interpretations (see, e.g., Lievers [104]). It is related to the general philosophical
debate between realism and anti-realism.

Besides being based on the association of meaning with its knowledge, the manifestation argument is also based on the ability to learn meanings, as well as the ability to
communicate them. Thus, in addition to the cognitive nature of language (including
logic, and mathematics in general), the *public* nature of language is involved in the
manifestation argument.

Since MTS cannot identify *uniquely* a model corresponding to the actual world, ver-
ifiability means *deciding*, given an arbitrary model, whether the truth-condition (constituting the MTS meaning of a sentence) obtains in the given model. In general, this
task is impossible even for the simplest sentences, involving only predication, as set
membership is not decidable[3] in general. It follows that *entailment*, a major concern
of all formal semantics definitions, is not effective either.

1.2.2   Explanatory power

Another kind of criticism of MTS questions its *explanatory power*. The received
wisdom regards MTS as a formalization of the relationship between language and the
world. Quine (in [158]) relates to this view as "the museum myth": NL expressions
are stuck on objects like labels in a great museum. The claim is that no theory can
succeed in directly relating language to the world. At most, language is related to
some meta-language (e.g., some set-theoretical language), used to specify models and
truth-conditions in them. This is true for MTS in general, but is particularly relevant
to the case of NL, which is its own ultimate meta-language. See Peregrin [140] for a
discussion of this issue.

Since I find this criticism a very compelling one, independent of any philosophical
stand on metaphysical issues, I want to elaborate more on it. Consider the usual MTS
definition of conjunction ('∧'), using the usual models in the form of *assignments g,*

---

[3]There is no precise statement by Dummett as to what is taken as "decidable". It is plausible, at
least in the context of computational linguistics, to identify this notion with *effectiveness* (i.e., *algorithmic
decidability*).

mapping each atomic sentence to a truth-value, and extending g to a full homomorphism.

$$g \vDash \varphi \wedge \psi \text{ iff } g \vDash \varphi \text{ and } g \vDash \psi \qquad (1.2.1)$$

How does such a clause define the meaning of the '$\wedge$' operator of the object language? Unless the meaning of 'and' (in the meta-language, here English) is *already* known, (1.2.1) does not define *meaning* at all! Otherwise, a similarly structured definition of a connective '*blob*' would be equally well-defined by

$$g \vDash \varphi \text{ blob } \psi \text{ iff } g \vDash \varphi \text{ blob } g \vDash \psi \qquad (1.2.2)$$

As argued by Tennant [219], even such meta-circular argument cannot be applied for "bad" connectives (lacking harmony in their I/E-rules – cf. Section 3.6).

1.2.3   Other criticisms of MTS and advantages of PTS

There are several other criticisms of MTS as a theory of meaning, as well as advantages of PTS as such a theory, which also are independent of cognitive and/or epistemic considerations, as well as from metaphysical ones.

Multiple truth-values: There are many well-known multi-valued logics that are based on more than two truth-values (sometimes, on infinitely many truth-values). For such logics, it is not obvious what an assignment of truth-conditions might consist of. Most often, a notion of a subset of *designated values* is brought into play in defining model-theoretic notions such as consequence or validity.
   This carries the model-theory further away from the pre-theoretic notion of 'meaning'.

Ontological richness: One may also add some dissatisfaction with the *ontological commitment* accompanying MTS, relating to various entities populating models: possible-worlds, events, times, degrees, kinds and many more. As emphasized by Paoli [136] when adhering to PTS, the definition of meaning need not appeal to any external apparatus; it can use the resources provided by the rules of the underlying deductive system. In other words, the deductive system is *meaning-conferring*; this will be also my own view throughout. A related issue, associated with entities in models, is the possibility of quantifying over "absolutely everything", accompanying MTS, see Rayo and Uzquiano [160].

Granularity of meaning: A notorious problem of MTS is its coarse granularity of meaning, where logically equivalent propositions are assigned the *same* meaning. As a simple example, in propositional Classical Logic, we have the equivalence

$$\varphi \wedge \psi \equiv \neg(\neg\varphi \vee \neg\psi) \qquad (1.2.3)$$

Both sides of the equivalence are assigned the same meaning (here, same truth-table). However, those two proposition do differ in several aspects involving

meaning, most notably in inference. While it is fairly natural to regard a transition from $\varphi \wedge \psi$ to φ as some kind of an "elementary" transition, a transition
from $\neg(\neg\varphi \vee \neg\psi)$ to φ, while model-theoretically valid, can hardly count as elementary, and its validity needs explanation by means of decomposition to more
elementary steps. A fortiori, the same applies to more complicated, less transparent logical equivalences. In particular, all logical validities are assigned the
same meaning. The granularity of meaning is further discussed in Section 5.2.4.
In natural language this discrepancy is even more salient. Identifying the meaning of every girl is a girl with that of every flower is a flower, and even with
the meaning of no non-bank is a bank, and is clearly inadequate.

Semantic distinctions: As pointed out by Paoli [136], Classical Logic conflates (at
least) two meanings of its connectives, which are conveniently distinguished
under PTS: connectives with *additive* rules and connectives with multiplicative
rules, abbreviated as additive vs. multiplicative connectives, respectively (see
Section 1.5.2). The existence of this ambiguity, as well as its formulation, are
very convenient once their different inferential roles are exposed via structural
rules (explained below). While they are expressible in a suitable (fairly complicated) model-theory, the appeal to proof-theory (in particular, sequent calculi)
is much more transparent and explanatory.

Exposing aspects of meaning: Appealing to PTS may allow exposing aspects of the
meaning of the defined expression not "visible" under their MTS definition.
A significant phenomenon of this type is discussed in Section 5.2.3, where
connectives are characterized as internalizations of structural properties of the
meaning-conferring proof-system.

Another important distinction is the one by Dummett between *assertoric contents* and *ingredient sense*, further elaborated upon in Section 1.4.

1.3   The idea of proof-theoretic semantics

The story of proof-theoretic semantics *in its current form* starts[4]. with the following
comment (to which we return in Section 3.4) by Gentzen (in [61], p. 80), one of the
fathers of modern (structural) proof-theory:

> ... The introductions represent, as it were, the 'definitions' of the symbol
> concerned, and the eliminations are no more, in the final analysis, than
> the consequences of these definitions. ...

Here 'introduction' and 'elimination' refer to rules of a natural-deduction proof-system
(see Section 1.5), to be abbreviated as *I*-rules and *E*-rules, respectively. This quotation expresses the view that the *meaning* of a logical constant (connective, quantifier,

---

[4]An early precursor can be seen in Sellars [199]

etc.) can be determined (or fixed) *solely* in proof-theoretic terms, which is the essence of PTS. The exact way the meaning of logical constants (and, beyond Gentzen's original intent, natural language expressions) is determined, and the proof-theoretical *semantic value* they become endowed with, constitute a major part of this book, and are presented in detail in the sequel. This revisionist[5] view contrasts with the then (and still ...) prevailing view, that meaning should be defined model-theoretically, appealing to notions like a model, entities, reference, denotation (in models) and, ultimately, truth. For example, anticipating the presentation of Intuitionistic Logic (Section 2.1), PTS maintains that the acquaintance with the following I/E-rules is all that is needed in order for one to maintain the understanding of the meaning of conjunction (the most "well-behaving" connective in this respect), denoted by '∧'.

$$\frac{\Gamma:\varphi \quad \Gamma:\psi}{\Gamma:\varphi\wedge\psi}\ (\wedge I) \quad \frac{\Gamma:\varphi\wedge\psi}{\Gamma:\varphi}\ (\wedge E_1) \quad \frac{\Gamma:\varphi\wedge\psi}{\Gamma:\psi}\ (\wedge E_2) \qquad (1.3.4)$$

(the notation is defined in Section 1.5). That is, inferring the conjunction from the two conjuncts, and inferring each conjunct from the conjunction, induces the *full meaning* of conjunction, and no appeal to its traditional truth-table is neither needed nor warranted in order to access its meaning.

In the meaning-conferring system, each operator has its *own* I/E-rules. Therefore, the determination of meaning is *localized*, a property of a theory of meaning known as *molecularism*. This may be contrasted with an axiomatic theory, which, while also of a proof-theoretical nature ("Hilbert system"), has its axioms relating all the operators to each other, a *holistic* view of meaning.

It is common to trace back the PTS view to Wittgenstein's dictum (see [237], VII, 30) 'meaning as use' (see de Queiroz [27] for a thorough bibliographic study of this matter).

> We can conceive the rules of inference [...] as giving the signs their meaning, because they are rules for the use of these signs.

See more about the way 'use' is captured by *ND* in Section 1.6.

While most proponent of PTS relate it to *ND*-systems, in particular after Prawitz ([146, 147]), there is also a possibility, pursued by some, to base PTS on a different presentation of rules, known as *Sequent Calculi (SC)* (also due to Gentzen.) I present those approaches in Section 6.5.

It is important to realize that PTS *does not* undermine the importance of model-theory as a discipline. Defining truth-conditions, and proving soundness and completeness of the proof-systems w.r.t. those truth-conditions have their own merit. PTS only rejects the view that model-theory is an adequate basis for a theory of meaning.

---

[5] A less revisionist version of PTS is presented in Chapter 4.

I note that in addition to the main controversy of what is an adequate theory of meaning, the rivalry between MTS and PTS generated a secondary controversy, as the disagreement over the right theory of meaning also involved a disagreement about the validity of some logical principles. The issue is, which is the preferred logic, mainly contrasting Intuitionistic Logic with Classical Logic, and to some extent Relevant Logic (known also as Relevance Logic), in a way seemingly orthogonal to the argument admitting only constructive means of proof in mathematics (for example, see Prawitz [150] among many other references). As observed by Marion [111], at the time of the emergence of this secondary dispute, those two logics were the only contenders. As PTS evolved, together with the emergence of a plentitude of alternative logics (see, for example, Van Benthem, Heinzmann, Rebusci and Visser [225]), more logics participate in this "race". As we shall see in the sequel, Intuitionistic Logic is viewed by some as better conforming to the PTS programme. This issue is also closely related to the philosophical dispute between realism and anti-realism (see Section 1.2.1) and, for example, Rahman, Primierro and Marion [159]). In particular, see also Bonnay and Cozie [15] for a recent discussion of the connection between the realism vs. anti-realism debate and the choice of a preferred logic. This paper actually presents a more radical version of anti-realism, leading to a rejection of *both* Classical and Intuitionistic logics, and a preference of Girard's *Linear Logic* [63]. However, since there are also a proof-theoretic justifications of classical logic, as described in Chapter 4, and of other logics, I will concentrate on the presentation of PTS without logical revisionism. The rivalry, if there is one, between Intuitionistic Logic, Classical Logic and Linear Logic (or even Relevance Logic that was mentioned in this context) should be resolved on other grounds than the availability of proof-theoretical justification. See Kürbis [95] for support of this view.

To summarize this point, PTS can be viewed as having two goals, to which I relate as primary and secondary.

Primary goal: Define the meaning of a logical operator *schematically* by means of self-justifying rules, that justify the other rules, when the operator is added to an *arbitrary* object language, in which this operator is not present, regardless of any other operator that *is* present.

Secondary goal: Define the meaning of a logical operator in the context of a *given* logic, with a known set of other operators, thereby providing a proof-theoretic justification to that logic.

This distinction becomes important, for example, in the discussion in Section 4.5.3, where the definition of the meaning of negation *in general* (i.e., schematically) is contrasted with its definition within Classical Logic, in the context of a proof-theoretic justification of the latter.

While PTS was originally conceived for logic, it was later suggested by philosophers of language such as Dummett and Brandom to apply also to meaning in natural language. It is worth noting, that a contrast similar to that between PTS and MTS exists also in the realm of programming languages, between *operational semantics*, that often takes the form of suitable proof-rules, and *denotational semantics*, that assigns abstract algebraic denotations to expressions in the language.

1.4   A bird's eye view of the PTS programme

In this section, I briefly present the central line of ideas of the PTS programme, to be unfolded in this book. In a nutshell, the PTS programme can be described as follows.

- *For (affirmative) sentences, replace the received approach of taking their meanings as* **truth-conditions** *(in arbitrary models) by an approach taking meanings to consist of* **canonical derivability conditions** *in the meaning-conferring* ND-*system (from suitable assumptions).* Two notions of canonicity of derivations are usually considered, to be further specified below. For logic (in part I), the usual natural-deduction (*ND*) proof systems are used. For natural language (part II), a "dedicated" natural-deduction proof-system is developed – on which the canonical derivability conditions are based. In a sense, the proof system should reflect the "use" of the sentences in the considered fragments, and should allow recovering pre-theoretic properties of the meanings of these sentences such as entailment, assertability conditions and consequence drawing.

- For *subsentential phrases*, down to lexical units (logical constants in logic, words in natural language), replace taking their denotations (extensions in arbitrary models) as meanings, by taking their **contributions** to the meanings (in our explication, canonical derivability conditions) of sentences in which they occur. This adheres to Frege's *context principle* [56], made more specific by the incorporation into a type-logical grammar for the fragment considered (see Moortgat [123]) .

Dummett introduces an important distinction between *assertoric content* and *ingredient sense* ([35], p. 48). The assertoric content of an (affirmative) sentence S is the meaning of S "in isolation", on its own. The ingredient sense of S is what S contributes to the meaning of an S', in which S occurs as a sub-expression, a component. A grasp of the ingredient sense always includes a grasp of the meaning, but not necessarily vice versa. This distinction is propagated to sub-sentential phrases too. This distinction features in a major way when meanings, both sentential and sub-sentential, are reified (discussed in detail in Chapter 5 for logic, and Chapters 8 and 9 for natural language). It is also related to the discussion in Section 4.5.3, where this distinction, under a specific interpretation, is alluded to in proof-theoretically justifying Classical Logic.

Following Stewart [210], one can view PTS as a three-tiered enterprise.

1. At the highest, most abstract tier, PTS is established as a more adequate theory of meaning than its MTS counterpart. To this tier belongs the criticism of MTS

as a theory of meaning, as well as the motivation and justification of PTS as such a theory. In particular, the PTS approach is bound to the 'meaning as use' paradigm. In this tier one can also include the philosophical arguments accompanying (and, as viewed by some, underlying) PTS, such as the realism/anti-realism controversy.

It is important to make a distinction between the rivalry of two competing approaches to meaning, and a disagreement about the actual meaning of some logical constant (or any other expression, for that matter) *within* some theory of meaning. Consider, for example, implication. One can disagree with its classical definition using the familiar truth-table, and claim that the entry assigning truth to the implication in case its antecedent is assigned falsehood does not really reflect its correct definition; preferring, for example, that the implication is assigned falsehood in such a case. Such a disagreement is only a disagreement about meaning of a specific expression – not about the method of assigning meaning itself. Such a disagreement may result in a different logic, for example in a *relevant logic*. An appeal to multi-truth-valued tables has the same status – it is still within the same MTS-based theory of meaning. On the other hand, a much deeper disagreement, the one I am concerned with in this book, is the rejection of the whole definition via truth-tables (or other model-theoretic means) as a definition of meaning, independently of which truth-table (or, more generally, which class of models) is proposed as the meaning of a given logical constant. Within PTS similar intra-method disagreements may arise. One can reject a certain I-rule (assuming natural-deduction is chosen as the vehicle for PTS – see the next item) for implication as really defining its correct meaning (i.e., its "correct use"), opting some other introduction rule. This will also result in a different meaning of an expression, but will not scrutinize the proof-theoretic approach itself.

2. The second tier presents the consideration of casting 'use' as a logical deductive system, studies in detail the candidate systems and presents arguments for locating meaning in whichever kind deductive system is chosen as meaning-conferring. When justifying the choice of a particular kind of system, one has to remember that it is the 'use' that induces meanings, and the chosen kind of deductive system is justified as serving a meaning-conferring kind of a deductive system inasmuch as it successfully reflects actual use. This has been formulated in Steinberger [207] as the following principle:

> *Principle of answerability*: only such deductive systems are permissible [to serve as meaning-conferring – N.F.] as can be seen to be suitably connected to our ordinary deductive inferential practices.

Note that this is an informal principle, and its satisfaction cannot be proved or disproved, only argued for/against informally. Obviously, once a deductive system answering this criterion is selected, individual rules within this system, associated with a particular expression, reflect the actual use of that expression, provided such an expression *is* in use in NL. When a *new* expression is introduced, then the rules *establish* use. Thus, the way I understand this principle is that it applies to deductive frameworks, and only occasionally to specific rules.

In addition to descriptive adequacy, Steinberger [207] also requires a well-chosen deductive system to exert some *normative force*, and adhere to general requirements such as finite presentability.

3. The lowest tier consists of a "tool-box" of techniques implemented in specific meaning-conferring deductive systems, formulated within the deductive framework chosen at the second tier. In addition, properties of meaning-conferring deductive systems are studied, qualifying them as such.

I shall briefly present some issues related to the first tier, but my main concern is in the two lower tiers, exhibiting a plentitude of technical details.

An important aspect of my presentation is an emphasis, not usually found in the PTS literature, on incorporating the meanings as determined by PTS into *grammars*. Meanings do not "hang in the air" – they are part of grammars (their semantic component) inasmuch as the syntax is part of grammar. In particular, the grammar formalism that underlies the presentation is *type-logical grammar (TLG)*, a strongly lexicalized formalism (see Moortgat [123]). This emphasis influences my view of *reification* (though in proof-theoretic terms) of meanings of sentences and words (cf. Section 1.6.3), as presented in Chapters 5 (for logic) and 9 (for natural language).

1.5 Natural-deduction proof-systems

The natural-deduction (*ND*) proof-systems were first presented by Gentzen in [61], as a presentation of (both classical and intuitionistic) logical inference. Similar ideas ere presented also in Jaśkowski [83]. For more on the history of natural-deduction (in particular, as it was presented in textbooks over the years) see Pelletier and Hazen [138]. The nature of such a system is derived from an attempt to characterize "natural reasoning", as done in informal Mathematics. We note here, as observed in Slater [202], that according to Gentzen's view, that became the prevailing view, *ND*-systems attempt to capture one specific relationship among propositions, namely *deducibility* of one from another (or a collection of others). As we shall find out when considering classical logic and negation, other relationships among propositions might be viewed as the target of *ND*-systems. At this stage, however, we proceed with Gentzen's view.

While there is no *unique* definition as to when a proof-system is to be viewed as an *ND* proof-system, there are several qualifications of a proof-system that are regarded as *essential* characterizations of natural-deduction.

- The rules of the system come in two main groups, *introduction rules* (*I*-rules) and *elimination rules* (*E*-rule(s)). Therefore, such systems are also called "int-elim" systems. This point is elaborated more below (p. 20).

- Derivations (proofs) in such system make extensive use of *assumptions*, possibly *discharged* during the derivation. The notion of a (varying) *dependence* on assumptions is central to *ND*-systems. This also is made more precise below. Because of their central role, assumptions will be discussed in more detail in Section 1.5.2.1.

- A preferred attribute of *ND*-systems is *purity*, whereby each logical connective has its own I/E-rules, not mentioning other operators explicitly. When an *ND*-system is used as meaning-conferring, purity of I-rules assure the independence of the meaning of the introduced operator from any other operator.

Because of the central role *ND*-systems play in the PTS programme, I present them in greater generality and in much more detail than found in the literature when specific *ND*-systems are presented. The *ND*-systems mostly concentrated upon are known as *unilateral single-conclusion ND*-systems ($SCND$). Two generalizations of *ND*-systems are presented in the sequel, where used.

- Multiple-conclusion *ND*-systems (see Section 4.2).

- Bilateral *ND*-systems (see Section 4.4).

1.5.1   Object languages, contexts and sequents

An *ND*-system[6], say $\mathcal{N}$, is defined over an *object language*, say L, the syntax of which is defined (usually, recursively) as a freely-generated[7] collection $\mathcal{F}$ of *formulas*, ranged over by (possibly indexed) φ, ψ, etc. When the object language is propositional, the basis of the recursion is usually a (countable) set $\mathcal{P} = \{p_i \mid i \geq 0\}$, referred to as *basic* (or *atomic*) sentences[8], closed under a (finite) collection of operators, *logical constants* (generating *compound sentences*). Sometimes, *propositional constants* are also included in the basis of the object language, most often '$\perp$' (falsum, absurdity) and its dual '$\top$' (verum). A more detailed discussion of the role of '$\perp$' in Intuitionistic Logic is presented in Section 2.1.3.   First-order logic, for example, has a more complicated basis for the recursion, where atomic sentences have their own structure. If the language contains variable-binding operators, the usual notions of *free* and *bound* variables are assumed, with $FV(\varphi)$ the collection of free variables in φ. The (capture-free) substitution of y for all occurrences of some $x \in FV(\varphi)$ is denoted by $\varphi[x := y]$. While in general an operator can have any arity, combining any number of formulas to form a new one, I shall employ as much as possible the use of binary (and unary) operators, facilitating the convenient infix notation. The *complexity* $|\varphi|$ of a formula φ is the number of operator occurrences it contains.

---

[6]While there exist *ND*-systems for multiple-valued logics in the literature, I shall not be concerned with them here.

[7]Freely-generated here means that, for example a ternary operator '$*$' will generate a formula of the form '$*(\varphi, \psi, \chi)$', allowing arbitrary sub-formulas, and not, say, '$*(\varphi, \psi, \varphi)$', restricting two sub-formulas to be identical.

[8]The base case in the recursive definition can also be viewed as *propositional variables*, amenable to substitution.

The formulas of the object language (and sub-expressions thereof) are the linguistic expressions the meaning of which are to be defined by PTS. In the second part of this book, where PTS is applied to natural language, a totally different form of an object language is considered.

There are two common variants of presentation modes of $SCND$-systems.

Simple presentation: The objects of $\mathcal{N}$ in its simple presentation, serving as premises and conclusion in a rule, are the formulas of the object language L themselves. The dependence of premises on undischarged assumptions is left implicit when formulating rules (see next section).

Logistic presentation: The objects of $\mathcal{N}$ in its logistic presentation, serving as premises and conclusion in a rule, are *sequents* of the form $\Gamma : \varphi$, where:

- Γ is a *context*, a (finite) list (i.e., ordered set[9]) of formulas of the object language, *assumptions* on which the derivation (defined below) of φ may depend. Γ is also referred to as the *antecedent* of the sequent. A context $\Gamma = \Gamma_1, \Gamma_2$ is a combination of two contexts into one, formed by concatenation; the comma is sometimes omitted. I will not distinguish between a singleton context, containing only one formula, say φ, and the formula φ itself.

- φ, the *succedent* of the sequent, is a formula of the object language.

- $\mathcal{L}$ is the logic defined by $\mathcal{N}$, the collection of all sequents provable in $\mathcal{N}$.

The informal interpretation of a sequent $\Gamma : \varphi$ is that φ *holds under the assumptions* Γ (known also as a *conditional assertion*). When $\mathcal{N}$ can be determined from context, or is immaterial, it is omitted from the sequent.

Below, I formulate the way a transition from one mode of presentation to the other can take place, thus allowing the choice of mode as a matter of convenience. Simple presentation will mostly be used for presenting examples, to cut short notational clutter.

Actually, this is a somewhat restricted form of a sequent, called a *single-conclusion sequent* ($SCND$). A more general notion of a sequent of the form $\Gamma : \Delta$, called a *multiple-conclusion sequent* ($MCND$), is more controversial as meaning-conferring in PTS, and is considered in Section 4.2. Much of the development is carried out in parallel, under *both* modes of presentation, as the distinction becomes more consequence bearing for multiple-conclusion systems.

As already mentioned, in part II of the book, I consider natural language (and slight extensions thereof) as the object language over which *ND*-systems are defined, giving rise to a much more complicated syntactic structure. This added complexity will bear

---

[9]For some logics, contexts may have a more complicated structure, for example trees; this will be specifically indicated if arising.

some consequences on the conception of *ND*-systems. At this stage, we remain with the simpler syntactic structure.

The interpretation of $\vdash_{\mathcal{N}} \Gamma : \varphi$ is that the holding of φ under the assumptions Γ can be established by the rules of $\mathcal{N}$. One should careful to distinguish this notion from that of implication '$\rightarrow$' (discussed below). In axiomatic presentations of some logics, e.g., Classical Logic, the two are related via what is known as the *deduction theorem*:

Theorem 1.5.1 (Deduction Theorem).

$$\vdash \Gamma, \varphi : \psi \text{ iff } \vdash \Gamma : \varphi \rightarrow \psi \tag{1.5.5}$$

However, the following two facts show that this connection need not be universal.

- The object language need not contain implication.

- Even if implication and conjunction are present in the object language, the deduction theorem need not hold, as is the case, for example, in some modal logics (depending on the definition of proof from assumptions in axiomatic presentations of logics, see Hakley and Negri [70]).

1.5.2   Rules

ND-systems consist of a collection of *rules*, in contrast to the axiomatic nature of Hilbert-style proof systems that preceded them. The latter had two main roles:

1. To "capture" in a syntactic, effective way the consequence relation of the logic as defined model-theoretically.

2. To justify the axioms (as correctly performing the task above) by means of soundness and (preferably) completeness w.r.t. the model-theoretically specified meaning.

There is an important methodological difference between Hilbert's axiomatic proof-theory and Gentzen's structural proof-theory (in both variants - natural-deduction and sequent-calculus). While the former regards proofs (forming *categorical assertions*, depending on no open assumptions), as of primary interest for logic, the latter attribute priority to *derivations* from open assumptions (forming *hypothetical arguments* for conditional assertions). Clearly, categorical arguments are a special case of the hypothetical ones, when the collection of open assumptions turns to be empty. Another way of expressing the same idea is to say the primary interest of logic, when carried out in an *ND*-setting, is *inference*, in contrast to *theoremhood*, when carried out axiomatically in a Hilbert system. This issue is further elaborated upon in Chapter 6.

In MTS terms, Hilbert systems capture *validities* (*tautologies* in the propositional case) – formulas yielding truth under every interpretation, while *ND*-systems captures *(logical) consequence*. This difference is reflected in several aspects of PTS, to be elaborated upon as I advance with the presentation.

For the formulation of rules, a meta-language is used, having additional symbols like the colon ':' and comma in sequents, the horizontal line, or the force markers (discussed in Section 4.4). A rule, say (R), has *premises* (usually, finitely many), usually presented as located over an horizontal line, and a *conclusion*, usually presented as located underneath the horizontal line. The premises and conclusions are objects depending on the presentation mode. A rule is *degenerate* iff its conclusion is identical to one of its premises. Throughout, I only consider non-degenerate rules. To distinguish the two presentations of (the same) rule (R), they are labelled (R_s) (simple presentation) and (R_l) (logistic presentation).

Simple presentation:

$$
\cfrac{
\begin{array}{ccc}
[\varphi_1^1, \cdots, \varphi_1^{m_1}]_1 & & [\varphi_p^1, \cdots, \varphi_p^{m_p}]_p \\
\vdots & & \vdots \\
\psi_1 & \cdots & \psi_p
\end{array}
}{\psi} \ (R_s^{\bar{i}})
\qquad (1.5.6)
$$

Here both the p premises and the conclusion are meta-variables over formulas of the object language L. Abusing the terminology, I refer to them as just formulas. A premise may (but need not) have discharged assumptions in L), wrapped in square brackets when present. The index $\bar{i} = 1 \cdots p$ on the rule's name (also formulas, called the *discharge label*, abbreviates the collection of indices of assumptions discharged by an application of the rule. I use $\Sigma_j = \{\varphi_j^1, \cdots, \varphi_j^{m_j}\}$ for the jth *block of assumptions*, for $1 \leq j \leq p$. When $m_j = 0$, the jth premise discharges no assumptions. In addition to the dischargeable assumptions Σ_j on which ψ_j depends, it may depend on additional (lateral) assumptions which are implicit.

Rules are classified into two families, *additive* and *multiplicative*.

Additive: There is a restriction that all the premises depend on the *same* collection of (implicit) lateral assumptions.

Multiplicative: Each premise may depend on different collections of (implicit) lateral assumptions.

Logistic presentation: In this presentation, both the premises and the conclusion are all sequents as above. The effect of a discharge of an assumption is exhibited here by the assumption present in the antecedent of a premise, but not in the antecedent of the conclusion.

The presentation also splits into two sub-cases, as follows.

Additive:

$$
\cfrac{\Gamma, \varphi_1^1, \cdots, \varphi_1^{m_1} : \psi_1 \quad \cdots \quad \Gamma, \varphi_p^1, \cdots, \varphi_p^{m_p} : \psi_p}{\Gamma : \psi} \ (R_l)
\qquad (1.5.7)
$$

The j'th logistically presented premise is abbreviated to $\Gamma, \Sigma_j \vdash_{\mathcal{N}} \psi_j$. The characteristic feature of additive rules is that all premises have the same context, Γ, hence they are also known as *context-sharing* rules.

Multiplicative:

$$\frac{\Gamma_1, \varphi_1^1, \cdots, \varphi_1^{m_1} : \psi_1 \quad \cdots \quad \Gamma_p, \varphi_p^1, \cdots, \varphi_p^{m_p} : \psi_p}{\Gamma_1 \cdots \Gamma_p : \psi} \ (R_l) \qquad (1.5.8)$$

Here the j'th logistically presented premise is abbreviated to $\Gamma_j, \Sigma_j \ : \psi_j$. The effect of a discharge of an assumption is exhibited here by the assumption present in the antecedent of a premise, but not in the antecedent of the conclusion. The characteristic feature of multiplicative rules is that each premise may have its own context, Γ_j, hence they are also known as *context-free* rules.

The difference between additive and multiplicative rules came to the fore most notably in Linear Logic [64].

Rules (actually, instances thereof) are said to be *applied* to their premises, *yielding* their conclusion(s). As mentioned above, one can mutually shift from one presentation to another; mostly, examples will be presented using the simple presentation.

Rules, in particularly as a tool for PTS, are to be read as *parametric* in the underlying object language. Thus, when a rule displayed as in (1.5.6), (1.5.7) and (1.5.8) is contained in an *ND*-system over object language L, φ ranges over *arbitrary L-formulas*, and for (1.5.7) and (1.5.8) Γ ranges over *arbitrary L-contexts*. So, such a rule can be freely included in different *ND*-systems. See the discussion on schematism in Section 1.6.4. Sometimes this schematism is replaced by closure under substitution of formulas. I will mainly adhere to the schematism via meta-variables. When the meta-variables are instantiated to actual formulas within a rule, an *instance* of the rule results. Note that by an instantiation different meta-variable may yield identical object-language formulas. For example, it may yield identical premises of a rule, like in the $(\wedge I)$ rule in (1.3.4). Such premises should not be collapsed to one. Consider an instantiation of both φ and ψ in $(\wedge I)$ to an atomic sentence p. The correct instantiated form of the rule (with Γ omitted) is

$$\frac{p \quad p}{p \wedge p} \ (\wedge I), \ \text{not} \ \frac{p}{p \wedge p} \ (\wedge I)$$

ND-systems for logic[10] contain the following kinds of rules.

Operational rules: For every *logical constant*[11] (connective, quantifier) of arity greater than 1, say[12] '$*$', of the object language, there are *two families of rules* (usually, both finite).

---

[10]In part II, *ND*-systems for natural language are proposed, of a certain different nature, as explained there.

[11]Yet another conception of the classification of rules is presented in Section 6.5.2, in discussing sequent calculi as a basis for PTS.

[12]As already mentioned, in general, logical constants can be of any arity. Recall that for readability, I most often present them as binary.

Elimination rules (E-rules): $(*E)$, determining which formulas can be deduced from $(\varphi_1 * \varphi_2)$ in the simple presentation, or which sequents can be deduced from $\Gamma : (\varphi_1 * \varphi_2)$ in the logistic presentations. The premise containing '$*$' is the *major premise*, while all other premises of the rule (if there are any) are *minor premises*. E-rules should reflect *a direct conclusion* of $(\varphi_1 * \varphi_2)$. Note that by the rules' non-degeneracy assumption, a rule like

$$\frac{\chi \quad \varphi * \psi}{\chi} \ (R)$$

does not qualify as an E-rule for '$*$'.

Introduction rules (I-rules): $(*I)$, determining from which formulas, in the simple presentation, can the conclusion $(\varphi_1 * \varphi_2)$ be deduced, or from which sequents, in the logistic presentations, can the conclusion $\Gamma : (\varphi_1 * \varphi_2)$ be deduced. I-rules should reflect *a direct derivation* of $(\varphi_1 * \varphi_2)$.

In both cases, $(\varphi_1 * \varphi_2)$ is the *principal formula* of the rule. Note that for 0-ary constants (to be encountered below), that have no principal operator, the I-rules introduce directly the constant. There is also a limit case of a constant that has no I-rules (cf. $\perp$ in NJ, Section 2.1.3). In a borderline case of a rule (R) of the form (in its simple presentation)

$$\frac{\varphi *_1 \psi \quad \cdots}{\varphi' *_2 \psi'} \ (R)$$

that could be considered either an I-rule for '$*_2$' or an E-rule for '$*_1$', it is the task of the ND-system to classify it. It cannot serve both purposes simultaneously. This property of a rule is called being *single ended* by Dummett [35] (pp.256-8).

Definition 1.5.1 (purity, simplicity). *An operational rule for an operator, say '$*$', is* pure *iff the rule does not mention any other connectives than '$*$'. The rule is* simple *iff it mentions '$*$' once only.*

Let $\mathcal{E}_\mathcal{N}$ denote the collection of all E-rules in $\mathcal{N}$, $\mathcal{I}_\mathcal{N}$ the collection of all I-rules in $\mathcal{N}$ and $\mathcal{O}_\mathcal{N} = \mathcal{I}_\mathcal{N} \cup \mathcal{E}_\mathcal{N}$. As usual, when $\mathcal{N}$ is clear from the context, it is omitted as a subscript.

Structural rules: These are rules not referring explicitly to any logical constant, and typically allow the manipulation of the context. However, in Sections 4.2 and 4.4 we shall also encounter structural rules that operate on the succedent of a sequent, the latter are important for drawing conclusions. The more typical structural rules control the order and the multiplicity of formulas in a sequent. Typically, a structural rule, say (S), is depicted as follows.

$$\frac{\Gamma : \varphi}{\Gamma' : \varphi} \ (S) \tag{1.5.9}$$

implying that the context Γ' can replace, in a derivation of φ (in $\mathcal{N}$), that of Γ. Occasionally, it may be convenient to use the following notation,

$$\frac{\Gamma(\Gamma_1) : \varphi}{\Gamma(\Gamma_2) : \varphi} \ (S)$$

where $\Gamma(\Gamma_1)$ refers to a context Γ containing a sub-context (sublist of assumptions) Γ_1, and $\Gamma(\Gamma_2)$ is obtained from the above by replacing Γ_1 by Γ_2. Let $\mathcal{S}_\mathcal{N}$ denote the collection of the structural rules of $\mathcal{N}$. $\mathcal{S}_\mathcal{N}$ is further divided as follows, where this classification is further discussed in Section 1.5.5.

S_a-**rules:** These are S-rules that only manipulate assumptions in a derivation, like Gentzen's original rules (for single conclusion) *Weakening*, *Contraction* and *Exchange*. Typically in *ND*-systems, S_a-rules are not taken as primitive; rather, they are absorbed into the identity axiom $\vdash\varphi : \varphi$ (or, sometimes, $\vdash\Gamma,\varphi : \varphi$).

S_i-**rules:** Those are structural rules that allow the introduction of a conclusion, like some *coordination rules* in the bilateral presentation of Classical Logic in [179], to be elaborated upon in Section 4.4, or right weakening in the multiple-conclusion presentation of classical logic, elaborated upon in Section 4.2.

S_e-**rules:** Those are structural rules that allow the elimination of a conclusion, again like the *coordination rules* in the bilateral presentation of Classical Logic in [179], or right contraction in the multiple-confusions presentation of classical logic.

There is a whole area of study, called *substructural logics*, dedicated to systems which contain only some (or none!) of the structural rules originally conceived by Gentzen. See, for example Restall [172] or Paoli [134].

Structural rules are relative to a logical system, not just to a logic, as the same logic may be defined using systems that differ w.r.t. the structural rules assumed primitive.

While typically *ND*-systems have the structural rules implicitly built into the operational rules, when not so, I will present them as explicit rules. More on the role of the structural rules in PTS is delineated in Section 1.6.2.

Sometimes, rules are defined so as to have *side-conditions* to their application. For example, the following I-rule for the universal quantifier (see Figure 2.3 below)

$$\frac{\Gamma : \varphi}{\Gamma : \forall x.\varphi} \ (\forall I)$$

has as side-condition $x \notin free(\Gamma)$, namely x does not occur free in any assumption of the context Γ. Similarly, the E-rule for necessitation in the modal logic S5

$$\frac{\Gamma : \varphi}{\Gamma : \Box\varphi} \ (\Box I)$$

has the side-condition that every assumption $\psi \in \Gamma$ is itself boxed, $\psi = \Box\psi'$ (for some ψ'). Note that this side-condition causes the rule to violate the context generality condition (defined in Section 1.6.4), making it inappropriate for serving as a meaning conferring rule for necessity. This violation was noted also in Humberstone [80] in the context of uniqueness of an operator. A possible solution is to change the form of the sequents over which rules for modalities are defined. I will not pursue this matter further here.

Below, when I speak of an application of a rule, it is always presupposed that the side-condition (if any) is satisfied.

Note that this kind of side-condition on a application of a rule should be distinguished from a different notion of a rule, found in axiomatic Hilbert-system, where the premises as well as the conclusion are assumed to be (formal) theorems of the logic, in contrast to the rules in *ND*-systems where premises are assumptions. More on this issue is presented in Section 1.5.2.1. Recall that no axioms are present in *ND*-systems (except for identity axioms), and that is why they can be considered meaning-conferring, exhausting the *full* meaning of whatever they define, as elaborated in Section 1.6.

1.5.2.1   On the role of assumptions

A typical characteristic of *ND* proof-systems is their use of *hypothetical reasoning*, which gives *ND* its force: A rule "temporarily" introduces *assumptions* (occurring in the antecedents of one or more premises), to be used in premises, and any number of their instances (including zero!) may be *discharged* by an application of (an instance of) the rule; if all instances are discharged, the assumed formula does not occur anymore in the antecedent of the conclusion sequent. The natural view of a discharged assumption conforms to the above mentioned interpretation of a sequent, reflecting a dependency of the succedent on the antecedent, in terms of "holding under assumption". A discharged assumption is also called *closed*, while an undischarged one is *open*.   In case of a discharge of zero occurrences, we refer to a *vacuous discharge*, *rejected by some logics*! In case of discharging more than one occurrence of the assumption, we refer to a *multiple discharge*, also rejected by some logics.

An important feature of reified meanings (defined below, Section 5.2) is, that the role of open assumptions (contexts) is different than the orthodox view of it, by which $\varphi \in \Gamma$ is *a placeholder* for an assumed proof of φ; rather, they are viewed as *grounds* for assertion (defined in Definition 12.2.79), or parts of *outcomes* of assertion . This change of view is particularly significant in an attempt to apply PTS to natural language; this is dealt by me more extensively in Part II of the book. It reflects a general recent trend to attribute priority of the *hypothetical* over the *categoric* (see, for example, Schroeder-Heister [192]). In a sense, this view is closer to the view of hypothetical reasoning in deductive databases (in computer science), where the database is extended temporarily, and after drawing some conclusion from the extended database, it is returned to its initial state (like an assumption discharge). In a sequent $\Gamma : \varphi$, the context Γ can be seen as resources, to be analyzed (particularly by applying E-rules to them) in order to derive φ. In contrast to the BHK-interpretation of the I-rule for implication in Intuitionistic Logic (in both simple and logistic presentations):

$$\begin{array}{c} [\varphi]_i \\ \vdots \\ \dfrac{\psi}{\varphi \to \psi} \ (\to I^i) \end{array} \qquad \dfrac{\Gamma, \varphi : \psi}{\Gamma : \varphi \to \psi} \ (\to I)$$

(1.5.10)

where by $(\to I)$ is viewed as a *proof-transformer*, transforming proofs of φ to proofs of ψ, my view is rather adding temporally φ to Γ as a resource, after the analysis of which (with Γ) it can be released, not usable anymore. This view also sheds light on Tennant's definition of *normal form*, where assumptions "stand proud" when serving as a major premise of an E-rule (see, for example, [215, 217]).

1.5.2.2   Rules classification

The following classification of *ND*-rules using some distinguishing criteria, will be useful for he upcoming discussion.

1. An *ND*-rule is *hypothetical* if it allows for at least one premise with *assumptions discharge*; otherwise, it is *categorical*. The latter seems to coincide with what Milne [114] calls a "*immediate inference*", mentioned on p. 10.

2. An *ND*-rule is *combining* if has more than one premise; otherwise it is non-combining.

3. An *ND*-rule is *parameterized* if (at least one of) its premises depends on a free variable; otherwise, it is non-parameterized.

4. an *ND*-rule is *conditional* if it has some *side-condition* on its applicability; otherwise – it is *unconditional*. I restrict myself here to side-conditions stating *freshness* of a variable for a context, meaning the variable does not occur *free* in any formula in the context. I assume there is an inexhaustible supply of fresh variables whenever a rule-application needs one. More general side-conditions may have to be considered in a more general setting.

Note that the classification by the above criteria is orthogonal to the I-rule vs. E-rule classification. Thus, the (logistically presented) conjunction-introduction rule

$$\dfrac{\Gamma : \varphi \quad \Gamma : \psi}{\Gamma : \varphi \wedge \psi} \ (\wedge I)$$

and the (logistically presented) conjunction-elimination rules

$$\dfrac{\Gamma : \varphi \wedge \psi}{\Gamma : \varphi} \ (\wedge E_1), \quad \dfrac{\Gamma : \varphi \wedge \psi}{\Gamma : \psi} \ (\wedge E_2)$$

are categorical, while the (logistically presented) implication-introduction rule

$$\dfrac{\Gamma, \varphi : \psi}{\Gamma : \varphi \to \psi} \ (\to I)$$

and the disjunction-elimination rule

$$\frac{\Gamma : \varphi \vee \psi \quad \Gamma, \varphi : \xi \quad \Gamma, \psi : \xi}{\Gamma : \xi} \ (\vee E)$$

are hypothetical. Also, the $(\wedge I)$ and $(\vee E)$ are combining, while $(\wedge E_i)$ and $(\rightarrow I)$ are non-combining. None of the above rules is parameterized (hence, none are conditional). A categorical, parameterized and conditional rule is the usual I-rule for existential quantification, that has the (simply presented) form

$$\frac{\varphi[x := t]}{\exists x.\varphi} \ (\exists I)$$

where t is safe for φ. The premise is here parameterized by an explicit substitution of the term t. Here $\varphi[x := t]$ stands for φ with the term t substituted for the free occurrences of x in φ. For simplicity, we do not consider function symbols here, so variables are the only terms.

1.5.3   Derivations

Derivations, like rules, are linguistic objects. Similarly to (an application of) a rule corresponding to a single step of reasoning, derivations correspond to complete arguments.

1.5.3.1   Defining derivations

$\mathcal{N}$-*Derivations*, ranged over by $\mathcal{D}$, are again defined separately for the two modes of presentations of $\mathcal{N}$. The form of derivation chosen here is Gentzen's tree-like format (see more below on this format).

Definition 1.5.2 ($\mathcal{N}$-Derivations).

Simple derivations: *Here a derivation $\overset{\mathcal{D}}{\psi}$ has an explicit (single) conclusion ψ, and* implicit *assumptions on which the conclusion ψ depends, denoted by $\mathbf{d}_\mathcal{D}$, defined in parallel to $\mathcal{D}$.*

- *Every assumption φ is a derivation $\overset{\mathcal{D}}{\varphi}$, with $\mathbf{d}_\mathcal{D} = \{\varphi\}$.*

$$\begin{array}{c} [\Sigma_j] \\ \mathcal{D}_j \end{array}$$

- *If* ψ_j *, for* $1 \le j \le p$*, are derivations with dependency sets* $\mathbf{d}_{\mathcal{D}_j}$ *with* $\Sigma_j \subseteq \mathbf{d}_{D_j}$*, and if*

$$\dfrac{\begin{array}{ccc} [\Sigma_1]_1 & & [\Sigma_p]_p \\ \vdots & & \vdots \\ \psi_1 & \cdots & \psi_p \end{array}}{\psi} \ (R_s^{\bar{i}})$$

is an instance of a (simply presented) rule in $\mathcal{R}_{\mathcal{N}}$ *with a* fresh *discharge label* $\bar{i}$*, then*

$$\mathcal{D} \atop \psi \ =^{\mathrm{df.}} \quad \dfrac{\begin{array}{ccc} [\Sigma_1]_1 & & [\Sigma_p]_p \\ \mathcal{D}_1 & & \mathcal{D}_p \\ \psi_1 & \cdots & \psi_p \end{array}}{\psi} \ (R_s^{\bar{i}}) \qquad\qquad (1.5.11)$$

is a derivation with $\mathbf{d}_{\mathcal{D}} = \cup_{1 \le j \le p} \mathbf{d}_{\mathcal{D}_j} - \cup_{1 \le j \le p} \hat{\Sigma}_j$*, where* $\hat{\Sigma}_j$ *is the collection of formulas actually discharged by this instance of the applied rule. The derivations* $\mathcal{D}_1, \cdots, \mathcal{D}_p$ *are the* direct sub-derivations *of* $\mathcal{D}$*, said to be* encompassed *by* $\mathcal{D}$*.*

$$\begin{array}{c} \varphi \\ \mathcal{D} \end{array}$$

It is convenient to use ψ*, when* $\varphi \in \mathbf{d}_{\mathcal{D}}$*, to focus on some assumption* φ *on which* ψ *depends.*

As for the additive/multiplicative distinction, the following[13] *holds.*

additive: *Recursively, the sub-derivations* $\mathcal{D}_1, \cdots, \mathcal{D}_p$ *are all additive,* $\mathbf{d}_{\mathcal{D}_1} = \cdots = \mathbf{d}_{\mathcal{D}_1}$*, and the rule* R_s *is additive. The resulting derivation* $\mathcal{D}$ *is additive too.*

multiplicative: *Recursively, the sub-derivations* $\mathcal{D}_1, \cdots, \mathcal{D}_p$ *are all multiplicative and the rule* R_s *is multiplicative. The resulting derivation* $\mathcal{D}$ *is multiplicative too.*

Logistic derivations: *Here the context* Γ *encodes explicitly, in a node of a derivation, the assumptions on which the succedent formula depends.*

- *Every instance of* $\varphi : \varphi$ *is a derivation.*

- **Additive:** *If* $\Gamma, \Sigma_1 : \psi_1, \cdots, \Gamma, \Sigma_p : \psi_p$*, for some* $p \ge 1$*, are (logistically presented) derivations, and if*

$$\dfrac{\Gamma, \Sigma_1 : \psi_1 \quad \cdots \quad \Gamma, \Sigma_p : \psi_p}{\Gamma : \psi} \ (R_l)$$

is an instance of a (logistically presented) additive rule in $\mathcal{R}_{\mathcal{N}}$*, then*

$$\mathcal{D} \atop \Gamma : \psi \ =^{\mathrm{df.}} \quad \dfrac{\begin{array}{ccc} \mathcal{D}_1 & & \mathcal{D}_p \\ \Gamma, \Sigma_1 : \psi_1 & \cdots & \Gamma, \Sigma_p : \psi_p \end{array}}{\Gamma : \psi} \ (R_l)$$

---

[13] For simplicity, I am ignoring here mixed derivations where additive and multiplicative rules are applied intermittently.

is a logistically presented derivation.

$$\mathcal{D}_1 \qquad\qquad \mathcal{D}_p$$

Multiplicative: *If* $\Gamma_1, \Sigma_1 : \psi_1, \cdots, \Gamma_p, \Sigma_p : \psi_p$, *for some* $p \geq 1$, *are (logistically presented) derivations, and if*

$$\frac{\Gamma_1, \Sigma_1 : \psi_1 \quad \cdots \quad \Gamma_p, \Sigma_p : \psi_p}{\Gamma_1 \cdots \Gamma_n : \psi} \ (R_l)$$

is an instance of a (logistically presented) multiplicative rule in $\mathcal{R}_\mathcal{N}$, *then*

$$\frac{\mathcal{D}}{\Gamma_1 \cdots \Gamma_n : \psi} =^{\mathrm{df.}} \frac{\overset{\mathcal{D}_1}{\Gamma_1, \Sigma_1 : \psi_1} \quad \cdots \quad \overset{\mathcal{D}_p}{\Gamma_p, \Sigma_p : \psi_p}}{\Gamma_1 \cdots \Gamma_n : \psi} \ (R_l)$$

is a logistically presented derivation.
The derivations $\mathcal{D}_1, \cdots, \mathcal{D}_p$ *are the* direct subderivations *of* $\mathcal{D}$.

Note that assumptions are introduced into a logistic derivation via identity derivations and remain as such as long as not discharged by some application of a rule (this is further elaborated in Section 1.5.3.2).

In cases where structural rules are kept in $\mathcal{N}$ as explicit rules, the definition of derivation has to incorporate their application too.

There is an easy mutual conversion between the the two presentation modes:

simple to logistic: Convert each node in $\overset{\mathcal{D}}{\psi}$ to $\overset{\mathcal{D}}{\Gamma : \psi}$, where $\Gamma = \mathbf{d}_{\psi, \mathcal{D}}$.

logistic to simple: Convert each node in $\overset{\mathcal{D}}{\Gamma : \psi}$ to $\overset{\mathcal{D}}{\psi}$, setting $\mathbf{d}_{\psi, \mathcal{D}} = \Gamma$.

Thus, the logistic presentation keeps track *explicitly* of the assumptions Γ on which ψ depends.

Definition 1.5.3 (derivability). ψ *is* derivable *from* Γ *in* $\mathcal{N}$ *(or, better,* $\Gamma : \psi$ *is* derivable *in* $\mathcal{N}$*), denoted by* $\vdash_\mathcal{N} \Gamma : \psi$, *iff there exist a simple* $\mathcal{N}$*-derivation* $\overset{\mathcal{D}}{\psi}$ *with* $\mathbf{d}_{\psi, \mathcal{D}} = \Gamma$ *(respectively, a logistic derivation* $\Gamma : \overset{\mathcal{D}}{\psi}$*).*

Note that $\vdash_\mathcal{N} : \psi$ indicates derivability of ψ from an empty context, in which case ψ is referred to as a *(formal) theorem* of $\mathcal{N}$. In such a case, the colon is most often omitted.

As noted by von Plato [145], in the original development of *ND*-systems by Gentzen there was an extra rule[14] $\frac{\varphi}{\varphi}$ (T), concluding an assumption, that was later abandoned.

---

[14]'T' is derived from 'tautological'.

The explanation of von Plato of this rule is to make the derivation of $\varphi \to \varphi$ look less awkward. Below is a presentation of such derivations with and without (T).

$$\frac{\dfrac{[\varphi]_i}{\varphi}\ (T)}{\varphi \to \varphi}\ (\to I^i) \qquad \frac{[\varphi]_i}{\varphi \to \varphi}\ (\to I^i)$$

Derivations in an *ND*-system are depicted here as trees[15] (of formulas, or of sequents, according to the presentation mode), where a node and its descendants are an instance (of an application of) one of the rules.

An advantage of the logistic presentation over the simple one is, that the tree-structure is more transparent. Consider the following two presentations of the same derivation.

$$\frac{\dfrac{[\varphi]_1 \quad \varphi \to \psi}{\psi}\ (\to E) \qquad \psi \to \chi}{\dfrac{\chi}{\varphi \to \xi}\ (\to I^1)}\ (\to E)$$

$$\frac{\dfrac{\varphi : \varphi \quad \varphi \to \psi : \varphi \to \psi}{\varphi, \varphi \to \psi : \psi}\ (\to E) \qquad \psi \to \chi : \psi \to \chi}{\dfrac{\varphi, \varphi \to \psi, \psi \to \chi : \chi}{\varphi \to \psi, \psi \to \chi : \varphi \to \xi}\ (\to I)}\ (\to E)$$

The logistic derivation is an actual tree, with three leaves, while the simply-presented derivation is "tree-like" and has two leaves only (since $[\varphi]$ is discharged, and "is not there really").

For a discussion of the advantages of the tree-form structure of *ND*-derivations over linearly structured *ND*-derivations, see von Plato [145]. In simply presented derivations, the discharged occurrences of the assumptions are enclosed in square brackets, and marked, $[...]_i$, for some index i, to match i on the applied rule-name, and are leaves in the tree depicting the derivation. Importantly, any *instance* of an assumption-discharging rule has a *unique* index! Rules of an *ND*-system operate on the succedent of a sequent *only*. I also use $\mathcal{D} : \vdash_N \Gamma : \varphi$ to indicate a specific derivation of that sequent. In such a derivation, Γ is also referred to as the *open assumptions* of $\mathcal{D}$. If Γ is empty, the derivation is *closed*. Since the simple presentation is less notationally cluttered, it will be used in the sequel to display specific derivations. As indicated by von Plato [145], the tree format enjoys several advantages over other, linear, presentations.

- Open (undischarged) assumptions are the leaves of the tree, making the dependence of the conclusion on them more transparent.

- The proof of closure under composition of derivations is simpler.

- The order of some rule applications can be "hidden" and easily permuted.

---

[15]For a *linear* depiction of *ND*-derivations, see, for example, Fitch [38].

In the PTS programme, derivations have their own role, beyond the study of derivability.

1.5.3.2 An analysis of derivations

The following characteristics of the traditional view of derivations emerge when expressed in a temporally metaphoric way, where derivations are viewed as evolving in time. Both the placement of an assumption and the application of a rule are seen as "instantaneous" events.

Future of an assumption : When an assumption is made it is open at the moment of placement. It is not yet determined at that moment whether the assumption will remain open or will be discharged (closed) by a later application of a rule discharging it.

Moment of rule-application : The moment of application of (a specific occurrence) of a rule (R) in a $\mathcal{N}$-derivation $\mathcal{D}$ is the moment at which its dischargeable assumptions (if any) *are discharged* and thereby closed.

Thus, there is something misleading in the way derivations are depicted as trees with some of the leaves bracketed (and indexed). The decorating bracket should "appear" on a leaf only when (and if) the assumption corresponding to that leaf is discharged. The misleading notation is an abuse of notation for convenience of depiction, while the intention is clear.

I propose to replace those characteristics to the following ones, that will turn useful for the definition of canonical derivations from open assumptions in Section 1.5.5:

Future of an assumption : When an assumption is made, it is open at that moment, and

- either it is intended to remain open throughout the derivation
- or it is potentially closed, associated with a rule the application of which will close it during the derivation.

Thus, it is determined at that moment whether it will remain open or be closed by a later application of a rule discharging it. A good notation should mark this distinction by some three-way notation for marking assumptions. I will not suggest one here.

It is possible to view the difference between assumptions intended to remain open and potentially closed assumptions by reading them as follows.

- An assumption φ placed in the intention of remaining open can be read as "assume φ" without further qualifications. If one wants to consider truth, it is like "assume φ true"; or, "assume φ possessed" by a cognitive agent via some other means. In the database analogy, φ is a part of the database.

- On the other hand, an assumption placed as potentially closed by applying (R) can be read as "assume φ towards applying (R)". In the database analogy, this is a temporary extension of the database for the sake of some hypothetical reasoning, where the current database is restored once the hypothetical conclusion has been drawn.

Moment of rule-application : For a rule (R) in an $\mathcal{N}$-derivation $\mathcal{D}$ that discharges assumptions, *there is no moment of application of (a specific occurrence) of it.* Rather, the rule application has a *duration* elapsing from the moment at which its assumptions are *placed* to the moment it discharges those assumptions. The limit case of application of rules which do not discharge assumptions can be still viewed as an instantaneous event.

Returning to the recursive specification of derivations, the duration of the application of an assumption-discharging rule is captured in (1.5.11) by the encompassed sub-derivations $\mathcal{D}_1, \cdots, \mathcal{D}_p$. Attributing to an assumption-discharging rule the property of being 'the last rule applied' in $\mathcal{D}$ is nothing more than a convenient parlance enabling to maintain the tree-shape of $\mathcal{D}$; it just binds together the sub-derivations it encompasses.

A central problem regarding derivations is their identity conditions: when are two derivations the same? The definition above is somewhat unusual in that the names of the rules applied in each step are included as part of a derivation. This evades a certain problem arising when names of applied rules are *not* included. Anticipating the presentation of the proof-system for Intuitionistic Logic (in Section 2.1), consider the E-rules for conjunction (with contexts omitted):

$$\frac{\varphi \wedge \psi}{\varphi} \ (\wedge_1 E) \qquad \frac{\varphi \wedge \psi}{\psi} \ (\wedge_2 E)$$

Now, consider a very simple 'derivation' in which applied rule names are not indicated for $\begin{array}{c} \varphi \wedge \varphi \\ \mathcal{D} \\ \varphi \end{array}$. In this case, $\mathcal{D}$ is ambiguous between

$$\mathcal{D}_1 : \frac{\varphi \wedge \varphi}{\varphi} \ (\wedge_1 E) \quad \text{and} \quad \mathcal{D}_2 : \frac{\varphi \wedge \varphi}{\varphi} \ (\wedge_2 E)$$

As another example (from Došen [30]), consider the following two 'derivations' for $\begin{array}{c} \varphi \wedge (\varphi \rightarrow \varphi) \\ \mathcal{D} \\ \varphi \end{array}$. Again, $\mathcal{D}$ is ambiguous between

$$\mathcal{D}_1 : \frac{\varphi \wedge (\varphi \rightarrow \varphi)}{\varphi} \ (\wedge_1 E) \quad \text{and} \quad \mathcal{D}_2 : \frac{\dfrac{\varphi \wedge (\varphi \rightarrow \varphi)}{\varphi} \ (\wedge_1 E) \quad \dfrac{\varphi \wedge (\varphi \rightarrow \varphi)}{\varphi \rightarrow \varphi} \ (\wedge_2 E)}{\varphi} \ (\rightarrow E)$$

Identifying $\mathcal{D}_1$ and $\mathcal{D}_2$ in those examples, and omitting names of applied rules from derivations in general, renders derivations *extensionalized*. While this issue is tangential to the PTS programme, it illustrates a general problem in proof-theory. See Došen [30] for a more thorough discussion of this issue.

Definition 1.5.4 (root of a derivation). *For a derivation $\mathcal{D}$ of φ (from any Γ), its* root *is $\rho(\mathcal{D}) = \varphi$.*

The function ρ is naturally extended to sets of derivations Δ by letting $\rho(\Delta) = \{\rho(\mathcal{D}) \mid \mathcal{D} \in \Delta\}$. The function ρ is further extended to contextualized functions by putting, for a function $\mathcal{F}$ from contexts Γ to sets of derivations from Γ, $\rho(\mathcal{F}) = \bigcup_{\Gamma \in \mathcal{P}(L_{prop})} \rho(\mathcal{F}(\Gamma)))$.

Definition 1.5.5 (separability). *An* ND-*systen $\mathcal{N}$ is* separable *iff whenever $\vdash_{\mathcal{N}} \Gamma : \varphi$,*

$$\begin{array}{c}\Gamma \\ \mathcal{D} \\ \varphi\end{array}$$

there is a derivation $\begin{smallmatrix}\Gamma\\\mathcal{D}\\\varphi\end{smallmatrix}$ in which the only operative rules used are those governing the operators occurring in Γ or in φ.

The significance of this property is discussed in section 3.1.

An important property of *ND*-derivations is their *closure under composition*. This closure establishes the *transitivity* of *ND*-derivability, namely '$\vdash_{\mathcal{N}}$'.

Definition 1.5.6 (closure under derivation composition).

Simple derivations *: Let $\begin{smallmatrix}\varphi\\\mathcal{D}\\\psi\end{smallmatrix}$ be any simply presented $\mathcal{N}$-derivation of ψ from φ (and assumptions $\mathbf{d}_{\mathcal{D}}$, left implicit), and let $\begin{smallmatrix}\mathcal{D}'\\\varphi\end{smallmatrix}$ be any simply presented $\mathcal{N}$-derivation of φ (from assumptions $\mathbf{d}_{\mathcal{D}'}$, left implicit). Then $\mathcal{N}$ is closed under derivation composition iff the result of replacing every occurrence of (a leaf) φ in $\mathcal{D}$ by the sub-tree $\mathcal{D}'$ is also a derivation, denoted $\begin{smallmatrix}\mathcal{D}''\\\psi\end{smallmatrix} =^{\mathrm{df.}} \begin{smallmatrix}\mathcal{D}[\varphi := \begin{smallmatrix}\mathcal{D}'\\\varphi\end{smallmatrix}]\\\psi\end{smallmatrix}$, with $\mathbf{d}_{\mathcal{D}''} = (\mathbf{d}_{\mathcal{D}} - \{\varphi\}) \cup \mathbf{d}_{\mathcal{D}'}$.*

Logistic derivations: *Let $\begin{smallmatrix}\mathcal{D}\\\Gamma, \varphi : \psi\end{smallmatrix}$ be any logistically presented $\mathcal{N}$-derivation (of ψ from Γ, φ), and let $\begin{smallmatrix}\mathcal{D}'\\\Gamma' : \varphi\end{smallmatrix}$ be any logistically presented $\mathcal{N}$-derivation (of φ, from Γ'). Then $\mathcal{N}$ is closed under derivation composition iff the result of*

1. *replacing every occurrence of (a leaf) $\varphi : \varphi$ in $\mathcal{D}$ by the sub-tree $\begin{smallmatrix}\mathcal{D}'\\\Gamma' : \varphi\end{smallmatrix}$.*

2. *replacing the antecedent of the sequent labelling any node in $\mathcal{D}$, say $\overline{\Gamma}$, by the new antecedent $\overline{\Gamma} - \varphi, \Gamma'$ (leaving the succedent of the sequent intact).*

is also a derivation, denoted $\begin{smallmatrix}\mathcal{D}''\\\Gamma - \varphi, \Gamma' : \psi\end{smallmatrix} =^{\mathrm{df.}} \begin{smallmatrix}\mathcal{D}[\varphi : \varphi := \begin{smallmatrix}\mathcal{D}'\\\Gamma' : \varphi\end{smallmatrix}]\\\Gamma - \varphi, : \psi\end{smallmatrix}$ (of ψ, from $\Gamma - \varphi, \Gamma'$).

Thus, in $\mathcal{D}''$, any use in $\mathcal{D}$ of the assumption φ is replaced by re-deriving φ according to $\mathcal{D}'$ (from possibly additional assumptions, on which the conclusion of $\mathcal{D}''$ (which is the same as the conclusion of $\mathcal{D}$) now depends). This conclusion, however, depends no more on the assumption φ. Note that if the assumption φ is not actually used in $\mathcal{D}$,

$$\mathcal{D}[\varphi := \begin{matrix}\mathcal{D}' \\ \varphi\end{matrix}] \equiv \mathcal{D}$$

then $\quad\quad\quad \psi \quad\quad$ (and similarly for the logistic counterpart).

Example 1.5.1. *Suppose $\mathcal{D}$ is the following simply presented derivation for $\varphi, \varphi{\to}\psi$, $\varphi{\to}\chi \vdash \psi{\wedge}\chi$, and $\mathcal{D}'$ is the following simply presented derivation for $\xi, \xi{\to}\varphi \vdash \varphi$ (both in Intuitionistic Logic).*

$$\mathcal{D}: \quad \dfrac{\dfrac{\varphi \quad \varphi{\to}\psi}{\psi}\,(\to E) \quad \dfrac{\varphi \quad \varphi{\to}\chi}{\chi}\,(\to E)}{\psi{\wedge}\chi}\,(\wedge I) \qquad\qquad \mathcal{D}': \quad \dfrac{\xi \quad \xi{\to}\varphi}{\varphi}\,(\to E) \qquad (1.5.12)$$

Then,

$$\begin{matrix}\mathcal{D}[\varphi := \begin{matrix}\mathcal{D}' \\ \varphi\end{matrix}] \\ \varphi{\wedge}\chi\end{matrix} = \dfrac{\dfrac{\dfrac{\xi \quad \xi{\to}\varphi}{\varphi}\,(\to E) \quad \varphi{\to}\psi}{\psi}\,(\to E) \quad \dfrac{\dfrac{\xi \quad \xi{\to}\varphi}{\varphi}\,(\to E) \quad \varphi{\to}\chi}{\chi}\,(\to E)}{\psi{\wedge}\chi}\,(\wedge I)$$

$$(1.5.13)$$

Note that both *occurrences of φ in $\mathcal{D}$ where replaced during the substitution.*

Example 1.5.2. *The composition of the above derivations when logistically presented is given in Equations 1.5.14 and 1.5.15.*

The definitions are naturally extended to multiple simultaneous substitution $\mathcal{D}[\varphi_1 := \begin{matrix}\mathcal{D}'_1 \\ \varphi_1\end{matrix}, \cdots, \varphi_m := \begin{matrix}\mathcal{D}'_m \\ \varphi_m\end{matrix}]$, for any $m \geq 1$. The satisfaction of the closure under derivation composition needs to be shown for the specific *ND*-systems used for meaning-conferring, as it underlies establishing harmony (see below). Most often, it is proved by induction on the structure of derivations.

Note that there is difference between this property of an *ND*-system $\mathcal{N}$ and the property of *admissibility* in $\mathcal{N}$ (cf. Section 1.5.4) of the *cut*-rule

$$\dfrac{\Gamma_1 : \varphi \quad \Gamma_2, \varphi : \psi}{\Gamma_1 \Gamma_2 : \psi}\,(cut)$$

also reflecting the transitivity of $\mathcal{N}$-derivability. The admissibility of (cut) establishes a property at the level of sequents: any sequent $\mathcal{N}$-derivable using (cut) has a *direct* $\mathcal{N}$-derivation without the use of (cut), which usually differs in form. The closure property above is a property of derivations, not of sequents (see more about this issue in Negri and von Plato [130]).

As a simple example of the failure of the closure under derivation composition, suppose that the definition of a derivation is modified, by adding a requirement that Γ, the collection of leaves in $\mathcal{D}$, is consistent. In such a case, $\begin{matrix}\Gamma \\ \mathcal{D} \\ \varphi\end{matrix}$ may have a consistent Γ, as will $\begin{matrix}\Gamma' \\ \mathcal{D}' \\ \varphi\end{matrix}$; however, this does not ensure the consistency of Γ, Γ', the leaves of

$$\mathcal{D}: \quad \dfrac{\dfrac{\varphi\vdash\varphi \quad \varphi\to\psi:\varphi\to\psi}{\varphi,\varphi\to\psi:\psi}(\to E) \qquad \dfrac{\varphi:\varphi \quad \varphi\to\chi:\varphi\to\chi}{\varphi,\varphi\to\chi:\chi}(\to E)}{\varphi,\varphi\to\psi,\varphi\to\chi:\psi\wedge\chi}(\wedge I)$$

$$(1.5.14)$$

$$\mathcal{D}': \quad \dfrac{\xi:\xi \quad \xi\to\varphi:\xi\to\varphi}{\xi,\xi\to\varphi:\varphi}(\to E)$$

Then,

$$\mathcal{D}[\varphi:=\xi,\xi\to\varphi:\varphi]$$

$$= \dfrac{\dfrac{\dfrac{\xi:\xi \quad \xi\to\varphi:\xi\to\varphi}{\xi,\xi\to\varphi:\varphi}(\to E) \quad \varphi\to\psi:\varphi\to\psi}{\xi,\xi\to\varphi,\varphi\to\psi:\psi}(\to E) \qquad \dfrac{\dfrac{\xi:\xi \quad \xi\to\varphi:\xi\to\varphi}{\xi,\xi\to\varphi:\varphi}(\to E) \quad \varphi\to\chi:\varphi\to\chi}{\xi,\xi\to\varphi,\varphi\to\chi:\chi}(\to E)}{\xi,\xi\to\varphi,\varphi\to\psi,\varphi\to\chi:\psi\wedge\chi}(\wedge I)$$

$$(1.5.15)$$

$\mathcal{D}[\varphi := \mathcal{D}']$, whereby the latter fails to be a legal derivation, causing failure of closure under derivation composition. Other examples of non-closure under composition with "blind substitution" of trees are presented in Section 2.4.1, Proposition 2.4.3. (for Relevant Logic), and in Section 2.5, example 2.5.6 (for Modal Logic). For yet another example of an *ND*-system not closed under derivation composition see Wadler [229].

There is another kind of operation on derivations, *term substitution*, the closure under which is needed once variables and quantifiers are present in the object language. The definition is only formulated for simply-presented derivations, the logistic case being a simple variation.

Definition 1.5.7 (closure of derivations under term substitution). *For a term t (of*
$$\varphi(y)$$
$$\mathcal{D}$$
the object language) and an $\mathcal{N}$- derivation ψ *, where $y \in FV(\varphi)$ is*[16] *fresh for (the*
$$\varphi(t)$$
$$\mathcal{D}[y := t]$$
suppressed) Γ and for ψ, $\mathcal{N}$ is closed under term substitution iff ψ *, the result of replacing (in the usual capture-free way) every free occurrence of y in any formula (including an open assumption) in $\mathcal{D}$ by t, is a legal $\mathcal{N}$-derivation.*

Again, this closure has to be separately established for specific *ND*-systems for specific logics.

1.5.4 Derived and admissible rules

It may be convenient, for various purposes, to augment a given *ND*-system $\mathcal{N}$ with additional rules, provided they do not change the derivability relation '$\vdash_{\mathcal{N}}$'. There are two major such augmentations. Consider a rule[17] (R) (in its simple presentation)

$$\frac{\varphi_1 \quad \cdots \quad \varphi_n}{\psi} \ (R)$$

(1.5.16)

Derived rules: The rule (R) is *derived* (in $\mathcal{N}$) provided $\vdash_{\mathcal{N}} \varphi_1, \cdots, \varphi_n : \psi$. That is, there is a $\mathcal{N}$-derivation (i.e., without using (R)) of the conclusion ψ from the premises $\varphi_1, \cdots, \varphi_n$. Derived rules are very useful to *shorten* $\mathcal{N}$-derivations. It preserves the original $\mathcal{N}$-derivability because in any derivation using R, any use of R can be replaced by its derivation.

Example 1.5.3. *Anticipating the presentation of Intuitionistic Logic in Section 2.1, with the following I/E-rules for implication ('→') (in simple presentation)*

---

[16]It is convenient to assume Barendregt's *variable convention* ([4], p. 26), by which no variable occurs both free and bound in the same formula.

[17]For simplicity, a non-discharging rule is shown, but the definition below applies to discharging rules as well.

$$\frac{\overset{[\varphi]_i}{\underset{\psi}{\vdots}}}{(\varphi \to \psi)} \ (\to I^i) \qquad \frac{\varphi \quad \varphi \to \psi}{\psi} \ (\to E)$$

consider the following rule ($\to^*$), embodying the transitivity of implication in Intuitionistic Logic.

$$\frac{\varphi \to \psi \quad \psi \to \chi}{\varphi \to \chi} \ (\to^*)$$

To show that '($\to^*$) is derivable in that logic, it suffices to show a derivation.

$$\frac{\dfrac{[\varphi]_i \quad \varphi \to \psi}{\psi} \ (\to E) \qquad \psi \to \chi}{\dfrac{\chi}{\varphi \to \chi} \ (\to I^i)} \ (\to E)$$

Example 1.5.4. *Anticipating the presentation of Classical Logic in Section 2.2, with its following alternative E-rules for negation ('$\neg$') (in simple presentation)*

$$\frac{\overset{[\neg\varphi]_i}{\underset{\bot}{\vdots}}}{\varphi} \ (RAA^i) \qquad \frac{\neg\neg\varphi}{\varphi} \ (DNE)$$

the following example shows an hypothetical derived rule that discharges an assumption by deriving (RAA) using (DNE).

$$\frac{\dfrac{\overset{[\neg\varphi]_i}{\underset{\bot}{\vdots}}}{\neg\neg\varphi} \ (RAA^i)}{\varphi} \ (DNE)$$

Note that a derived (R) *is not* one of the I/E-rules, just an auxiliary rule.

Admissible rules: The rule (R) is *admissible* (in $\mathcal{N}$) whenever $\vdash_{\mathcal{N}\cup(R)} \Gamma : \varphi$ implies $\vdash_{\mathcal{N}} \Gamma : \varphi$. That is, for any derivation using (R), there is an equivalent derivation (of the same conclusion from the same assumptions) *without* using (R).

Note that there need not be a derivation of the conclusion of (R) from its premises. Admissibility of a rule means that whenever each premise of the rule is itself derivable, so is the conclusion of the rule. The way to establish admissibility is by some *global* consideration, usually transforming derivations with (R) to derivations without (R). Examples will be presented when the need arises.

1.5.5 Canonicity of derivations

There is a family of derivations that plays a central role in the PTS programme, as being the vehicle through which meaning is conferred by the operational rules, called *canonical derivations*, defined below. In the literature, canonicity is seen as a property of *proofs* (derivation from an empty context).

Definition 1.5.8 (canonical proof). *A proof $\mathcal{D}$ (in $\mathcal{N}$) is canonical iff it ends with an application of an I-rule.*

In a canonical proof of (a compound) φ from Γ, the conclusion φ is the conclusion of an application of an I-rule of the main operator of φ. This is viewed as the *most direct* (though not always the shortest) way to derive φ, a derivation *according to the meaning of its principal operator*. I extend canonicity here, following Francez [49], from proofs to arbitrary derivations from open assumptions, preserving the idea of being the most direct conclusion, but adapted to the presence of open assumptions.

I distinguish between two kinds of canonicity, I-canonicity and E-canonicity, serving the I-view and E-view of PTS (as explained in detail in Section 1.6.1), respectively. Usually, when 'canonicity' is used in the literature on PTS, I-canonicity is meant.

There is natural temptation to define an I-canonical derivation from open assumptions by simply adopting the same condition as that of canonical proofs, namely A derivation from open assumptions $\mathcal{D}$ (in $\mathcal{N}$) is I-*canonical* iff it ends with an application of an I-rule. As will be argued in Section 5.2, this definition would be too restrictive in terms of the reified meaning it induces. A more adequate definition is the following.

Definition 1.5.9 (canonical derivations from open assumptions). *Let ψ be a compound sentence.*

- *A $\mathcal{N}$-derivation $\mathcal{D}$ for $\Gamma : \psi$ is I-canonical iff it satisfies one of the following two conditions.*

    - *The last rule applied in $\mathcal{D}$ is an I-rule (for the main operator of ψ).*
    - *The last rule applied in $\mathcal{D}$ is an assumption-discharging E-rule, the major premise of which is some φ in Γ, and its encompassed sub-derivations $\mathcal{D}_1, \cdots, \mathcal{D}_n$ are all canonical derivations of ψ.*

    Denote by $\vdash_{\mathcal{N}}^{I_c}$ I-canonical derivability in $\mathcal{N}$.

- *A derivation $\mathcal{D}$ (in $\mathcal{N}$) for $\Gamma, \varphi : \psi$ (for any ψ) is E-canonical (w.r.t a compound φ) iff it starts with an application of an E-rule (with φ the major premise).*

    Denote by $\vdash_{\mathcal{N}}^{E_c}$ E-canonical derivability in $\mathcal{N}$.

For Γ empty, the definition of I-canonicity reduces to the above definition of a canonical proof. Note the recursion involved in this definition. The important observation regarding this recursion is that it always terminates via the first clause, namely by an application of an I-rule. In particular, the attribute of canonical proofs that there are none for an *atomic* sentence is preserved by the definition of canonical derivations from open assumptions too (but, see Section 1.6.3). Henceforth, I will refer to such derivations as *essentially-ending* with an application of an I-rule.

The observation mentioned above justifies the definition of canonicity for derivations from open assumptions Γ by preserving the *directness* of the inference, but the conclusion ψ, inferred via an I-rule, needs to be *propagated* through application of E-rules to open assumptions in Γ. This also justifies viewing Γ as grounds for assertion of ψ. The significance of the second clause in the definition is exemplified in Section 5.2.

An important observation is that a derivation of a compound φ from itself by means of the *identity axiom* in every ND-system, namely $\vdash\varphi : \varphi$ (or more generally $\vdash\Gamma, \varphi : \varphi$), *is not* canonical, and will not count as deriving φ according to its meaning once the definition of meaning is given.

In an E-canonical derivation (say of ψ) from φ, φ is the major premise of an application of an E-rule of the main operator of φ. This is viewed as the most direct conclusion of φ, according to the meaning of its principal operator. Note that the application of the E-rule to φ in an E-canonical derivation from φ need not be the first rule application in the derivation. It has only to be the first rule *applied to* φ, the latter being a major premise. This can be seen from the following example.

Example 1.5.5. *Anticipating the E-rule for implication, the traditional modus ponens, consider the following derivation, which is E-canonical w.r.t.* $(\varphi_1{\rightarrow}\varphi_3){\rightarrow}\chi$.

$$\cfrac{\cfrac{\cfrac{\cfrac{[\varphi_1]_1 \quad \varphi_1{\rightarrow}\varphi_2}{\varphi_2}\ (\rightarrow E) \quad \varphi_2{\rightarrow}\varphi_3}{\cfrac{\varphi_3}{\varphi_1{\rightarrow}\varphi_3}\ (\rightarrow I^1)}\ (\rightarrow E) \quad (\varphi_1{\rightarrow}\varphi_3){\rightarrow}\chi}{\chi}\ (\rightarrow E) \quad \chi{\rightarrow}\psi}{\psi}\ (\rightarrow E)$$

$$(1.5.17)$$

The role of canonicity in the determination of proof-theoretic meanings is further discussed and exemplified in Section 5.2, page 184.

It is important to note the following two attributes of canonical derivations of a compound sentence from open assumptions:

- A sub-derivations of an I-canonical derivation according to the first clause of the definition need *not* be canonical itself. Only the second clause of the definition requires canonicity of (encompassed) sub-derivations.

- A sub-derivations of an E-canonical derivation also need *not* be canonical itself.

These observations plays an important role in the reification of meaning, and are discussed further in Chapter 5.

In Section 1.6.2, after identifying the role of the structural rules in conferring meaning, a more general notion of canonicity is defined (in Definition 1.6.10).

Canonical derivations figure centrally in the definition of (reified) meanings. I say more about the role of canonicity of derivations in PTS in Section 5.2.

The distinction between two kinds of canonicity of derivations reflects two aspects of 'use' embodied in *ND*-rules, related to two speech-acts.

assertion: φ can be *asserted* if there is context Γ, a *warrant for assertion*, and an *I*-canonical derivation of φ from Γ.

concluding: ψ can be *concluded* from φ if there is a context Γ, a *warrant for conclusion*, and an *E*-canonical derivation w.r.t. φ, of ψ from Γ, φ.

The following notation for (possibly empty) collections of derivations in an *ND*-system $\mathcal{N}$ (understood from the context) will be used in the sequel. Let φ be any formula in the object language, and Γ a context (for that language).

- $[[\varphi]]_\Gamma^*$: the collection of all derivations (if any) of φ from Γ.

- $[[\varphi]]_\Gamma^{Ic}$: the collection of all *I*-canonical derivations (if any) of φ from Γ.

- $[[\varphi]]_\Gamma^{Ec}$: the collection of all *E*-canonical derivations (if any) from φ, Γ.

1.5.6   The Fundamental Assumption

Dummett [35] requires the following property of *ND*-systems, called *the fundamental assumption (FA)*, to always[18] hold for every meaning-conferring $\mathcal{N}$.
FA: (in my own notation) For every complex formula φ in the object language, if $[[\varphi]]_\varnothing^* \neq \varnothing$, then also $[[\varphi]]_\varnothing^{Ic} \neq \varnothing$.
Namely, if there is *any* proof of φ, there is also an *I*-canonical proof of φ. Thus, Dummett turns a mathematical result about formal systems, namely *normalizability* (cf. Section 3.6.1.2), known to hold, for example, for the *ND*-system NJ (for Intuitionistic Logic (Section 2.1)), into a *semantic condition*. For example, for conjunction this amounts to the following.

$$
\begin{array}{c}
\mathcal{D} \\
\varphi \wedge \psi
\end{array}
\Longrightarrow
\begin{array}{c}
\mathcal{D}_1 \quad \mathcal{D}_2 \\
\varphi \quad \psi \\
\hline
\varphi \wedge \psi
\end{array}
(\wedge I)
\tag{1.5.18}
$$

(for some $\mathcal{D}_1, \mathcal{D}_2$). In general, the sub-derivations may use extra assumptions, if those are discharged by the application of the *I*-rule.

The Fundamental Assumption forms the basis of explaining *any* use of some logical constant by means of a meaning-constitutive use of that constant. See Moriconi [124]

---

[18] Dummett himself did doubt the general holding of **FA**, e.g., in [35] (Ch. 12).

and Kürbis [96] for detailed critical discussions of the role of **FA** within the PTS programme. As emphasized by Kürbis, The imposition of FA is biased to start with in favor of anti-realism and Intuitionistic Logic as is evident by its application to disjunction. Such an application, requiring any proof of $\varphi \vee \psi$ to end with an application of an I-rule, requires the availability of a proof for φ or a proof of ψ.

Some caution is needed here, not to apply **FA** to assumptions with free variables (open formula assumptions). Otherwise, wrong conclusions may be reached. To see that, suppose $\forall x.\varphi(x) \vee \psi(x)$ has been derived using $\forall I$:

$$\frac{\Gamma : \varphi(x) \vee \psi(x)}{\Gamma : \forall x.\varphi(x) \vee \psi(x)} \; (\forall I) \text{, } x \text{ free in } \Gamma$$

Thus, the premise $\varphi(y) \vee \psi(y)$ has beed derived. If we now assume, according to **FA**, that this premise was derived by means of $(\vee I)$, than we had derivations either for $\varphi(y)$ or for $\psi(y)$. Hence $\forall x(\varphi(x)) \vee \forall x(\psi(x))$ would be (wrongly!) derivable.

Note, however, that in contrast to proofs (i.e., derivations from an empty context), derivations (from arbitrary contexts) cannot be assumed in general to satisfy the analogue of **FA**, if the latter is taken only as the existence of a I-canonical derivation from the same assumptions, interpreted only as the first clause in Definition 1.5.9 (pace Dummett [32], p. 254). For example, in Intuitionistic (propositional) Logic (see Section 2.1), certainly $\vdash \varphi, (\varphi \rightarrow \psi) : \psi$, but $\nvdash^{Ic} \varphi, (\varphi \rightarrow \psi) : \psi$. Also, $\vdash \varphi \vee \psi : \varphi \vee \psi$ (an axiom); but to get $\vdash^{Ic} \varphi \vee \psi : \varphi \vee \psi$, one of $\vdash \varphi \vee \psi : \varphi$, $\vdash \varphi \vee \psi : \psi$ needs to hold, but neither does hold in Intuitionistic Logic. This was also observed by Kürbis [96], who concludes that **FA** should not be imposed on derivations from open assumptions at all. However, the full definition of canonical derivation from open assumptions (Definition 1.5.9) does extend the scope of applicability of **FA**.

1.5.7   Complexity and Non-Circularity

Since the I-rules are meaning-conferring, they must satisfy a condition that will prevent a *circularity* in this meaning determination. Otherwise, they will not be capable of forming an explication of the cognitive process of understanding (or grasping) the determined meanings. Thus, the premises of an I-rule (including the discharged ones) have to be of *lower* complexity than that of its conclusion. Traditionally[19], by complexity it is understood the *syntactic complexity*, depending on depth of the recursive structure of the formula. Note that this *does not* mean that in an I-rule for some operator, say $(*I)$, the premises cannot contain '$*$'; they certainly can, provided such an occurrence of '$*$' forms a formula of lower complexity than the conclusion. In other words, any such apparent circularity should be *recursive*, leading to some atomic formulas in a finite number[20] of steps. The danger of apparent circularity can be clearly

---

[19] An exception to this view of complexity with regards to negation is discussed in sub-section 4.5.4.

[20] I ignore here the issue of *tractability* of this process. This aspect of PTS has not been investigated in the literature.

seen by inspecting the rule $(\to I)$ for forming an implication (see Section 2.1).

$$\frac{\Gamma, \varphi : \psi}{\Gamma : \varphi \to \psi} \ (\to I)$$

(1.5.19)

This rule *determines* the meaning of implication by alluding to *all* derivations of ψ from the augmentations of the context Γ by the assumption φ (discharged by this rule application); however, this collection of derivations in the premise may well contain other applications of $(\to I)$, even though this rule is added to some underlying system that does not contain implication yet. So, it is crucial for the circularity not being vicious that any such 'inner' appeal to the rule is with a conclusion lower in complexity than that of $\varphi \to \psi$. The avoidance of (vicious) circularity and the wish to have premises of meaning-conferring rules with lower complexity than their conclusion underlies the appeal to canonical derivations as the basis of proof-theoretic meaning.

1.6   Using Natural-Deduction for Conferring Meaning

In this section, I summarize the role *ND*-systems play within PTS as a theory of meaning. Recall that *ND*-systems are conceived as reflecting the '*use*' of whatever needs its meaning to be defined. The notion of 'use' intended to be captured by *ND*-systems has two layers, both reflecting the actual inferential practices.

- The practice of deduction itself: this layer of use is captured by the structural rules (with their schematism of presentation, Section 1.6.4). Those rule define '⊢' itself, before any specific operator is considered. This layer is the one that is more closely associated with the answerability principle.

- The practice of use of specific operators: this layer of 'use' is captured by the *ND* operational rules. In case the operator is an abstraction of (one facet of) an NL particle, its operative rules will reflect, in some sense, this use. But if the operator does not originate in NL, the operational rules prescribe its intended inferential use. For example, Sheffer's stroke (known also as *nand*), expressing falsity of at least one of its arguments, is not part of any NL, and has no 'natural use' to be captured by its rules. See, for example, Read [163].

As emphasized in Došen [30], *ND*-systems are a formalization of the informal notion of *deduction*, which is what gets to be 'used' in our inferential practices. In logic, what is being defined are logical constants (part I). In natural language (part II), the meanings of more complicated expressions are defined. In the latter case, there is always a "natural use" to be captured by operational rules.

In much of the literature on PTS, the scope of this approach is confined to logical constants. An immediate question arising is, *what is a logical constant?* For an early discussion of this question, see Peacocke [137]. See also Hacking [69]. For a more recent approach, see Došen [28], and for a survey see Gómez-Torrenta [68]. I will not concern myself with this question. According to my approach, PTS is applicable to any *compound* expression, whether logical or not (in particular, NL expressions) for which deduction rules can be formulated, endowing it with a proof-theoretic meaning relative to some primitive expressions.

1.6.1   Views of PTS

There are several major views of PTS as based on *ND*-systems. Each of them induces a different notion of reified sentential meaning, as discussed in Section 5.2. The classification was first presented in Francez [46].

Inferentialism: According to the inferentialism view of PTS, referred to also as the *I*-view, the meaning of a sentence containing a proof-theoretically defined expression as its main operator is characterized as follows.

- *I*-rules (for the defined expression) are *self-justifying*, and *E*-rules are justified by the *I*-rules by means of a *justification procedure* (explained in Section 3.4). This approach to PTS is championed by (in addition to Gentzen, implicitely) Prawitz (e.g., [152]), Tennant (e.g., [215]), Read (e.g., [164]) and others.
- The meaning of a sentence with the defined expression as its main operator is based on its *I*-canonical derivations from its grounds for warranted assertion.

Pragmatism: According to the pragmatism view of PTS, referred to also as the *E*-view, the meaning of a sentence with a proof-theoretically defined expression as its main operator is characterized as follows.

- *E*-rules (for the defined expression) are *self-justifying*, and *I*-rules are justified by the *E*-rules by means of a *justification procedure*.
- The meaning of a sentence with the defined expression as its main operator is based on its *E*-canonical derivations to its warranted consequences.

  Dummett, in [35], mentions both of the above two views, but relates mainly to the inferentialism view. I refer to those two approaches as the *I*-view and the *E*-view of PTS. Generalizing the terminology of Tennant [214] (p. 94), according to the *I*-view the *I*-rules (for some operator, say '***') are *constitutive* of the meaning of '***', while the *E*-rules are *explicative* of this meaning. On the *E*-view, the roles of the *I/E* rules are reversed. As emphasized by Humberstone ([80], Section 4.14), when a constitutive collection of rules is involved in determining the meaning of an operator, not only *all* rules in that collection take part in meaning determination, but *only* rules in that collection do so. This observation is embodied in the notions of canonicity (cf. Section 1.5.5), where *I*-canonical derivation may essentially end with *any I*-rule for the principal operator, and *only* in such a rule, and, mutatis mutandis, for *E*-canonicity.

Combined view: There is also a less common view (e.g., in Tennant [219]), that might be called the *combined* view (*C*-view), that regards both *I*-rules and *E*-rules *jointly* conferring meaning, no group needing a justification by the other. De Queiroz [26, 27] can also be seen as supporting the *C*-view. For a variant of the *C*-view that also includes the reduction/expansion transformations (discussed below), *in addition* to the *I/E*-rules, as determining meaning, see Moriconi [124]. I shall not pursue this view further here.

Non-uniform view: What is common to the three views above is that all operators are defined[21] in a *uniform* way, where the three views differ on the preferred way. There is also a *non-uniform*, hybrid, also less common approach, letting each operator determine *individually* which rules for it are self-justifying, and which other rules are justified by this choice. In general, it is not clear what *objective* basis can underline such a determination. One such view is presented by Milne in [114], where there is seemingly a preference to take as self-justifying rules that express an *immediate inference*, where there is no dependence of the rule on the way each premise was derived. Another such view is presented by Kürbis [98], based on the number of *I*-rules or *E*-rules a constant has. The *I*-view is attributed to constants having exactly one *I*-rule, while the *E*-view is attributed to constants having exactly one *E*-rule. It is stipulated that each constant is of one of those two types. More on the justifications later on in Section 3.4.

None of the above views, however, needs to appeal to model-theoretic entities for their justification. In particular, the usual view of soundness and completeness w.r.t. some model-theoretic definition, *is not* taken as a basis for justifying rules.

In [19], Contu rejects both of the first two views, with attributing a certain advantage to Pragmatism (*E*-view), lacking a circularity he finds in the former. He favors the combined approach, argued for by the non-eliminability of either the *I*-rules or the *E*-rules. I return to this criticism in Section 5.4.

As noted by Kürbis [96], *FA* (the fundamental assumption) fits only the *I*-view; no analog requirement fitting the *E*-view has been proposed (though [96] hints toward such a condition).

1.6.2   The role of structural rules in PTS

However, neither of the views of PTS defines meaning "from scratch". According to all those views, the definition is always on the basis of some *background*. This background is that of the *structural rules*, that, implicitly, take part in the proof-theoretic definition of meaning. They determine the meaning of the *derivability relation* itself, before any specific operator of the object-language is considered. They embody the

---

[21]MTS also treats all operators uniformly, where truth-conditions is the common way to define meaning for all operators.

part of 'use' that is the inferential practice itself, before any use of a specific operator is accounted for. The dependence of PTS on this background was hinted to by Nieuwendijk (in [131], pp. 136-7) and was explicitly observed in Hjortland [76] (p. 138). For a more recent discussion of the role of the structural rules in the PTS programme, see Hjortland [77]. While Gentzen, who was interested only in Intuitionistic Logic and Classical Logic, did not relate to any variation on the structural rules, later developments did. For example, additional structural rules are considered in Section 4.4, in the context of Bilateralism.

One particularly important consequence of this view that structural rules participate in meaning-conferring emerges as the need to revise the notion of canonicity of derivations (see Section 1.5.5). Recall that a canonical derivation (under both kinds of canonicity considered) is a derivation *according to meaning*! So, if structural rules participate in conferring meaning, their application should also be considered canonical. This observation is due to Murzi [125].

Definition 1.6.10 (structural canonicity). *A derivation $\mathcal{D}$ in an* ND-*system $\mathcal{N}$ is* structurally canonical (*S-canonical*) *iff the last rule applied in $\mathcal{D}$ is either essentially an I-rule or an S_i structural rule.*

Thus, structural canonicity (according to the I-view) includes the structural rules among the meaning-constitutive rules. During most of the current presentation, only the standard (Gentzen's) structural rules (or a subset thereof) are assumed (mostly, as admissible, not necessary primitive rules). So, canonicity can be maintained in its narrower sense, as defined in Section 1.5.5. A notable example resorting to structural canonicity is when Bilateralism is used for justifying Classical Logic, *adding* structural rules (see Section 4.4.3). This definition has also a direct impact notion of harmony (cf. Section 3.6.1.3) and on the fundamental assumption (**FA**), extending its role beyond the limit of Dummett's view of it.

I will point out the effect of structural rules whenever it emerges.

An even more radical view of the elements of an *ND*-system taking part in the proof-theoretic definition of meaning is presented in Moriconi [124] (as mentioned in the discussion of the *C*-view, p. 42).

1.6.3 The reification of meaning

This section is based on Francez [46]. As stated in the introduction, ever since Gentzen's casual remark, *I*-rules are taken, according to the PTS inferentialism view, to *determine* the meanings of the logical constants, but they do not *constitute* those meanings by providing some *explicit* definition of the determined meanings. Similarly for the *E*-based pragmatism PTS. Thus, the general conception in the inferentialism literature is that PTS provides an *implicit definition* of the meanings of the logical constants. Let me note that one can approach (and either support or oppose) PTS from (at least) two different points of view.

Philosophical: From this point of view (probably held by most of its proponents and opponents), PTS is rooted in the philosophical argument between realism and anti-realism concerning the possibility of verification-transcendent truth. As a consequence, much of the deliberation regarding PTS concerns the rivalry between Classical Logic and intuitionistic logic, the latter conceived as better amenable to a proof-theoretic justification (see Chapter 3). An implicit definition of meaning certainly meets the needs of such a philosophical goal.

Computational: From this point of view (which is my own point of view), the main issue is not necessarily related to any philosophical stance, or to any rivalry between specific logics, but to the use of meaning in *grammars*, tools for linguistic specification (of either formal languages like those employed by logic or of natural languages). To meet this end, PTS needs to provide an *explicit definition* of meanings; the latter do not "hang in the air" but reside (as abstract entities) within a grammar. I refer to such meanings, explicitly defined, as '*reified meanings*'. Those take the form of a *proof-theoretic semantic value* as determined by the I-rules or the E-rules of the meaning-conferring ND-system; something of the form

$$[[*]] = \cdots \tag{1.6.20}$$

where '$*$' is any operator the meaning of which is proof-theoretically determined. There are several reasons why such a definition of a reified meaning is needed. For example:

- facilitating a discussion of the compositionality of the meaning of formulas having the defined operator (e.g., a logical constant or an NL expression such as a determiner) as their main operator as determined by the rules at hand.

- designing a *lexicon* for a grammar (specifying both syntax and semantics) for a (logical or natural) language of which the defined expression is a part, for computational purposes. This fits smoothly the approach of Type-Logical grammar (TLG) (see Moortgat [123]), my preferred tool for grammar specification.

- allowing a definition of synonymity (sameness of meaning) and translation from one language to another (more important for natural language, though).

- in the case of NL sentences, having their semantic value propositionalize sentential meanings, to serve as arguments for *propositional attitudes*.

But what exactly is explicitly defined as the result of the determination of meaning via the meaning-conferring rules?

In spite of the vast literature on the subject of PTS, there is hardly an explicit definition for a *reification* of proof-theoretic meaning. The issue was hinted at already in Dummett [35]:

The meaning of the logical constants cannot [...] *consist* in their role in deductive inference: they must have meanings of a more general kind whereby they contribute to the meanings of sentences containing them just as other words do.

According to one prevailing view of what is being determined by the meaning-conferring rules (e.g., Engel [37], Hjortland [76], Garson [58, 59]) PTS is viewed as the ability to reconstruct (or recover) the truth-tables (also for multi-valued logics) from the *ND*-rules. Thus, we are back to truth in a model as the meaning determined – actually back to MTS. I consider this approach as defeating the purpose of PTS and therefore to be rejected by PTS proponents, and I am not going to relate to it any further.

Rather, I present an alternative answer to the question what *does* constitute those reified meanings. In Francez and Ben-Avi [51] and Francez, Dyckhoff and Ben-Avi [55], such a definition is provided (described in detail in subsequent chapters), for two strata of meaning. A central claim there, adopted here too (as already described in Section 1.1.1), is that it is the explicit definition of *sentential* meaning of compound sentences that is determined directly by the *ND*-rules; the meanings of sub-sentential phrases, in particular of the logical operators themselves are *derived* from the sentential meanings of sentences formed using those operators. This contrasts the usual MTS compositional approach (especially as manifested in TLG tradition) where lexical meanings are assumed to be given, and sentential meanings are generated compositionally from those word meanings. Note that in spite of viewing meanings as reified, seemingly being denotations (a feature usually attributed to MTS), this view *does not* contradict PTS! The denotations (i.e., semantic values) are proof-theoretic objects (functions into collections of canonical derivations), as elaborated in the appropriate chapters to follow, and there is no notion of satisfaction in a model, the main characteristic of MTS. The main line of presentation follows the *I*-based inferentialism PTS, with occasional indications about the other approaches.

The *reification* of meaning, referred to as a (proof-theoretic) *semantic value*, is obtained by adhering to the following principles.

Sentential meanings: There is a distinction here between atomic and compound sentences.

    Atomic sentences: Meanings of atomic sentences are usually viewed in PTS as *given* (possibly from outside the meaning-conferring *ND*-system), and the meaning of compound sentences are obtained *relatively* to the given meanings of atomic sentences. In order to overcome the indeterminacy arising from this relativization, I will adhere to the following view:

      Atomic canonicity: For an atomic sentence p, the axiom $p : p$ is viewed as a canonical derivation.

      Thereby, more uniformity in the presentation is achieved, and the sentential atomic meaning falls under the same definition as that for compound sentences, below. This view[22] plays a role in the definition of proof-theoretic consequence in Section 5.2.2.

Compound sentences: For compound sentences, sentential meanings are defined as the (contextualised) *collection of canonical derivations* (the canonicity depending on the chosen view of PTS). This is very much in the spirit of the modern approach "propositions as types" (for example, Martin-Löf [105]), the inhabitants of a type being the the proofs[23] of the proposition. For logic, this is presented in Chapter 5, and for natural language – in Chapter 8.

Subsentential meanings: Subsentential meanings, and in particular the meanings of the expressions with which the I/E-rules are associated, are defined by certain *abstractions* over the sentential meanings. For logic, this is presented in detail in Chapter 5, and for natural language in Chapter 9.

A possible reification of sentential meanings in MTS is via the *satisfaction* (of a formula, in a model) denoted by $\mathcal{M} \vDash \varphi$. This leads to the definition

$$[[\varphi]] = \{\mathcal{M} \mid \mathcal{M} \vDash \varphi\}$$

Namely, the (reified) meaning of a sentence is the collection of all its models (of the appropriate kind). This definition, however, does not lend itself to a natural generalization of meanings of sub-sentential phrases, which are of crucial importance for grammars. For example, in 1st-order logic, we do not have anything to qualify as the meaning of the universal quantifier, namely $[[\forall]]$. In Chapter 5, PTS is shown to yield such reified subsentential meanings.

A methodological principle manifested by the way meanings are reified, based on canonical derivations, is the precedence in the order of explanation, of deduction over assertion. By this view, the assertion (of a compound sentence) is justified only if a certain deduction can take place. This theme is repeated continuously throughout the presentation.

1.6.4   Schematism

While the operative I/E-rules of an *ND*-system are applied in the context of a fully-specified object language, when viewed as meaning-conferring, the rules are viewed *fully schematic*, applicable to *any* object language containing the operators explicitly mentioned in the rule. For that to serve its purpose, the following *generality* properties (cf. Humberstone [80]) should hold, in particular when the *ND*-system is to be used for conferring meaning, as described below:

---

[22]This idea was suggested to me by Andreas Fjellstad.

[23]Actually, the proof-objects; however, I do not allude here to *proof-terms*, mainly because they are not well-developed for NL sentences.

formula generality: Formulas should appear in their most general form allowed by L. For example, in the conclusion, if φ has as its principal operator a binary operator, say '$*$', then φ should be presented as '$\psi * \chi$', allowing ψ and χ to differ, and not as, say, $\psi * \psi$, forcing the the two sub-formulas to be the same.

context generality: Contexts variables should be present in every premise and conclusion, ranging, as mentioned above, over arbitrary object language contexts.

Consider, for example, intuitionistic implication (forming *conditionals*). Anticipating the development in the sequel, its I/E-rules are the following.

$$\frac{\Gamma, \varphi : \psi}{\Gamma : \varphi \rightarrow \psi} \ (\rightarrow I) \qquad \frac{\Gamma : \varphi \rightarrow \psi \quad \Gamma : \varphi}{\Gamma : \psi} \ (\rightarrow E) \qquad\qquad (1.6.21)$$

The schematism of the rules, when viewed as meaning-conferring, means that the formulas φ and ψ are viewed as formulas of *any* object language containing implication. Similarly, the context Γ spans over collections of formulas in *any* object language. Consequently, implication, and in general *any proof-theoretically defined* expression obtains its meaning independently of any specific object language including that expression. This property is also known as the *molecularity* of the theory of meaning. However, in order to facilitate such a schematic presentation, the meaning-conferring rules *have to be pure*! The issue of purity of the *ND*-rules is further elaborated upon in Section 3.1.

Violation of the context-generality criterion and its effect are further considered in Section 3.6.2.1 in connection with quantum disjunction, and in Section 3.7.5 in connection with Modal Logic.

1.6.5   Summary

This chapter introduced the idea of PTS, delineated it very grossly, and presented the basic machinery needed for its development. The main technical device used for meaning-conferring is (single conclusion) natural-deduction proof systems. Several criteria were specified to render such *ND*-systems qualified for meeting the desired task: conferring proof-theoretically determined meaning. As the book unfolds, those detailed will be refined and explained. In particular, as we will see in Chapter 3, *ND*-systems will be shown to be in need of meeting certain additional qualifications in order to "deserve" being viewed as meaning-conferring.

In this book, I am only interested in the theory of meaning, both for logic and for natural language. For logic, much of the actual logical theory, studying the properties of the traditionally defined operators, can be found in Humberstone [80], as well as in many standard textbooks. I will only consider those properties of logic and NL needed for the task of PTS.

Part I

Proof-Theoretic Semantics in Logic

Chapter 2

A variety of logics

This chapter is dedicated to the presentation of several logics, for which a PTS will be defined, and which will be evaluated according to the criteria set forth in Chapter 3. For each logic considered, only its object language and an *ND*-system, a candidate for serving as the meaning-conferring *ND*-system, are presented. Readers familiar with all of these logics may skip this chapter and continue with Chapter 3.

2.1 Intuitionistic Logic

2.1.1 Introduction

In this section, I present the (first-order) Intuitionistic logic (IL), that is considered by many PTS adherents as "well-behaved". For an introductory exposition of IL, with many suggestions for further reading, see van Dalen [226].

The standard natural-deduction proof-system for IL is Gentzen's *NJ* [61], presented below, in stages. First, *minimal logic* is considered (Johansson [84]), and then *falsum* and negation are added.

2.1.2 Propositional minimal logic

I consider an object language $L_{propmin}$ (for propositional minimal logic), containing the following operators: *implication*: $\rightarrow$, *conjunction*: $\wedge$, *disjunction*: $\vee$. *formulas*, ranged over by metavariables φ, ψ (possibly subscripted), are defined[1] recursively, starting with a countable collection P of *propositional variables* (also called atomic propositions – I will use the two terms synonymously), ranged over by metavariables p, q (also possibly subscripted).

$$\varphi := p \mid (\varphi_1 \to \varphi_2) \mid (\varphi_1 \wedge \varphi_2) \mid (\varphi_1 \vee \varphi_2)$$

Occasionally, outermost parentheses are omitted. Contexts here are finite (possibly empty) lists (i.e. ordered multi-sets, with possible repetition).

2.1.3   Intuitionistic propositional calculus

The language $L_{propint}$ extends $L_{propmin}$ with the 0-ary connective '$\perp$' (known as *falsun*). Typically, *negation* in Intuitionistic Logic is not taken as a primitive operator. Negation is defined by

$$\neg\varphi =^{df.} (\varphi \to \perp) \tag{2.1.1}$$

Formulas are extended accordingly.

There are several conceptions of the nature of the falsum '$\perp$', not all naturally leading to the above interpretation of intuitionistic negation.

- According to Prawitz ([151]), '$\perp$' is a propositional constant *underivable* from any consistent context. As such a constant, it is eligible to be arbitrary embedded in formulas (in particular, to form a consequent of an implication). It is the proof-theoretic analogue of the model-theoretic view as being false in every model, under any assignment of truth-values to the propositional variables. Hence, it is characterized by having no I-rule (cf. Section 2.1.5).

- According to Dummett ([35], p. 295), with analogy to quantification, '$\perp$' does have an I-rule, albeit an *infinitary* one.

$$\frac{p_0 \quad p_1 \quad p_2 \quad \cdots}{\perp} (\perp I) \tag{2.1.2}$$

  where the countably-many premises are all the propositional variables (atomic sentences). Using this approach calls for a revision of the definition of a derivation. Furthermore, the fact that the (infinite) conjunction of all atomic propositions is contradictory is a mere stipulation as was noted by Dummett himself ([35], p. 295).

- Another common view of '$\perp$' is an *abbreviation* of some (fixed, but arbitrary) *contradiction*, like $p_0 \wedge \neg p_0$ (or $0 = 1$ in a first-order language with identity over arithmetic). Obviously, such a view cannot admit the definition of negation in (2.1.1) on pain of circularity. This view, in the first-order case, has been

---

[1] Sometimes, those formulas are referred to as *well-formed formulas (wffs)*. Since no ill-formed formulas are ever considered, I stick to just 'formulas'.

criticized (for example, by Tennant [216]) as rendering negation as depending on the language to which it is added (arithmetic, in the above example). There does not seem to exist a universal canonical absurd sentence, especially in a non-mathematical discourse. Furthermore, as observed by Cook and Cogburn [21], this definition is circular, because 0 = 1 is a contradiction only in virtue of Peano's axiom, that themselves involve negation (in the form of inequality), which is already assumed to behave as expected.

- Yet another view is the one expressed by Tennant ([216]), according to which ⊥ is not a propositional constant, not belonging at all to the object language (hence, not eligible for embedding in sentences). A fortiori, it has no operational rules. Rather, it is a meta-linguistic *punctuation mark* of the structural language, occurring only in derivations, indicating, when occurring in a derivation, that a contradiction has been reached at that stage. Analogously, it is like an empty succedent of a sequent in the sequent-calculus presentation of Intuitionistic Logic. It is more convenient in *ND*-system, in order not to have empty conclusions.

For presenting Intuitionistic Logic here, I shall mostly follow Prawitz's view.

2.1.4   First-order intuitionistic logic

The language L_{foint} (for first-order Intuitionistic Logic) has, instead of the propositional variables, predicate symbols $\mathcal{P}$ (of arbitrary arity) and a countable collection $\mathcal{V}$ of individual variables[2], ranged over by $x, y, ...$, possibly indexed, for forming *atomic formulas*. In addition to the sentential connectives of $IL_{minprop}$, there are two *quantifiers*: *universal*: '∀' and *existential*: '∃', also used to form compound formulas.

$$\varphi := P(x_1, \cdots, x_n), x_i \in \mathcal{V}, i = 1, n, P \in \mathcal{P} \text{ (atomic formulas)}$$

$$\varphi \wedge / \vee / \rightarrow \psi, \ \neg\varphi, \ \ \forall x.\varphi, \exists x.\varphi$$

The dot marks the *scope* of the quantifier. An occurrence of a variable x is *bound* if it falls under the scope of a quantifier – *free* otherwise. I use $Fr(\varphi)$ for the collection of all individual variables having a free occurrence in φ. This notation is naturally extended to $Fr(\Gamma)$. The notation $\varphi[x := y]$ means the formula obtained from φ by substituting y for all free occurrences of x. For technical reasons, it is convenient to regard $\varphi[x := y]$ as a sub-formula of both of $\forall x.\varphi, \exists x.\varphi$, deviating from the strictly recursive definition of sub-formulas.

It is convenient to enhance readability of formulas by assuming *Barendregt's convention* ([4], p.26): no variable occurs both free and bound in the same formula. For example, $(\forall x.\varphi(x)) \wedge \psi(x)$ is presented more clearly in its equivalent form

---

[2]Sometimes, individual constants and function symbols are assumed too. We'll leave it to context to determine which version is used.

$$\frac{}{\varphi : \varphi} \ (Ax)$$

$$\frac{\Gamma, \varphi : \psi}{\Gamma : (\varphi \rightarrow \psi)} \ (\rightarrow I) \qquad \frac{\Gamma : \psi \quad \Gamma : (\psi \rightarrow \varphi)}{\Gamma : \varphi} \ (\rightarrow E)$$

$$\frac{\Gamma : \varphi \quad \Gamma : \psi}{\Gamma : (\varphi \wedge \psi)} \ (\wedge I) \qquad \frac{\Gamma : (\varphi \wedge \psi)}{\Gamma : \varphi} \ (\wedge E_1) \qquad \frac{\Gamma : (\varphi \wedge \psi)}{\Gamma : \psi} \ (\wedge E_2)$$

$$\frac{\Gamma : \varphi}{\Gamma : (\varphi \vee \psi)} \ (\vee I_1) \qquad \frac{\Gamma : \psi}{\Gamma : (\varphi \vee \psi)} \ (\vee I_2) \qquad \frac{\Gamma : (\varphi \vee \psi) \quad \Gamma, \varphi : \chi \quad \Gamma, \psi : \chi}{\Gamma : \chi} \ (\vee E)$$

Figure 2.1: The propositional *NJ*-system: operational rules for minimal logic

$(\forall y.\varphi(y)) \wedge \psi(x)$. Since there is an infinite supply of fresh variables, no generality is lost by this restriction. I adhere to it throughout the book.

Sometimes some of the connectives and quantifiers are not taken as primitive, as we have seen with intuitionistic negation, but are viewed as *defined* in terms of the primitive connectives/quantifiers. There may be an impact of such a practice on PTS, a topic not extensively discussed in PTS. Unless otherwise stated, the operators referred to in the *ND*-system are assumed primitive.

2.1.5   The Natural-Deduction proof-system *NJ*

The standard *ND*-system for Intuitionistic Logic is *NJ*, originating from Gentzen's work. The operational rules of *NJ* for the propositional minimal logic are presented in Figure 2.1.

Remarks:

- The rule for implication $(\rightarrow I)$ mimics *conditional proof*, where the antecedent of an implication is temporarily assumed, the conclusion derived (using the assumption any number of times needed), and then *any number of instances* of the assumption are discharged. This rule is non-combining (having a single premise) and hypothetical (discharging an assumption). It is unconditional (not having any side-conditions).

  The rule $(\rightarrow E)$ is the familiar *modus ponens* rule (MP). It is combining, categorical and unconditional.

- The disjunction $(\vee E)$ rules mimic *proof by cases*. In order to derive *an arbitrary conclusion* ξ from $\varphi \vee \psi$, ξ has to be derived separately from each disjunct, possibly using auxiliary assumptions. It is combining, hypothetical and unconditional.

$$\frac{\Gamma : \psi}{\Gamma, \varphi : \psi} \ (W) \qquad \frac{\Gamma, \varphi, \varphi : \psi}{\Gamma, \varphi : \psi} \ (C) \qquad \frac{\Gamma_1, \varphi_2, \varphi_1, \Gamma_2 : \psi}{\Gamma_1, \varphi_1, \varphi_2, \Gamma_2 : \psi} \ (E)$$

Figure 2.2: The propositional *NJ*-system: structural rules

- Note that all the rules of propositional minimal *NJ* are pure.

Much of the presentation can be developed in terms of minimal logic. The extension for obtaining the rules for I_{propil} is the addition of

$$\frac{\Gamma : \bot}{\Gamma : \varphi} \ (\bot E)$$

(2.1.3)

Remark: The rule $(\bot E)$ is one way to effect '*ex falso quodlibet*' (everything is derivable from a contradiction). Note the absence of an $(\bot I)$ rule - contradiction cannot be introduced.

The above definition of negation (as implication of absurdity) induces the following I/E-rules for negation, that are special cases of those for implication.

$$\frac{\Gamma, \varphi : \bot}{\Gamma : \neg \varphi} \ (\neg I) \qquad \frac{\Gamma : \varphi \quad \Gamma : \neg \varphi}{\Gamma : \bot} \ (\neg E)$$

(2.1.4)

One may wish to regard negation as primitive, not relying on $\bot$. In this case, the above rules can be formulated as follows, using an explicit contradiction.

$$\frac{\Gamma, \varphi : \psi \quad \Gamma, \varphi : \neg \psi}{\Gamma : \neg \varphi} \ (\neg I) \qquad \frac{\Gamma : \varphi \quad \Gamma : \neg \varphi}{\Gamma : \psi} \ (\neg E)$$

(2.1.5)

The structural rules of *NJ* (common to both Minimal Logic and Intuitionistic Logic, both propositional and first order) are presented in Figure 2.2.

Remarks:

- Note that in the identity axiom (Ax), an open formula is possible too, namely $\vdash_{\mathbf{NJ}} \varphi(x) : \varphi(x)$. I will still retain the term 'open assumption' to mean an undischarged assumption, and use, if needed, 'open-formula assumption' for an assumption with free variables. Similarly for 'closed'. The need for open-formula assumptions can be seen in the following derivation of $\vdash_{NJ}$: $\forall x. \varphi(x) \rightarrow \varphi(x)$.

$$\frac{\dfrac{\varphi(x) : \varphi(x)}{: \varphi(x) \rightarrow \varphi(x)} \ (\rightarrow I)}{: \forall x. \varphi(x) \rightarrow \varphi(x)} \ (\forall I)$$

(2.1.6)

Note that the application of $(\forall I)$ satisfies the freshness side-condition on x stated below.

$$\frac{\Gamma : \varphi(x)}{\Gamma : \forall x.\varphi} \ (\forall I), \ x \notin Fr(\Gamma) \qquad \frac{\Gamma : \varphi[x := y]}{\Gamma : \exists x.\varphi} \ (\exists I)$$

$$\frac{\Gamma : \forall x.\varphi}{\Gamma : \varphi[x := y]} \ (\forall E) \qquad \frac{\Gamma : \exists x.\varphi \quad \Gamma, \varphi[x := y] : \psi}{\Gamma : \psi} \ (\exists E), \ x \notin Fr(\Gamma, \psi)$$

Figure 2.3: The first-order *NJ* system

- The structural rule (W) (*weakening*) allows the addition of "superfluous" (unused) assumption, thereby manifesting the *monotonicity* of derivability in *NJ*.

- The structural rule (C) (*contraction*) allows reduplication of assumptions, expressing multiple use of an assumption.

- The structural rule (E) (*exchange*) allows changing the order of assumptions in a context.

Actually, the effect of (W) and of (C) in *NJ* can be achieved via the assumptions discharge policy, allowing vacuous – and multiple – discharge. However, these rules became independent of assumptions discharge in the literature, and it is common, and convenient, to present them as separate rules. As a result of the presence of those three structural rules, we can assume henceforth that context for *NJ* are just sets of formulas.

The *ND*-rules for L_{foint} are presented in Figure 2.3.

Remarks:

- The rule $(\forall I)$ captures the idea of proving a universal generalization by showing that an *arbitrary* individual, not mentioned in the context (the open assumptions) satisfies the the property expressed by φ. The variable x is also called the *eigenvariable* of the rule. The rule $(\forall E)$ captures the idea that a universal generalization holds indeed for *every* instance.

  It is sometimes claimed that the rule $(\forall I)$ does not capture the inference of the universal generalization resulting from surveying a *finite* domain. Thus, it is claimed, that from $\varphi(y_1), \cdots, \varphi(y_n)$ one can infer $\forall x.\varphi(x)$. However, such an argument is really an enthymeme, as there is no explicit premise stating that $y_1, \cdots, y_n$ *exhaust* the domain of quantification. The full argument is captured by the following derivation, employing also the identity rules (in Section 2.3).

$$\cfrac{\cfrac{\cfrac{\forall x.x = y_1 \vee \cdots \vee x = y_n}{x = y_1, \vee \cdots \vee x = y_n} \ (\forall E) \quad \cfrac{[x = y_1]_1 \quad \varphi(y_1)}{\varphi(x)} \ (= E) \quad \cdots \quad \cfrac{[x = y_n]_n \quad \varphi(y_n)}{\varphi(x)} \ (= E)}{\varphi(x)} \ (\vee E^{1, \cdots, n})}{\forall x.\varphi(x)} \ (\forall I)$$

(2.1.7)

- The rule $(\exists I)$ allows to infer the existence of an instance of φ from the presence of a *witness* for such an instance. The rule $(\exists E)$ allows inferring an arbitrary conclusion from the existential generalization, provided this conclusion can be drawn by assuming temporarily an instance of φ. This instance of φ uses a free variable as an (individual) *parameter*. Such a use of parameters is heavily relied upon in Part II of the book, when treating natural language. For a detailed discussion about the difficulties encountered when attempting to interpret the use of parameterized discharged assumptions, in particular in $(\exists E)$, in contrast to their proof-theoretic interpretation as meaning-conferring on the existential quantifier, see Milne [116].

Note that, in contrast to MTS, quantifier rules (and hence the quantifiers' meanings conferred by the rules) do not relate to a "domain of quantification" and its relationship to the universe. It does evade the problems involved in the concept of *quantifying over everything*, known also as *absolute generality* (mentioned in Section 1.2), from the set-theoretic point of view.

2.1.5.1   Some example Derivations

Below are some examples of derivations in *NJ*, manifesting some of its properties.

1. $\vdash_{NJ} : (\varphi \rightarrow \varphi)$: Such a derivation is usually more complicated in an axiomatic proof-system **H** for IL.

$$\frac{\vdash_{NJ} \varphi : \varphi}{\vdash_{NJ} : (\varphi \rightarrow \varphi)} \; (\rightarrow I) \tag{2.1.8}$$

2. $\vdash_{NJ} : ((\varphi \rightarrow (\psi \rightarrow \varphi)))$: An axiom of **H**.

$$\frac{\dfrac{\vdash_{NJ} \varphi : \varphi}{\vdash_{NJ} \varphi : (\psi \rightarrow \varphi)} \; (\rightarrow I)}{\vdash_{NJ} (\varphi \rightarrow (\psi \rightarrow \varphi))} \; (\rightarrow I) \tag{2.1.9}$$

Note the possibility of $\varphi = \psi$. This is why the labelling on the simply presented $(\rightarrow I)$ is important. *Vacuous discharge* of ψ, objected to by some alternative systems (see Section 2.4 on Relevant Logic). Another derivation for $\vdash_{NJ} : (\varphi \rightarrow (\psi \rightarrow \varphi))$, using (W) (Weakening), is shown below.

$$\frac{\dfrac{\dfrac{\vdash_{NJ} \varphi : \varphi}{\vdash_{NJ} \varphi, \psi : \varphi} \; (W)}{\vdash_{NJ} \varphi : (\psi \rightarrow \varphi)} \; (\rightarrow I)}{\vdash_{NJ} : (\varphi \rightarrow (\psi \rightarrow \varphi))} \; (\rightarrow I) \tag{2.1.10}$$

Weakening is also rejected by some logics, notably Relevant Logic (Section 2.4).

3. $\vdash_{NJ} (\varphi \rightarrow (\varphi \rightarrow \psi)) : (\varphi \rightarrow \psi)$:

$$\frac{\vdash_{NJ} \varphi \to (\varphi \to \psi)) : (\varphi \to (\varphi \to \psi)) \quad \vdash_{NJ} \varphi : \varphi}{\vdash_{NJ} \varphi \to (\varphi \to \psi)) \ \varphi : (\varphi \to \psi)} \ (\to E) \qquad \vdash_{NJ} \varphi : \varphi}{\dfrac{\vdash_{NJ} (\varphi \to (\varphi \to \psi)) \ \varphi : \psi}{\vdash_{NJ} : (\varphi \to (\varphi \to \psi)) : (\varphi \to \psi)} \ (\to I)} \ (\to E) \tag{2.1.11}$$

Multiple (double, here) discharge! Note that weakening was omitted for simplicity.

The example above looks as follows in its simple presentation.

$$\frac{\dfrac{(\varphi \to (\varphi \to \psi)) \quad [\varphi]_1}{(\varphi \to \psi)} \ (\to E) \qquad [\varphi]_1}{\dfrac{\psi}{(\varphi \to \psi)} \ (\to I^1)} \ (\to E) \tag{2.1.12}$$

Example with quantifiers:

4. Assume $x \notin FV(\psi)$.

$$\frac{[\exists x \varphi(x)]_j \qquad \dfrac{\dfrac{[\forall x(\varphi(x) \to \psi)]_l}{\varphi(x) \to \psi} \ (\forall E) \qquad [\varphi(x)]_i}{\psi} \ (\to E)}{\dfrac{\dfrac{\psi}{\exists x \varphi(x) \to \psi} \ (\to I^j)}{\forall x(\varphi(x) \to \psi) \to (\exists x \varphi(x) \to \psi)} \ (\to I^i)} \ (\exists E^i) \tag{2.1.13}$$

5.

$$\frac{[\exists y \forall x \varphi(x,y)]_j \qquad \dfrac{\dfrac{[\forall x \varphi(x,y)]_i}{\varphi(x,y)} \ (\forall E)}{\exists y \varphi(x,y)} \ (\exists I)}{\dfrac{\dfrac{\exists y \varphi(x,y)}{\forall x \exists y \phi(x,y)} \ (\forall I)}{\exists y \forall x \varphi(x,y) \to \forall x \exists y \varphi(x,y)} \ (\to I^i)} \ (\exists E^i) \tag{2.1.14}$$

Next, non-derivability is considered. The most famous intuitionistically non-derivable sentence is $\not\vdash_{NJ}: \varphi \lor \neg \varphi$, the law of excluded middle (*tertium non datur*). Had such a proof (from an empty context) existed, it had to end with $(\lor I)$, either from a derivation of φ from an empty context, or from a derivation of $\neg \varphi$. However, since φ can be atomic (and $\neg \varphi$ a negated atomic sentence), and no such literals are derivable from an empty context, no such derivation can exist.

Another famous formula, underivable from the empty context, is Peirce's law $(((\varphi \to \psi) \to \varphi) \to \varphi)$, discussed further below.

Proposition 2.1.1 (closure of *NJ* under derivation composition). NJ *is closed under derivation composition.*

Proof: By induction on the derivation $\mathcal{D}$.

- If $\mathcal{D}$ is an axiom $\vdash_{NJ} \varphi : \varphi$, then $\mathcal{D}[\varphi := \overset{\mathcal{D}'}{\varphi}]$ is just $\mathcal{D}'$, which is a derivation by assumption.

- Otherwise, the assumption φ is an assumption of some of the immediate sub-derivations of $\mathcal{D}$, each of which is closed under derivation composition by the inductive derivation. The result is then obtained by applying the rule last applied in $\mathcal{D}$ to the modified sub-derivations. Note that some renaming of the discharge labels of the sub-derivations might be needed before composing, to preserve the uniqueness of those labels in the resulting derivation[3].

This proposition is the analogue, by the Curry-Howard correspondence, of the 2nd substitution lemma for derivations in Hindley [74]. The proposition will be used in Section 3.7.1 where *NJ* will be shown to be "well behaving" w.r.t. the PTS programme.

Proposition 2.1.2 (closure of *NJ* under term substitution). NJ-*derivations are closed under (capture free) term-substitutions.*

The proof is again by induction on the derivation.

2.2   Classical Logic

In this section, I present *classical logic (CL)*, that is considered by many PTS adherents problematic and inferior to IL. Readers familiar with this logic may skip this section.

The object language of propositional classical logic is the same like that of its intuitionistic counterpart, but with negation (usually) assumed primitive. Similarly for the first-order Classical Logic.

The *CL ND*-system may be viewed as a strengthening of its intuitionistic counterpart, disregarding the latter's insistence on the constructive character of the admissible inference modes, hence of the allowed I/E-rules in its $SCND$ presentations. This strengthening is sometimes referred to as "achieving classical strength". There are several ways of doing this, presented in the following sub-sections. All of them are fiercely rejected by Intuitionists.

---

[3]This is similar to renaming bound variables.

2.2.1 Double-Negation elimination

Double-negation elimination is, as its name suggests, an E-rule. However, it differs from the *NJ* other E-rules in that it does not eliminate a single occurrence of an operator, but two occurrences of one.

$$\frac{\Gamma : \neg\neg\varphi}{\Gamma : \varphi} \; (DNE) \tag{2.2.15}$$

As is clear on immediate inspection, this rule violates the complexity restriction, in that the premise $\neg\neg\varphi$ is of higher complexity that the conclusion φ. It is disharmonious (to be explained in the next chapter) since it introduces grounds for the assertion of φ that are not related to the consequences of φ. As a meaning-conferring rule, it would cause a circularity in the meaning determination. This rule causes, via the following derivation of Peirce's law (2.2.16), a non-conservative extension of the positive implicational fragment.

$$\cfrac{[((\varphi\to\psi)\to\varphi)]_3 \quad \cfrac{\cfrac{[\neg\varphi]_1 \quad [\varphi]_2}{\psi}(\neg E)}{(\varphi\to\psi)}(\to I_2)}{\cfrac{\cfrac{\varphi}{\cfrac{\neg\neg\varphi}{\varphi}(DNE)}(\to E) \quad [\neg\varphi]_1}{(((\varphi\to\psi)\to\varphi)\to\varphi)}(\neg I_1)}(\to I_3) \tag{2.2.16}$$

Another indication of the power of this rule is the following proof of (LEM) (from no open assumptions), that is not provable in Intuitionistic Logic.

$$\cfrac{\cfrac{\cfrac{[\neg(\varphi\vee\neg\varphi)]_3 \quad \cfrac{[\varphi]_1}{(\varphi\vee\neg\varphi)}(\vee_1 I)}{\neg\varphi}(\neg I_1) \quad \cfrac{[\neg(\varphi\vee\neg\varphi)]_3 \quad \cfrac{[\neg\varphi]_2}{(\varphi\vee\neg\varphi)}(\vee_2 I)}{\neg\neg\varphi}(\neg I_2)}{\neg\neg(\varphi\vee\neg\varphi)}(\neg I_3)}{(\varphi\vee\neg\varphi)}(DNE) \tag{2.2.17}$$

It is not accidental that (DNE) is used only in the *last* step in the derivation, where the sub-derivation above its application is intuitionistically acceptable. This feature gives rise to general translations of classical theorems to their intuitionistic "double-negation counterparts", e.g., Kolmogorov ([90]), Glivenko [66], and Gödel [67].

Note that a "dual" rule (DNI), namely double-negation introduction, is derivable even in IL.

$$\cfrac{\cfrac{\varphi \quad [\neg\varphi]_i}{\bot}(\neg E)}{\neg\neg\varphi}(\neg I^i) \tag{2.2.18}$$

2.2.2 Indirect Proof

The following rule mimics the principle of indirect proof commonly used in mathematics, where one proves a proposition by assuming its negation and deriving a contradiction. It is known as *reductio ad absurdum*, hence its name.

$$\frac{\Gamma, \neg\varphi : \perp}{\Gamma : \varphi} \ (RAA)$$

(2.2.19)

The rule is formulated with primitive negation. If we keep the definitional view of negation (i.e., implying $\perp$), the rule can be seen as an additional $(\to E)$-rule, eliminating '$\to$' differently in the special case of the consequent being $\perp$. This rule is also disharmonious since it introduces grounds for asserting φ which are not related to the consequences of φ. So, it is not surprising to realize its non-conservativity over the *positive* implicational fragment, as seen by the derivation of Peirce's law (2.2.20) below.

Clearly, this rule also violates the complexity restriction. Furthermore, it is not pure as it refers to two operators: negation and absurdity. A major source of antagonism towards this rule is its applicability to a conclusion $\exists x.\psi$, allowing the establishment of an existence claim without having a witness, a construction of an x satisfying ψ. Following is a derivation, using this rule, of Peirce's law.

$$\frac{[\neg\varphi]_2 \quad \dfrac{[(\varphi \to \psi) \to \varphi]_3 \quad \dfrac{\dfrac{\dfrac{[\varphi]_1 \quad [\neg\varphi]_2}{\perp}\,(\perp I)}{\psi}\,(\perp E)}{\varphi \to \psi}\,(\to I^1)}{\varphi}\,(\to E)}{\dfrac{\dfrac{\perp}{\varphi}\,(RAA^2)}{((\varphi \to \psi) \to \varphi) \to \varphi}\,(\to I^3)}$$

(2.2.20)

2.2.3 Excluded Middle

Another option to get classical strength is to add the law of *excluded middle (LEM)* as an additional I-rule for disjunction.

$$\frac{}{\Gamma : \varphi \vee \neg\varphi} \ (LEM)$$

(2.2.21)

As an example, consider the derivation of (DNE) using (LEM).

$$\frac{\dfrac{}{\varphi \vee \neg\varphi}\,(LEM) \quad [\varphi]_1 \quad \dfrac{\dfrac{\neg\neg\varphi \quad [\neg\varphi]_2}{\perp}\,(\to E)}{\varphi}\,(\perp E)}{\varphi}\,(\vee E^{1,2})$$

(2.2.22)

As observed by von Plato ([145], p. 333), using (LEM) together with $(\vee E)$ causes the "disappearance" of φ and, consequently, a loss of the sub-formula property, stated as a reason to move to a SC-system for Classical Logic by Gentzen.

2.3   Identity

Sometimes, the identity relation '=' is added as a relation interpreted as identity, with its own I/E-rules. I adopt a version of the rules in Read [166], as presented in [169].

$$\frac{\Gamma, \varphi(x) : \varphi(y)}{\Gamma : x = y} \ (= I) \qquad \frac{\Gamma : x = y \quad \Gamma : \varphi(x) \quad \Gamma, \varphi(y) : \psi}{\Gamma : \psi} \ (= E) \tag{2.3.23}$$

where φ does not occur in Γ. By choosing the arbitrary conclusion ψ in $(= E)$ as $\varphi(y)$ itself, we again have the following simplified rule $(= \hat{E})$.

$$\frac{\Gamma : x = y \quad \Gamma : \varphi(x)}{\Gamma : \varphi(y)} \ (= \hat{E}) \tag{2.3.24}$$

From these, one can derive rules for reflexivity $(= -refl)$, symmetry $(= -sym)$ and transitivity $(= -tr)$. For shortening the presentation of derivations, combinations of these rules are still referred to as applications of $(= E)$.

$$\frac{}{\Gamma : x = x} \ (= -refl) \quad \text{derived by} \quad \frac{\dfrac{}{\Gamma, \varphi(x) : \varphi(x)} \ (Ax)}{\Gamma : x = x} \ (= I) \tag{2.3.25}$$

$$\frac{\Gamma : x = y}{\Gamma : y = x} \ (= -sym) \quad \text{derived by} \quad \frac{\dfrac{}{\Gamma : x = x} \ (= -refl) \quad \Gamma : x = y}{\Gamma : y = x} \ (= \hat{E}) \tag{2.3.26}$$

$$\frac{\Gamma : x = y \quad \Gamma : y = z}{\Gamma : x = z} \ (= -trans) \quad \text{derived by} \quad \frac{\Gamma : y = z \quad \Gamma : x = y}{\Gamma : x = z} \ (= \hat{E}) \tag{2.3.27}$$

2.4   Relevant Logic

Relevant Logic (or, better, *Relevance Logic*) is a generic name for a family of logics, that arose as a reaction to what became known as the *paradoxes of material implication*, and *Lewis' paradox*, related to the rule *ex falso quodlibet*, allowing *any* conclusion to be drawn from a contradiction.

The two main tautologies (of both Intuitionistic and Classical Logic) manifesting the paradoxes of material implication and Lewis's paradox are delineated below,

$\vdash : \varphi \to (\psi \to \varphi)$: This tautology expresses a "fact", that a true proposition φ is implied by *any* other proposition ψ, without any connection in content between φ and ψ. An *NJ* derivation of it was presented in (2.1.9), using vacuous discharge.

$\vdash : \neg\varphi \to (\varphi \to \psi)$: This tautology expresses another "fact", that a false proposition φ implies *any* other proposition ψ, again without any connection in content between φ and ψ.

$$\frac{}{\vdash_{\mathbf{R}}\varphi_1 : \varphi_{\{1\}}} \ (Ax)$$

$$\frac{\varphi_\alpha \quad \psi_\alpha}{(\varphi \wedge_{\mathbf{R}} \psi)_\alpha} \ (\wedge_{\mathbf{R}} I) \qquad \frac{(\varphi \wedge_{\mathbf{R}} \psi)_\alpha}{\varphi_\alpha} \ (\wedge_{\mathbf{R},1} E) \qquad \frac{(\varphi \wedge_{\mathbf{R}} \psi)_\alpha}{\psi_\alpha} \ (\wedge_{\mathbf{R},2} E)$$

$$\frac{\begin{array}{c}[\varphi]_i\\ \vdots\\ \psi_\alpha\end{array}}{(\varphi \to \psi)_{\alpha - \{i\}}} \ (\to_{\mathbf{R}} I) \qquad (i \in \alpha) \qquad \frac{\varphi_\alpha \quad (\varphi \to \psi)_\beta}{\psi_{\alpha \cup \beta}} \ (\to_{\mathbf{R}} E)$$

Figure 2.4: The conjunction-implication fragment of **R**

On the structural level, the first tautology amounts to admitting the weakening structural rule '(W)', and the second can be seen as the rejection of the property $\varphi, \neg\varphi : \psi$ (or its equivalent formulation using '$\bot$').

The "relevant logician" insists that in an implication $\varphi \to \psi$, φ must be "relevant" in some sense to ψ in order to imply it. One such sense of relevance, typical to *ND* presentations of Relevant Logic, is that φ must be *used* in the derivation of ψ to allow for $(\to I)$ to apply. As a result, both vacuous discharge and the weakening structural rule (W) are rejected.

A common property of Relevant Logics is the inadmittance of *unrestricted transitivity*, which in *ND*-system is manifested by the failure of closure under derivation composition (and failure of the admissibility of unrestricted applications of '(CUT)' in sequent calculi formulations of Relevant Logics).

2.4.1 The relevant logic R

The relevant logic **R** was introduced in Anderson and Belnap [1] (see also Mares [110]). The way its *ND*-presentation enforces use of an assumption as a condition for its discharge is by introducing an explicit indexing of formulas in the derivation, recording the assumptions used in the sub-derivation above this node. The operational rules for the conjunction-implication fragment of **R** (in tree-style format) are (simply) presented in Figure 2.4, and its structural rules in Figure 2.5. Lower case Greek letter range over sets of indices. To formulate the definition of *relevant derivability*, I use the following convention: a context is always indexed with consecutive indices, starting from 1; namely $\Gamma = \varphi_1, \cdots, \varphi_n$. Let $i(\Gamma) = \{1, \cdots, n\}$.

Definition 2.4.11 (relevant derivability). $\vdash_{\mathbf{R}} \Gamma : \varphi_\alpha$ *iff* $\vdash \Gamma : \varphi$ *and* $\alpha = i(\Gamma)$.

Here $\vdash \Gamma : \varphi_\alpha$ is the usual recursively defined notion of an *ND*-derivation (it can be thought of as intuitionistic derivability).

$$\frac{\Gamma, \varphi, \varphi : \psi}{\Gamma, \varphi : \psi} \ (C) \qquad \frac{\Gamma_1, \varphi_2, \varphi_1, \Gamma_2 : \psi}{\Gamma_1, \varphi_1, \varphi_2, \Gamma_2 : \psi} \ (E)$$

Figure 2.5: The structural rules of **R**

Remarks:

1. Assumptions are always introduced with a *fresh* index (not used elsewhere). When C (contraction) is applied, the two copies of φ carry the same index, the index of φ in the premise. Actually, there is no need to assume (C) as primitive, as it is admissible due to the possibility of multiple discharge. Note that (W) (weakening) is *not* present, rendering **R** a sub-structural logic.

2. Conjunction can be introduced only if both conjuncts depend on the same set of used assumptions. This is known as *extensional conjunction*. For an *intensional conjunction* with unshared used assumptions see Mares [109].

3. Relevant implication can be introduced only if its discharged assumption has been used in the sub-derivation in the premise (indicated by the side condition $i \in \alpha$). The discharge is indicated by the removal of the index from the conclusion's dependence set. The application of $(\to_{\mathbf{R}} E)$ constitutes such a use.

4. Note that the derivation of, say, the first paradox of material implication above, presented in (2.1.9), is blocked, as no vacuous discharge is possible. The side condition on the unused assumption $[\psi]_2$ is violated, as $2 \notin \{1\}$.

$$\frac{\dfrac{\varphi_1 : \varphi_{\{1\}} [Ax]}{\vdash_{\mathbf{R}}^{?} \varphi_1 : (\psi \to \varphi)_{\{?\}}} \ (\to_{\mathbf{R}} I)}{\vdash_{\mathbf{R}}^{\{??\}} : (\varphi \to (\psi \to \varphi))} \ (\to_{\mathbf{R}} I) \qquad\qquad (2.4.28)$$

Similarly, the other derivation of the first paradox of material implication in (2.1.10) is also blocked, as the structural rule of weakening is not present.

5. Note that

$$\nvdash_{\mathbf{R}} \varphi_1, \psi_2 : (\varphi \wedge_{\mathbf{R}} \psi)_{1,2} \qquad\qquad (2.4.29)$$

6. Had we used the following version $(\wedge_{\mathbf{R}}^{i})$ of $(\wedge_{\mathbf{R}} I)$, called *intensional conjunction*, that *does* support (2.4.29)

$$\frac{\varphi_\alpha \quad \psi_\beta}{(\varphi \wedge_{\mathbf{R}}^{i} \psi)_{\alpha \cup \beta}} \ (\wedge_{\mathbf{R}^i} I) \qquad\qquad (2.4.30)$$

The above paradox would become relevantly derivable again, with the following derivation (omitting contexts).

$$\frac{\dfrac{\dfrac{\dfrac{\varphi_1 \quad \psi_2}{(\varphi \wedge_{\mathbf{R}}^i \psi)_{\{1,2\}}} \; (\wedge_{\mathbf{R}}^i I)}{\varphi_{\{1,2\}}} \; (\wedge_{\mathbf{R}}^i E)}{(\psi \to_{\mathbf{R}} \varphi)_{\{1\}}} \; (\to_{\mathbf{R}} I)}{(\varphi \to (\psi \to \varphi))_{\varnothing}} \; (\to_{\mathbf{R}} I) \qquad (2.4.31)$$

See Mares [109] for a discussion.

7. Below are GE-rules for $\to_{\mathbf{R}}$ and $\wedge_{\mathbf{R}}$.

$$\frac{(\varphi \to_{\mathbf{R}} \psi)_\alpha \quad \varphi_\beta \quad \chi_\gamma \quad \overset{[\psi]_i}{\overset{\vdots}{}}}{\chi_{\alpha \cup \beta \cup \gamma - \{i\}}} \; (\to_{\mathbf{R}} E^i) \; (i \in \gamma)$$

$$\frac{(\varphi \wedge_{\mathbf{R}} \psi)_\alpha \quad \chi_\beta \quad \overset{[\varphi]_i, [\psi]_j}{\overset{\vdots}{}}}{\chi_{\alpha \cup \beta - \{i,j\}}} \; (\wedge_{\mathbf{R}} E^{i,j}) \; (\{i,j\} \subseteq \beta) \qquad (2.4.32)$$

Let $\mathbf{R}_\to$ denote the implicational fragment of $\mathbf{R}$. For the rest of the discussion, it suffices to consider $\mathbf{R}_\to$.

Proposition 2.4.3. $\mathbf{R}_\to$ *is* not *closed under derivation composition.*

Proof: Consider the following $\mathbf{R}_\to$ derivations.

$$\mathcal{D}: \frac{\varphi_1 \quad (\varphi \to_{\mathbf{R}} \psi)_2}{\psi_{1,2}} \; (\to_{\mathbf{R}} E) \qquad \mathcal{D}': \frac{\chi_1 \quad (\chi \to_{\mathbf{R}} \varphi)_2}{\varphi_{1,2}} \; (\to_{\mathbf{R}} E) \qquad (2.4.33)$$

and

$$\mathcal{D}[\varphi := \overset{\mathcal{D}'}{\varphi}] = \frac{\dfrac{\chi_1 \quad (\chi \to_{\mathbf{R}} \varphi)_2}{\varphi_{1,2}} \; (\to_{\mathbf{R}} E) \quad (\varphi \to_{\mathbf{R}} \psi)_2}{\psi_{1,2}} \; (\to_{\mathbf{R}} E) \qquad (2.4.34)$$

Clearly, (2.4.34) is *not* a legal $\mathbf{R}_\to$ derivation. There are two violations.

1. The first, more trivial, violation is, that in Γ, Γ' here are two assumptions indexed '2', contrary the definition. This can be amended easily, by modifying the definition of derivation composition so that the indices in Γ and in Γ' are renamed before the transformation as follows: change $(\varphi \to \psi)_3 \in \Gamma$ to $(\varphi \to \psi)_3$ (and keep Γ' unchanged in this special case). Doing this to (2.4.34), changing in Γ $(\varphi \to_{\mathbf{R}} \psi)_2$ to $(\varphi \to_{\mathbf{R}} \psi)_3$ results in

$$\mathcal{D}[\varphi := \overset{\mathcal{D}'}{\varphi}] = \frac{\dfrac{\chi_1 \quad (\chi \to_{\mathbf{R}} \varphi)_2}{\varphi_{2,3}} \; (\to E) \quad (\varphi \to_{\mathbf{R}} \psi)_3}{\psi_{1,2}} \; (\to E) \qquad (2.4.35)$$

The resulting Γ, Γ' is now properly indexed.

Note that there is no problem in assumptions occurring not according to the order of their indices. This is still incorrect, as the dependency of the final conclusion is wrong!

2. The more essential change is in the "body" of $\mathcal{D}$, where all the dependancies have to be modified, to be consistent with the renaming above. What we want for the example is

$$\mathcal{D}[\varphi := \overset{\mathcal{D}'}{\varphi}] = \cfrac{\cfrac{\chi_1 \quad (\chi \to_{\mathbf{R}} \varphi)_3}{\varphi_{1,2}} \ (\to_{\mathbf{R}} E) \quad (\varphi \to_{\mathbf{R}} \psi)_1}{\psi_{1,2,3}} \ (\to E) \qquad (2.4.36)$$

which a legal $\mathbf{R}_\to$ derivation.

The observations about "correcting" the derivation composition can be generalized to the following definition.

Definition 2.4.12 (relevant derivation composition). *Let $\overset{\Gamma_1, \varphi_i, \Gamma_2}{\underset{\psi_\gamma}{\mathcal{D}}}$ be a derivation of ψ_γ from $\Gamma_1, \varphi_i, \Gamma_2$ (in $\mathbf{R}_\to$), having φ_i as one of its leaves, the rest of the leaves constituting Γ_1, Γ_2. Let $\overset{\Gamma'}{\underset{\varphi_\alpha}{\mathcal{D}'}}$ be a derivation (in $\mathbf{R}_\to$) of φ_α from Γ'. Then $\mathcal{D}[\varphi := \mathcal{D}']^{\mathbf{R}}$ is the tree resulting by replacing the leaf φ_i of $\mathcal{D}$ with the sub-tree $\mathcal{D}'$ (identifying the root of the latter with the leaf of the former), subject to the following modifications.*

1. *Let*

$$\hat{j} = \begin{cases} j & \chi_j \in \Gamma_1 \\ j + |\Gamma_1| & \chi \in \Gamma' \\ j + |\Gamma'| & \chi_j \in \Gamma_2 \end{cases} \qquad (2.4.37)$$

Now, replace every $\chi_j \in \Gamma_1, \Gamma', \Gamma_2$ by $\chi_{\hat{j}}$.

2. *For a collection of indices γ, let $\hat{\gamma} = \{\hat{j} \mid j \in \gamma\}$. Replace, in the resulting tree, every occurrence χ_γ by*

$$\begin{cases} \chi_{\hat{\gamma}} & i \notin \gamma \\ \chi_{\hat{\gamma} \cup \hat{\alpha} - \{i\}} & i \in \gamma \end{cases} \qquad (2.4.38)$$

The following proposition follows directly from the definition of relevant derivation composition.

Proposition 2.4.4 (closure of $\mathbf{R}_\to$ under relevant derivation composition). *For $\mathcal{D}, \mathcal{D}'$ as in Definition 2.4.12, $\mathcal{D}[\varphi := \overset{\mathcal{D}'}{\varphi}]^{\mathbf{R}}$ is a derivation (in $\mathbf{R}_\to$).*

2.5   Modal Logic

This section surveys *Modal Logic (ML)*, a family of logics extending the object language of propositional logics with (unary) connectives expressing *necessity* ($\Box$) and *possibility* ($\Diamond$). Logics in this family differ in properties of those operators. I will only consider here the propositional fragment. A formula in this language is called *modal* if it is of one of the forms $\Box\varphi$, $\neg\Diamond\varphi$ and *co-modal* if it is a negation of a modal formula. By Γ being modal (co-modal) is meant that each φ in Γ is modal (co-modal), respectively. When negation (either primitive or defined as implying $\bot$) is present, sometimes (e.g., in CL) only one modal operator is taken as primitive, the other defined as its dual.

$$\Diamond\varphi =^{\mathrm{df.}} \neg\Box\neg\varphi, \qquad \Box\varphi =^{\mathrm{df.}} \neg\Diamond\neg\varphi \tag{2.5.39}$$

Initially, the various logics in this family were presented axiomatically, in Hilbert-style. Early *ND*-systems for ML were proposed by Curry [22] and Fitch [38]. Curry's I/E-rules for the ML known as **S4** are the following (both simply and logistically presented, for emphasis on the side-condition on the open assumptions).

$$\begin{array}{c} \Box\varphi_1,\cdots,\Box\varphi_n \\ \vdots \\ \dfrac{\varphi}{\Box\varphi}\ (\Box I) \end{array} \qquad \dfrac{\Gamma:\varphi}{\Gamma:\Box\varphi}\ (\Box I), \ (\text{modal } \Gamma) \tag{2.5.40}$$

$$\dfrac{\Box\varphi}{\varphi}\ (\Box E) \qquad \dfrac{\Gamma:\Box\varphi}{\Gamma:\varphi}\ (\Box E) \tag{2.5.41}$$

For the case that $\Diamond$ is considered primitive, Fitch provides the following rules (notationally adjusted to Prawitz-like notation).

$$\dfrac{\varphi}{\Diamond\varphi}\ (\Diamond I) \qquad \dfrac{\Gamma:\varphi}{\Gamma:\Diamond\varphi}\ (\Diamond I) \tag{2.5.42}$$

$$\begin{array}{c} [\varphi]_i \\ \vdots \\ \dfrac{\Diamond\varphi \quad \Diamond\psi}{\Diamond\psi}\ (\Diamond E^i) \end{array} \qquad \dfrac{\Gamma:\Diamond\varphi \quad \Gamma,\varphi:\Diamond\psi}{\Gamma:\Diamond\psi}\ (\Diamond E)\ (\text{modal } \Gamma) \tag{2.5.43}$$

By conflating the modal and co-modal formulas the stronger ML **S5** is obtained, and the weaker ML **T** is obtained if the side-condition on ($\Box I$) replaced by the conclusion being $\Box\Gamma:\Box\varphi$ (where $\Box\Gamma = \{\Box\varphi_1,\cdots,\ \Box\varphi_m\}$ for $\Gamma = \{\varphi_1,\cdots,\ \varphi_m\}$). Finally, other weak MLs can be obtained by omitting ($\Box E$) altogether, and taking $\Diamond$ as defined.

By weakening the side condition on ($\Box I$) so that a modal formula is present in every path from an open assumption to the conclusion, all those MLs were shown to normalize by Prawitz [146].

As noted by Bierman and de Pavia [13], the **S4** system is not closed under derivation composition, the latter possibly causing a violation of the side condition, as shown in the following example.

Example 2.5.6 (Bierman and de Pavia).

$$\mathcal{D}_1 : \quad \frac{\begin{array}{c} \Box\varphi_1, \cdots, \Box\varphi_n \\ \vdots \\ \psi \end{array}}{\Box\psi} \; (\Box I) \tag{2.5.44}$$

Clearly, the side-condition on the open assumption is satisfied. Consider now composing $\mathcal{D}_1$ with the following $\mathcal{D}_2$.

$$\mathcal{D}_2 : \quad \frac{\chi \to \Box\varphi_1 \quad \chi}{\Box\varphi_1} \; (\to E) \tag{2.5.45}$$

resulting in

$$\mathcal{D}_3 : \quad \frac{\begin{array}{c} \dfrac{\chi \to \Box\varphi_1 \quad \chi}{\Box\varphi_1} \; (\to E), \cdots, \Box\varphi_n \\ \vdots \\ \psi \end{array}}{\Box\psi} \; (\Box I) \tag{2.5.46}$$

clearly violating the side-condition of all open assumptions being modal.

To fix this problem, Bierman and de Pavia prefer the following alternative $(\Box\hat{I})$-rule (both logistically and simply presented), having been considered before (see their [13] for more on this rule and its history) with no connection to closure under composition.

$$\frac{\Gamma : \Box\varphi_1 \quad \cdots \quad \Gamma : \Box\varphi_n \quad \Box\varphi_1, \cdots, \Box\varphi_n : \psi}{\Gamma : \Box\psi} \; (\Box\hat{I}) \qquad \frac{\Box\varphi_1 \quad \cdots \quad \Box\varphi_n \quad \begin{array}{c} [\Box\varphi_1, \cdots, \Box\varphi_n]_i \\ \vdots \\ \psi \end{array}}{\Box\psi} \; (\Box\hat{I}^i) \tag{2.5.47}$$

This rule does not cause trouble regarding closure under composition, as the minor premises can be derived arbitrarily, while the boxed premises are discharged by applying $(\Box\hat{I})$, thus not affected by composition. Note that Prawitz's relation of the side-condition also leads to closure under derivation composition. A similar problem is posed by the $(\Diamond E)$-rule, reformulated as

$$\frac{\Gamma : \Diamond\psi \quad \Gamma : \Box\varphi_1 \quad \cdots \quad \Gamma : \Box\varphi_n \quad \Diamond\psi, \Box\varphi_1, \cdots, \Box\varphi_n : \Diamond\chi}{\Gamma : \Diamond\chi} \; (\Diamond\hat{E}^i)$$

$$\frac{\Diamond\psi \quad \Box\varphi_1 \quad \cdots \quad \Box\varphi_n \quad \begin{array}{c} [\Diamond\psi, \Box\varphi_1, \cdots, \Box\varphi_n]_i \\ \vdots \\ \Diamond\chi \end{array}}{\Diamond\chi} \; (\Diamond\hat{E}^i) \tag{2.5.48}$$

Traditionally, MTS defines the meaning of the modal operators by alluding to Kripke's *possible-worlds semantics* [93]. See Read [167] for a persuasive argument against the fitting of possible-worlds semantics as a tool for defining the meaning of the modal operators.

2.6 Summary

In this Chapter, several well-known logics were briefly presented, with the intention to be judged subsequently for meeting the criteria (those already mentioned, and those to be presented in the next chapter) for their *ND*-systems to be adequate for conferring meaning. The issue of "rivalry" among those logics (and others) is not of concern here.

Chapter 3

The basis of acceptability of ND-rules as meaning-conferring

3.1 Introduction

This section is devoted to a formulation of criteria that ND-systems, and in particular I/E-rules, have to meet in order to qualify as meaning-conferring. First, I mentioned already the informal criterion of answerability, originating from the wish to connect the meaning-conferring rules to "use", taken as some actual inferential practice. The following criteria have already emerged from the discussion in the Introduction chapter.

purity: The property of *purity* of ND-rules (cf. Definition 1.5.1) is seen by many as a criterion for rules to be considered as meaning-conferring, originating from the mere concept of 'defining the meaning of an operator'. If an operational rule, say for '$*$', is not pure, it violates the formula generality property, and it may not be clear which connective is being introduced/eliminated, and some explicit marking of the principal formula would be needed. As for other violations of formula generality, an impure rule cannot be expected to confer the *full* meaning of the operator.

Example 3.1.7. *Consider an object language having only conjunction and disjunction, and letting the mutual distributivity rules serve as the respective as I/E-rules.*

$$\frac{(\varphi\wedge\psi)\vee(\varphi\wedge\chi)}{\varphi\wedge(\psi\vee\chi)} \ (\wedge I_d) \qquad \frac{\varphi\wedge(\psi\vee\chi)}{(\varphi\wedge\psi)\vee(\varphi\wedge\chi)} \ (\wedge E_d) \qquad (3.1.1)$$

$$\frac{(\varphi\lor\psi)\land(\varphi\lor\chi)}{\varphi\lor(\psi\land\chi)} \ (\lor I_d) \qquad \frac{\varphi\lor(\psi\land\chi)}{(\varphi\lor\psi)\land(\varphi\lor\chi)} \ (\lor E_d) \qquad\qquad (3.1.2)$$

Clearly, there is a circular mutual dependency of those operators induced by the rules, and many essential properties of the operators are left underivable. In particular, the standard I/E-rules are not derivable, nor are commutativity and associativity of the operators (cf. derivations (5.2.11) and (5.2.12)). In addition, it is hard to consider those rules as self-justifying; they are even to accepted by quantum logic.

In Bonnay [14], the purity property of the rules pertaining to some operator is viewed as a translation of logical expressions into structural expressions. In Milne [115], a different notion of an I-rule is proposed, whereby '$*$' need not be introduced into the top-level, i.e., be the dominant operator, implying impurity of the rule, as another connective is mentioned, the dominant one. This is discussed in more detail in Section 4.5.3.

A different approach regarding purity, seemingly endorsed by Dummett too ([35], pp. 256-7), is that rules may generate *dependence* of an operator, say '$*$', on others, mentioned explicitly in the $*$-rules (as long as no circularity arises in the dependence relation). This means that groups of operators might be *jointly added and understood*. An example of such an approach, due to Milne [115], is presented in Section 4.5.3, as a way to proof-theoretically justify Classical Logic (cf. Section 2.2), as well as explaining a difference between Classical Logic and Intuitionistic Logic (cf. Sections 2.1 and 2.2).

separability: An ND-system enjoying this property, when used to confer meaning, endows each operator its own inferential power. When $\mathcal{N}$ lacks this property, this is most notably manifested when $\mathcal{N}$ is obtained as a non-conservative extension of some underlying sub-system, as is the case when (DNE) is added to Intuitionistic Logic, where it is used to prove Peirce's Rule, not containing negation. In other words, If $\mathcal{N}$ requires rules of some operator not occurring in φ when deriving $\vdash_{\mathcal{N}}\Gamma : \varphi$, then the rules of that operator do not confer the *whole* meaning of that operator. See also Bendall [11] for a discussion of the necessity of separability of meaning-conferring ND-systems.

generality: An ND-system enjoying the generality properties discussed under 'schematism' (cf. Section 1.6.4) allow the introduction of an operator with a proof-theoretically conferred meaning independently of the object language into which the operator is introduced.

Additional criteria emerge out of the attack on PTS as described below in Section 3.2.

3.2   Tonk – an attack on PTS

An attack on the mere idea of PTS in general, and of the idea that rules can be "self-justifying" in particular, was presented by Prior in [156], with the introduction of a sentential connective, called *tonk*, the *I*-rule of which is that of *disjunction*, but the *E*-rule of which is that of *conjunction*.

$$\frac{\varphi}{(\varphi \; tonk \; \psi)} \; (tonk \; I) \qquad \frac{(\varphi \; tonk \; \psi)}{\psi} \; (tonk \; E) \tag{3.2.3}$$

Clearly, those rules are pure and satisfy separability. Under some minimal assumptions about the derivability relation (its reflexivity and transitivity), the effect of adding this connective to any logic is devastating, almost trivializing the derivability relation so that for *any* two formulas φ, ψ:

$$\vdash \varphi : \psi \tag{3.2.4}$$

Here is the proof.

$$\frac{\dfrac{\varphi}{(\varphi \; tonk \; \psi)} \; (tonk \; I)}{\psi} \; (tonk \; E) \tag{3.2.5}$$

Thus, adding *tonk* to a "bare" ND-system (over an object language with no other connectives) changes the meaning of the background derivability relation '$\vdash$', assumed to satisfy the structural rules as the *only* schematic rules (cf. p. 42). In [20], Cook shows a non-trivial logic with a non-transitive derivability relation (alas, by using truth-tables, related to Relevant Logic), to which adding *tonk* does not cause trivialization. Even more radically, in [39] Fjellstad considers a non-reflexive derivability relation to "salvage" *tonk*. In the rest of this section, I assume that the underlying derivability relation *is* reflexive and transitive.

Note that the effect of *tonk*, on its own, when added to a "bare" ND-system, is *not* the generation of inconsistency, neither under definition of having both $\vdash : \psi$ and $\vdash : \neg\psi$ (negation need not be present in the language at all), nor under the definition of Post, namely $\vdash : \psi$ for *every* ψ in the object language (actually, for *no* ψ do we have $\vdash : \psi$, as no *tonk*-rule discharges an assumption, and the initial context can never "be emptied"). However, as observed by Tennant [219], under the modest assumption of the presence of '$\rightarrow$' with its assumption-discharging *I*-rule (or any other discharging *I*-rule), inconsistency *does* arise, as evident by the derivation of $\vdash : \psi$ below.

$$\frac{\dfrac{\dfrac{[\varphi]_i}{\varphi \rightarrow \varphi} \; (\rightarrow I^i)}{(\varphi \rightarrow \varphi) \; tonk \; \psi} \; (tonk \; I)}{\psi} \; (tonk \; E) \tag{3.2.6}$$

Even worse, the effect of adding *tonk* to a "non-bare" language is a possible *change of meaning* of other connectives. Suppose that the language already had the defining rules (say *I*-rules) for conjunction and implication. By continuing the derivation in (3.2.5), we obtain

$$\frac{\varphi \qquad \dfrac{\dfrac{\varphi}{(\varphi \; tonk \; \psi)} \; (tonk \; I)}{\psi} \; (tonk \; E)}{\varphi \wedge \psi} \; (\wedge I) \qquad \text{and} \qquad \frac{\dfrac{\dfrac{[\varphi]_1}{(\varphi \; tonk \; \psi)} \; (tonk \; I)}{\psi} \; (tonk \; E)}{\varphi \rightarrow \psi} \; (\rightarrow I^1) \tag{3.2.7}$$

causing a change in the meaning of '∧', allowing the inference of any conjunction from its left conjunct only, and "internalizing" the degenerate derivability into a degenerate implication, where any formula implies any other formula (cf. the discussion of internalization in Section 5.2.3).

The claimed conclusion was two-fold, stating that if adequacy of the proof-system for conferring meaning should obtain:

- rules cannot be self-justifying.

- meaning cannot be conferred merely by rules.

The situation is even graver. In Wansing [232], it is shown that even for logics where *tonk* can be admitted harmlessly (like the 4-valued logic presented in Cook [20]), it is always possible to construct rules for "connectives stranger that *tonk*", that cause similar disastrous effects.

Note that, as sharply stated by Read ([162], p.162), the problem *tonk* raises is not merely that there is no *truth-functional* connective. It is a much wider attack on the whole PTS approach.

In [209], Stevenson was quick to concede to the (original) *tonk*-attack, and admit that rules for a connective need to induce a truth-table in order to confer meaning. Another more or less concessive reply is in Wagner [230], where, by appealing to the sense-denotation distinction, Wagner also sees rules for a connective as having to conform to some external meaning (again, truth-tables) to be acceptable.

Several responses to the *tonk*-attack, trying to overcome it in the *spirit* of the PTS programme, have been proposed, and I will survey them in the coming sections.

3.3   Additional criteria for meaning-conferring rules

Clearly, a certain revision of the PTS programme is needed, so that not *any* ND-rules should be viewed as meaning-conferring – only rules meeting certain adequacy criteria. Merely conventional acceptance cannot underlie a theory of meaning. However, those criteria themselves are *proof-theoretically grounded*, appealing in their formulation to inferential considerations, without resorting to model-theoretic considerations incorporated via the "back door". These criteria have two tasks to accomplish.

1. To ascertain that the self-justifying rules indeed justify the other rules.

2. To ascertain suitability of the rules to serve as meaning-conferring.

The first task can be seen as a realization, for the inferentialism approach to PTS, of Gentzen's comment, mentioned at the beginning of Section 1.3, where the E-rules "being a consequence" of the I-rules can be interpreted as a claim that the former are justified by the latter. Of course, a dual realization for the pragmatism approach to PTS is needed too. A stronger claim about the interpretation of Gentzen's comment for inferentialism PTS is presented in Section 3.6.3.

Obviously, any criterion for being a meaning-conferring rule should not be satisfied by the above rules for *tonk*, both explaining why the *tonk*-rules do not qualify as meaning-conferring, and eliminating its inclusion in meaning-conferring proof-systems. Below, I review the main criteria proposed for amending the PTS programme.

3.4   Rule justification

I first attend to the first task, that of justification of the non-self-justifying rules. Recall that in MTS, the standard way to justify rules is by showing their soundness and completeness w.r.t. whichever model-theoretic definition constituting their MTS meaning. For a PTS justification, the basic idea is to divide the operational rules into two families, that are treated differently in terms of justification. The one group, referred to as the *base* group, is regarded as *self-justifying*. It is important to note that the base rules are taken as self-justifying not merely by some convention, nor by an appeal to self-evidence. Rather, their being self-justified originates in the role these rules play in the notion of derivation (and proof) as underlying the determination of meaning. The other group is justified by the base rules if they *preserve availability of canonical derivations*. I treat separately the I-based and E-based approaches.

Recall that by the I-based approach, the I-rules are taken as the base, and meaning is determined via I-canonical derivations. The actual way this is done is presented in Chapter 5 for logic, and in Chapter 8 for NL. Thus, there is a need to justify the E-rules. The following definition provides a procedure (called by Dummett the *upward* justification procedure) that is taken as embodying the required justification in the I-based approach.

Definition 3.4.13 (upward rule justification). *A rule* (R) *is justified w.r.t. the base of I-rules iff the existence of I-canonical derivations for the premises of (R) implies the existence of an I-canonical derivation of the conclusion of (R).*

In Section 3.6 I show how this justification can be carried out.

Next, suppose the base consists of the E-rules, requiring the justification of the I-rules. The following definition provides a procedure (called by Dummett the *downward* justification procedure) that is taken as embodying the required justification in the E-based approach.

Definition 3.4.14 (downward rule justification). *A rule* (R) *is justified w.r.t. the base of E-rules iff the existence of E-canonical derivation for any consequence* ψ *of the conclusion of* (R) *implies the existence of an E-canonical derivation of* ψ *from the major premise of* (R).

As an example (Nieuwendijk [131]), considers the following rule of *distributivity* of conjunction over disjunction.

$$\frac{\Gamma : \varphi \wedge (\psi \vee \chi)}{\Gamma : (\varphi \wedge \psi) \vee (\varphi \wedge \chi)} \ (dist)$$

This rule does not serve as as an I-rule for $\vee$. It is not pure, mentioning also $\wedge$, thereby it would introduce $\vee$ only in a specific context. Also, its conclusion is of higher complexity than its premise. However, it is a useful rule, that can be justified both upwards and downwards.

upward: Suppose there is an I-canonical derivation of the premise. Thus, one of the following derivations establishes this assumption.

$$\frac{\Gamma : \varphi \quad \dfrac{\Gamma : \psi}{\Gamma : \psi \vee \chi} \ (\vee_1 I)}{\Gamma : \varphi \wedge (\psi \vee \chi)} \ (\wedge I) \qquad \frac{\Gamma : \varphi \quad \dfrac{\Gamma : \chi}{\Gamma : \psi \vee \chi} \ (\vee_2 I)}{\Gamma : \varphi \wedge (\psi \vee \chi)} \ (\wedge I)$$

From that, we can construct the following I-canonical derivation for the conclusion of $(dist)$ for the first case.

$$\frac{\dfrac{\Gamma : \varphi \quad \Gamma : \psi}{\Gamma : \varphi \wedge \psi} \ (\wedge I)}{\Gamma : (\varphi \wedge \psi) \vee (\varphi \wedge \chi)} \ (\vee_1 I)$$

A similar construction exists in the second case.

downward: An arbitrary E-canonical derivation of the conclusion of $(dist)$) has the following form.

$$\frac{\Gamma : (\varphi \wedge \psi) \vee (\varphi \wedge \chi) \quad \Gamma, \varphi \wedge \psi : \xi \quad \Gamma, \varphi \wedge \chi : \xi}{\Gamma : \xi} \ (\vee E)$$

From the two minor premises we get that either $\Gamma, \varphi : \xi$ using only E-rules, or both $\Gamma, \psi : \xi$ and $\Gamma, \chi : \xi$ using only E-rules. From those sub-derivation we can construct an E-canonical derivation, using only E-rules, from the major premise of $(dist)$, namely $\Gamma, \varphi \wedge (\psi \vee \chi) : \xi$.

3.5   Conservative Extension

One of the first responses to the *tonk*-attack was presented in Belnap [9]. Let us recall some basic definitions.

Definition 3.5.15 (extension). *An ND-system $\mathcal{N}_2$, based on object language L_2, extends an ND-system $\mathcal{N}_1$, based on object language L_1 iff*

1. *$L_1 \subseteq L_2$ and $\mathcal{F}_1 \subseteq \mathcal{F}_2$, namely the object language and class of formulas are extended.*

2. *$\vdash_{\mathcal{N}_1} \subseteq \vdash_{\mathcal{N}_2}$, namely derivability is extended.*

The issue now is, is the extension of derivability also covering sequents purely in the language L_1, not involving any of the "new" operators in L_2?

Definition 3.5.16 (conservativity of extensions). *An extension $\mathcal{N}_2$ of $\mathcal{N}_1$ is conservative iff for any Γ, φ in L_1, $\vdash_{\mathcal{N}_2} \Gamma : \varphi$ implies $\vdash_{\mathcal{N}_1} \Gamma : \varphi$.*

Thus, in an conservative extension, derivability in the "old vocabulary" is preserved: no new "old" sequents become derivable.

Note that conservativity of extension is a relative concept. In Section 2.2, it is shown that adding classical negation extends the implicational fragment of Intuitionistic (propositional) Logic non-conservatively.

Belnap's basic observation was, that adding *tonk* to the most basic system, the "bare" structural derivability relation having only reflexivity and transitivity, is non-conservative. Even stronger, for any $\mathcal{N}$ such that there are two formulas φ, ψ s.t. $\nvdash_{\mathcal{N}} \varphi : \psi$, adding *tonk* to $\mathcal{N}$ results in a non-conservative extension. This brings him to propose that only such ND-rules for a connective that extend conservatively the "bare" derivability relation are admissible as meaning-conferring (Belnap does not relate to the issue of justifying one group of rules by another). Obviously, this criterion bans *tonk*, as desired. Note, however, that this criterion bans also, besides *tonk*-like trivializing connectives, definitions of connectives by rules that do not trivialize the underlying "bare" deducibility relation, like in the case of adding double-negation elimination to Intuitionistic Logic.

When considering conservative extension as a criterion for the constitutive rules of an operator to qualify as meaning-conferring, the underlying assumptions about '$\vdash_{\mathcal{N}}$' need not necessarily assume all three of Gentzen's structural rules. Thus, adding an operator to a substructural logic may be judged by this criterion too, as can other assumptions about the underlying '$\vdash$' (an example to be encountered later, when bilateralism is considered).

Belnap's view draws on an analogy to *definitions*, that require a proof that the defined object exists and is unique. In Read's words ([162], p. 163),

one cannot define things into existence

(no "ontological proofs" ...). Belnap views conservative extension as a proof of existence (uniqueness is discussed below). However, as Read (op. cit.) notes, this analogy is not perfect. Clearly defining an extension by rules for a new operator does not carry with it a claim that the new operator is defined in terms of the old ones.

According to this view, defining the meaning of a new operator is conceived as an incremental step, always extending an underlying deducibility relation[1] that already has some definite meaning, a meaning that should be preserved by any extension with a "new" operator (cf. the remark on p. 42); at the very least, the underlying derivability relation is the one based on the structural rules only (before *any* operator has been defined), but the underlying derivability relation can already have any number of operators, to which the new one is added. In case of a conservative extension, the previously determined meanings are preserved, while new ones are added. It is important to notice that the preserved meanings of the "old" operators are themselves assumed to have been conferred via some set of rules. Otherwise, if there is an external notion of meaning involved (like in MTS) it is possible that a newly operator added non-conservatively still has well-defined meaning. Of course, such a justification of non-conservativity defeats the purpose of a PTS.

Such an incremental view raises additional questions about dependency on the order of introduction. Garson [58] proposes a stronger notion of conservativity (though in a MTS context, using *natural semantics*, whereby rules are interpreted as preserving validity, not truth), that is intended to overcome the problem of dependence on the order of extension of a collection of operators.

Definition 3.5.17 (strong conservativity). $\mathcal{N}$ *(over L) is* strongly conservative *iff* $\mathcal{N}$ *conservatively extends* every *sub-system* $\mathcal{N}'$ *(over some sub-language L' of L)*.

If strong conservativity holds, the meaning of any single operator c does not depend after which other operators it is being incorporated into a language. Any further incorporation of another operator will preserve the meaning of c. Strong conservativity prevents a particular situation, whereby each of two connectives, when added to a base language, yields a conservative extension, but the *joint* addition of both connectives to the base language is non-conservative. This property is essentially equivalent to separability of $\mathcal{N}$ (cf. Definition 1.5.5).

Actually, Belnap adjoins to the requirement of the rules conservatively extending a given consequence relation, viewed by him as securing the *existence*[2] of the defined operator, another requirement, that of *uniqueness*. In his formulation, if $*_1$ and $*_2$ are operators *subject to identical rules*, then for every φ, ψ (in a language containing

---

[1] Importantly, what is being extended is the derivability relation (via the defining rules), not merely the object language (via the connective).

[2] This is a kind of misnomer; this is evident, in particular, from a result in McKay [113], where two operators are presented, each separately conservatively extending positive Intuitionistic Logic, but jointly non-conservatively extend it (actually, leading to inconsistency). It does not make sense to call an operator "existing" on it own, but "non-existent" when considered with another one. I will not use this term any further.

both), the following should hold.

$$(\mathbf{u}) \quad \vdash \varphi *_1 \psi : \varphi *_2 \psi \quad \text{and} \quad \vdash \varphi *_2 \psi : \varphi *_1 \psi \tag{3.5.8}$$

Namely, both operators play the same deductive role. This requirement is justified by the wish not to be able to draw distinctions between such operators if the rules determine (the whole) meaning. Consider the following example.

Example 3.5.8. *Assume the rules* '$(\wedge I)$' *and* '$(\wedge E)$', *and their "copies"* '$(\wedge' I)$' *and* '$(\wedge' E)$'. *The mutual derivability of* $\varphi \wedge \psi$ *and* $\varphi \wedge' \psi$ *is given by*

$$\dfrac{\dfrac{\varphi \wedge \psi}{\varphi} \,(\wedge E) \quad \dfrac{\varphi \wedge \psi}{\psi} \,(\wedge E)}{\varphi \wedge' \psi} \,(\wedge' I) \qquad \dfrac{\dfrac{\varphi \wedge' \psi}{\varphi} \,(\wedge E') \quad \dfrac{\varphi \wedge' \psi}{\psi} \,(\wedge' E)}{\varphi \wedge \psi} \,(\wedge I) \tag{3.5.9}$$

Note, however, that the notion of 'being subject to the same rules' is not so clear. For example, as noted by Humberstone (in [80], p.504), if the rules governing an operator, say '$*$', are impure (mentioning also, say, another auxiliary operator '$@$', should the rules of '$@$' also have "copies"? If not, this leads to a *relativization* of uniqueness, i.e., uniqueness *in terms* of the auxiliary operator. If '$@$' is endowed with its own copy of rules, this leads to a notion of uniqueness of *the pair* $\langle *, @ \rangle$. For another example, how are side-conditions on a rule treated? In a "copy" of the rule $(\forall I)$ (cf. Figure 2.3), should the same eigenvariable be used? Similarly, for the formulation of $(\Box I)$, should $\Box'$ also count as modal? See Restall [177] for another critical study of uniqueness, relating also to $(\forall I)$. A comprehensive discussion of both existence and uniqueness of a connective can be found in Humberstone [80]. A recent discussion of uniqueness of a connective in comparison to another property of proof-systems is by Naibo and Petrollo [126].

Another point left unsettled by Belnap is a specification of which rules can be used in order to show the inter-derivability (**u**): only the rules of $*_1, *_2$, or all the rules in the system?

There were several criticism of conservative extensibility as the ultimate sole criterion for the admissibility of an ND-rules for a logical constant as meaning-conferring on that constant. One of them, that of being stronger than needed to ban trivializing connectives, has already been mentioned above, the ban of classical negation added to Intuitionistic Logic. While indeed being non-conservative, it does not seem as harmful as *tonk* is.

Another question raised by adopting conservative extension as the criterion for conferring meaning is why "blame" the extending operator for the failure of conservativity? It may well be that the rules of the original system, *before the extension*, are problematic in some sense, for example confer a too weak meaning on one or more of the "old" operator. We will encounter such a view regarding negation (see page 111).

Another problem with the adoption of conservative extension as the sole criterion for a set of rules to qualify as meaning-conferring is that it takes care of part of the problem only. If the rules for an operator are non-conservative, this means that the E-rules are too strong w.r.t. the I-rules, allowing deducing "too much". Indeed, this is the case for *tonk*, which is indeed eliminated by this criterion. However, as is emphasized in Section 3.6, the E-rules can be also to weak w.r.t. I-rules, a deficiency not prevented by the conservative extension criterion.

3.6   Harmony and stability

According to Dummett [35] (and mentioned previously in [32]), what is needed for a collection of I/E-rules to confer meaning, is some *balance* between introduction and elimination rules, that renders them as kinds of "inverses" of each other and constitutes an "harmonious relationship" between them. Whenever such a balance holds, the E-rules can extract the *full meaning* (in the inferentialism PTS) conferred by the I-rules, in being able to exploit *all and only* consequences of a compound φ as induced by its meaning. Most importantly, no atomic sentences can be proved (under no open assumptions) solely on the basis of logic. So, this is certainly a required property of rules to be counted as meaning-conferring, and will ban, as is evidently recognized, rules like those of *tonk* above from counting as meaning-conferring.

One can distinguish two "generic" properties of ND-systems in this respect. Those properties can be each characterized in two equivalent phrasings, depending whether the I-view or the E-view is meant. Let $\mathcal{CON}$ denote proof-theoretic consequence.

Harmony: This property[3] is intended to ban a situation in which, for some logical constant, the E-rules are *too strong* w.r.t. the corresponding defining I-rules, or that the I-rules are *too weak* w.r.t. the defining E-rules. This is the case of *over-generation* of conclusions of some φ by the E-rules, not in accord with the ground for the introduction of φ (by the I-rules of its principal logical constant). That is, for every suitable context Γ and any formula φ:

$$\mathcal{CON}[[\varphi, \Gamma]] \subseteq \mathcal{CON}[[G[[\varphi]], \Gamma]]$$

This property is elaborated upon in Section 3.6.1, and shown to be violated by *tonk*. A different approach to harmony is presented in Section 6.6.

Stability: This is a dual property, intended to ban the situation in which, for some logical constant, the E-rules are *too weak* w.r.t. the defining I-rules, or the I-rules are too strong w.r.t. the defining E-rules. This is the case of *under-generation* of conclusions of some φ , not able to draw some conclusion from φ that does follow from its grounds for introduction. That is, or every suitable context Γ and any formula φ:

$$\mathcal{CON}[[G[[\varphi]], \Gamma]] \subseteq \mathcal{CON}[[\varphi, \Gamma]]$$

---

[3]According to Schroeder-Heister [191], the first use of this term in this sense is by Tennant (in [213] p. 74).

This property is elaborated upon in Section 3.6.2. An "artificial" connective to be shown to violate this property is[4] $tunk$, with the following I/E-rules.

$$\frac{\varphi \quad \psi}{\varphi \, tunk \, \psi} \; (tunk I) \qquad \frac{\varphi \, tunk \, \psi \quad \overset{[\varphi]_i}{\underset{\vdots}{\chi}} \quad \overset{[\psi]_j}{\underset{\vdots}{\chi}}}{\chi} \; (tunk E^{i,j}) \qquad (3.6.10)$$

Thus, if both properties are satisfied by an ND-system, we have that for every context Γ and every formula φ

$$\mathcal{CON}[\![\varphi, \Gamma]\!] = \mathcal{CON}[\![G[\![\varphi]\!], \Gamma]\!] \qquad (3.6.11)$$

Sometimes, 'harmony' is taken to refer to both properties holding. I will reserve it, unless otherwise stated, to the first one only.

3.6.1 Harmony

One can discern several notions[5] of harmony that might be intended for the purpose for which harmony is needed: a criterion for I/E-rules to confer meaning.

intrinsic harmony: This is a local property of the I/E-rules for a logical constant, say '$*$', that do not depend on the ND-system as a whole.

harmony in form: This also is a local property that supposedly depends *only on the form* of the I/E-rules, independent of any underlying considerations about derivations in $\mathcal{N}$.

symmetric harmony: This notion of harmony is not based on the view of a partition of the rules into a self-justifying base and and a justified group of rules. Rather, it views the I/E rules *both* being meaning constitutive, jointly justified in a symmetric way.

3.6.1.1 Intrinsic harmony

In [147], going back to his original formulation of [146], Prawitz formulates the following principle[6], that came to be known [7] as the *inversion principle*, which supposedly captures Gentzen's remark that the E-rules should be "read off" the I-rules.

---

[4]Presented by Hjortland (in [76], p. 43), attributed there to Crispin Wright.
[5]Those names are mine.
[6]The quotation is notationally modified, to fit the notation used here.
[7]The origin of this term is from Lorentzen's operative logic, [106].

Let ρ be an application of an elimination rule that has ψ as conclusion. Then, the derivation that justifies the sufficient condition [...] for deriving the major premiss of of ρ, when combined with the derivations of the minor premises of ρ (if any), already "contain" a derivation of ψ; the derivation of ψ is thus obtainable directly from the given derivations without the addition of ρ.

In a sense, this principle embodies the old conception of a valid argument, one the conclusions of which are "contained" in its assumptions. For a historic survey of this principle, see Tesconi [221].

Dummett basically adheres to the same, coining its obtaining as '*harmony*'.

the corresponding introductions and eliminations are inverses of each other, in the sense that the conclusion obtained by an elimination does not state anything more than what must have already been obtained if the major premise of the elimination was inferred by an introduction. In other words, a proof of the conclusion of an elimination is already "contained" in the proofs of the premises when the major premiss is inferred by introduction.

This principle can be made more precise as follows.

Definition 3.6.18 (maximal formula). *An occurrence of a formula φ in a derivation $\mathcal{D}$ is maximal iff it is the conclusion of an application of an I-rule (of its principal operator), and the major premise of an application of an E-rule (for the same operator).*

Thus, $\mathcal{D}$ has the following form:

$$\frac{\dfrac{\mathcal{D}}{\varphi}\;(...I) \quad \{\mathcal{D}_i\}}{\psi}\;(...E)$$

The presence of a maximal formula (also called a *local peak*) in a derivation is a 'detour' in the derivation, constituting an indirect derivation of the final conclusion of $\mathcal{D}$. Below are to examples of maximal formulas in IL_{prop} derivations.

$$\frac{\dfrac{\mathcal{D}_1 \quad \mathcal{D}_2}{\dfrac{\varphi \quad \psi}{(\varphi\wedge\psi)}\;(\wedge I)}}{\varphi}\;(\wedge E) \qquad \frac{\dfrac{\dfrac{[\varphi]_i}{\mathcal{D}_1}}{\dfrac{\psi}{(\varphi\rightarrow\psi)}\;(\rightarrow I^i)} \quad \dfrac{\mathcal{D}_2}{\varphi}}{\psi}\;(\rightarrow E) \qquad (3.6.12)$$

Note that the following derivation *does not* display a maximal formula. The $(\rightarrow E)$ elimination rule, while indeed applied to a formula introduced by an application of an

I-rule, is not applied to that formula *as a major premise*, as required by the definition of a maximal formula.

$$\frac{\dfrac{[\varphi]_1 \atop \vdots \atop \psi}{\varphi \to \psi}\ (\to I^1) \qquad ((\varphi \to \psi) \to \chi)}{\chi}\ (\to E) \qquad\qquad (3.6.13)$$

The harmony requirement for the principal operator of φ can be stated as the ability to 'level' all local peaks, namely, remove any maximal formula, by means of a *reduction* of $\mathcal{D}$ to an equivalent $\mathcal{D}'$, having the same open assumptions and conclusion, but without that occurrence of the maximal formula. This is denoted by $\mathcal{D} \leadsto_r \mathcal{D}'$. This process of reduction embodies the requirement that the E-rules (for the operator in question) are not too strong. Any conclusion drawn by an application of an E-rule to a formula that was "just introduced", can be drawn, via the reduced derivation $\mathcal{D}'$, without the need of this introduction. As an example, consider the following reduction in the intuitionistic propositional logic, corresponding to the derivations in (3.6.12). simply presented.

$$\frac{\dfrac{[\varphi]_i \atop \mathcal{D}_1 \atop \psi}{(\varphi \to \psi)}\ (\to I^i) \quad \dfrac{\mathcal{D}_2}{\varphi}}{\psi}\ (\to E) \quad \leadsto_r \quad \dfrac{\mathcal{D}_2}{\mathcal{D}_1[\varphi := \varphi]} \atop \psi \qquad \dfrac{\dfrac{\mathcal{D}_1 \quad \mathcal{D}_2}{\varphi \quad \psi}\ (\wedge I)}{(\varphi \wedge \psi)}\ (\wedge E)}{\varphi} \quad \leadsto_r \quad \dfrac{\mathcal{D}_1}{\varphi}\ ,$$

$$(3.6.14)$$

The first reduction is well-defined due to the closure under derivation composition of IL.

Definition 3.6.19 (harmony).   *1. The rules for a logical constant (in $\mathcal{N}$), say $*$, are harmonious iff for every $\mathcal{N}$-derivation $\mathcal{D}$ and every occurrence of a maximal formula φ with $*$ as a principal operator, there is a reduction of $\mathcal{D}$ eliminating φ.*

   2. An ND-system $\mathcal{N}$ is harmonious iff all its rules are harmonious.

In the context of *computational logic*, the property of availability of reduction for every operator is called[8] *local-soundness* by Pfenning and Davies ([24, 143]).

Note that a reduction need not necessarily take place at the root of a derivation, but anywhere where a maximal formula is present. For example

$$\frac{\dfrac{\dfrac{\mathcal{D}_1 \quad \mathcal{D}_2}{\varphi \quad \psi}\ (\wedge I)}{\varphi \wedge \psi}\ (\wedge E_1)}{\dfrac{\varphi}{\mathcal{D}_3 \atop \chi}} \quad \leadsto_r \quad \dfrac{\mathcal{D}_1 \atop \varphi \atop \mathcal{D}_3}{\chi} \qquad\qquad (3.6.15)$$

---

[8]Note that the interpretation of this term, similarly to that of 'local-completeness' (in Section 3.6.2) reflects a bias towards the I-view.

which is a shorthand notation for

$$\mathcal{D}_3[\varphi := \begin{matrix} \mathcal{D}_1 \\ \varphi \\ \chi \end{matrix}]$$

Example 3.6.9. *As an example of disharmony, suppose that the E-rule for $\rightarrow$ is chosen "mistakenly" as follows.*

$$\frac{\varphi \rightarrow \psi}{\psi} \ (\rightarrow E^*)$$

while ($\rightarrow I$) is kept intact. In this case, the following derivation has an irreducible maximal formula, and the pair of rules violates harmony (local-soundness).

$$\frac{\begin{matrix} [\varphi]_i \\ \vdots \\ \psi \end{matrix}}{\dfrac{\dfrac{\psi}{\varphi \rightarrow \psi} \ (\rightarrow I^i)}{\psi} \ (\rightarrow E^*)}$$

Thus, following Prawitz, Dummett, Tennant and others, researchers in the mainstream PTS community see harmony as a requirement for ND-rules to be meaning-conferring. Obviously, the rules for *tonk* are disharmonious: the maximal formula φ *tonk* ψ in (3.2.5) cannot be removed by any reduction. In addition, they view harmony also as the justification of the E-rules when the I-rules are taken as self-justifying (the "upward justification" in Dummett's terms, see Section 3.4). In Section 3.7, I compare Intuitionistic and Classical Logic w.r.t. meeting the harmony criterion.

3.6.1.2 Harmony and normalizability

Occasionally, there are in the literature confusing references to intrinsic harmony as 'normalizability' (e.g., Read [168], Došen [29]). In this section, I would like to clarify this confusion and stress the differences. Normalizability is a more demanding condition on ND-systems than intrinsic harmony.

Definition 3.6.20 (normal derivation). *A derivation is* normal *iff it has* no maximal occurrences *[no "detours"].*

There are two properties of ND-systems that fall under the term 'normalization', a term related to *iterated* applications of reduction steps:

Weak normalization: For every derivation *there exists* of a terminating sequence of iterated reductions leading to an equivalent normal derivation.

Strong Normalization: For every derivation *there exists no infinite sequence* of iterated reductions. In other words, every iterated sequence of reduction steps terminates

The reduction step involved in showing local-soundness is the inductive step in a proof of both weak normalization and strong normalization, but its availability does not guarantee either form of normalization. These are global-soundness properties, having important consequences when holding for some $\mathcal{N}$. In Prawitz [146], normalizability was established for intuitionistic logic.

Note that it may be possible to have *more than one normal derivation* for the same sequent. Here is an example.

Example 3.6.10 (multiple normal derivations). $: ((\varphi \to \varphi) \to (\psi \to (\varphi \to \varphi)))$.

$$\dfrac{\dfrac{\dfrac{[\varphi]_i}{(\varphi \to \varphi)} \ (\to I^i)}{(\psi \to (\varphi \to \varphi))} \ (\to I)}{((\varphi \to \varphi) \to (\psi \to (\varphi \to \varphi))))} \ (\to I) \qquad\qquad (3.6.16)$$

Note: all discharges except first *are vacuous!*

$$\dfrac{\dfrac{[(\varphi \to \varphi)]_i}{(\psi \to (\varphi \to \varphi))} \ (\to I)}{((\varphi \to \varphi) \to (\psi \to (\varphi \to \varphi))))} \ (\to I^i) \qquad\qquad (3.6.17)$$

Why is strong normalization a problem? There are two main difficulties.

1. The elimination of a maximal formula may result in creating new maximal formulas in $\mathcal{D}'$, not present in $\mathcal{D}$. Below is an example in intuitionistic logic (in Section 2.1). Let $\mathcal{D}$ be the following derivation.

$$\dfrac{\dfrac{\dfrac{\dfrac{\mathcal{D}^1}{\alpha \to \gamma} \ \dfrac{[\alpha \wedge \beta]_1}{\alpha} \ (\to E)}{\gamma} \ (\to E) \quad \dfrac{\dfrac{\mathcal{D}^2}{\beta \to \delta} \ \dfrac{[\alpha \wedge \beta]_1}{\beta} \ (\wedge E)}{\delta} \ (\to E)}{\dfrac{\gamma \wedge \delta}{\alpha \wedge \beta \to \gamma \wedge \delta} \ (\to I^1)} \ (\wedge I) \quad \dfrac{\dfrac{\mathcal{D}^3}{\alpha} \ \dfrac{\mathcal{D}^4}{\beta}}{\alpha \wedge \beta} \ (\wedge I)}{\dfrac{\gamma \wedge \delta}{\delta} \ (\wedge E)} \ (\to E)$$

$\mathcal{D}$ has one maximal formula, $\alpha \wedge \beta \to \gamma \wedge \delta$. Eliminating it produces the following $\mathcal{D}'$, in which three new maximal formulas were generated: $\alpha \wedge \beta$ twice, and $\gamma \wedge \delta$.

$$\dfrac{\dfrac{\dfrac{\mathcal{D}^1}{\alpha \to \gamma} \ \dfrac{\dfrac{\mathcal{D}^3}{\alpha} \ \dfrac{\mathcal{D}^4}{\beta}}{\alpha \wedge \beta} \ (\wedge I)}{\dfrac{\alpha}{\gamma} \ (\to E)} \ (\to E) \quad \dfrac{\dfrac{\mathcal{D}^2}{\beta \to \delta} \ \dfrac{\dfrac{\mathcal{D}^3}{\alpha} \ \dfrac{\mathcal{D}^4}{\beta}}{\alpha \wedge \beta} \ (\wedge I)}{\delta} \ (\wedge E)}{\dfrac{\gamma \wedge \delta}{\delta} \ (\wedge E)} \ (\wedge I)$$

2. If a substituted sub-derivation itself contains another maximal formula, and if the assumption for which this sub-derivation is substituted is used more than once, the number of residual maximal formulas in $\mathcal{D}'$ grows. Consider the following example.

$$\dfrac{\dfrac{\dfrac{[\varphi]_1 \quad [\varphi]_1}{\varphi \wedge \varphi}\,(\wedge I)}{(\varphi \to (\varphi \wedge \varphi))}\,(\to I^1) \quad \dfrac{\dfrac{\varphi \quad \chi}{\varphi \wedge \chi}\,(\wedge I)}{\varphi}\,(\wedge E)}{\varphi \wedge \varphi}\,(\to E)$$

When removing the maximal formula $(\varphi \to (\varphi \wedge \varphi))$ by substituting the derivation of the minor premise for the assumption (that occurs *twice*), the maximal formula $\varphi \wedge \chi$ of this derivation of the minor premise doubles itself in the resulting derivation after reduction.

$$\dfrac{\dfrac{\dfrac{\varphi \quad \chi}{\varphi \wedge \chi}\,(\wedge I)}{\varphi}\,(\wedge E) \quad \dfrac{\dfrac{\varphi \quad \chi}{\varphi \wedge \chi}\,(\wedge I)}{\varphi}\,(\wedge E)}{\varphi \wedge \varphi}\,(\wedge I)$$

However, a careful inductive arguments shows that for Intuitionistic Logic there is a certain mapping of the derivation to a well-founded set that decreases after *every* reduction, so this process must eventually stop.

Why is strong normalizability a desirable property of ND-systems?

Definition 3.6.21 (sub-formula property). *An ND-system $\mathcal{N}$ has the* sub-formula *property iff in every normal derivation $\mathcal{D}$ for $\vdash_{\mathcal{N}} \Gamma : \varphi$, the only formulas used in nodes of $\mathcal{D}$ (in the simple presentation) are sub-formulas of formulas in Γ or of φ.*

The sub-formula property guarantees the availability of terminating proof-search procedures, hence to the decidability of derivability. For example, the implicative fragment of IL enjoys this property. However, being normal is not always sufficient for having the sub-formula property. A well-known example involves disjunction (even in IL).

$$\dfrac{\varphi \vee \psi \quad \dfrac{\dfrac{\substack{[\varphi]_1, [\chi]_3 \\ \mathcal{D}_1 \\ \xi}}{\chi \to \xi}\,(\to I^3) \quad \dfrac{\substack{[\psi]_2, [\chi]_3 \\ \mathcal{D}_2 \\ \xi}}{\chi \to \xi}\,(\to I^3)}{\chi \to \xi}\,(\vee E^{1,2}) \quad \substack{\mathcal{D}_3 \\ \chi}}{\xi}\,(\to E) \tag{3.6.18}$$

This derivation is normal as it has no maximal formula, as can be seen by inspection. Still, there is no reason to expect that either χ or $\chi \to \xi$ are sub-formulas of χ or the open assumptions on which χ depends, violating the sub-formula property. But, the following is an equivalent derivation.

$$\cfrac{\varphi\vee\psi \qquad \cfrac{[\varphi]_1,\ \overset{\mathcal{D}_3}{\chi}}{\cfrac{\mathcal{D}_1}{\xi}} \qquad \cfrac{[\psi]_1,\ \overset{\mathcal{D}_3}{\chi}}{\cfrac{\mathcal{D}_2}{\xi}}}{\xi}(\vee E^{1,2}) \tag{3.6.19}$$

Clearly, here the sub-formula property *does* hold. The trouble is, that there is no way to reduce the derivation in (3.6.18) to that in (3.6.19) only by reductions removing maximal formulas.

The solution here is to add another group of reductions, called *permutative reductions*, that will do the job. For disjunction, the reduction is the following.

$$\cfrac{\cfrac{\varphi\vee\psi \quad \cfrac{[\varphi]_i}{\overset{\mathcal{D}_1}{\chi}} \quad \cfrac{[\psi]_j}{\overset{\mathcal{D}_2}{\chi}}}{\chi}(\vee E^{i,j}) \quad \mathcal{D}}{\xi}(E) \quad\rightsquigarrow_r\quad \cfrac{\varphi\vee\psi \quad \cfrac{\cfrac{[\varphi]_i}{\overset{\mathcal{D}_1}{\chi}} \quad \mathcal{D}}{\xi}(E) \quad \cfrac{\cfrac{[\psi]_i}{\overset{\mathcal{D}_2}{\chi}} \quad \mathcal{D}}{\xi}(E)}{\xi}(\vee e^{i,j}) \tag{3.6.20}$$

where (E) is any E-rule. Adding permutative reductions to IL preserves normalizability of NJ, though the proof is more complicated. Note that permutative reductions are not related to harmony.

Another property of normalizability is expressed by the following theorem.

Theorem 3.6.2 (normalizability implies conservativity). *If the rules for an operator normalize than adding them to any $\mathcal{N}$ forms a conservative extension $\mathcal{N}'$ of $\mathcal{N}$.*

Proof. Suppose $\mathcal{D}$ established $\vdash_{\mathcal{N}'} \Gamma : \varphi$, and uses an $(*I)$-rule. If $\mathcal{D}$ ends with an essential application of $(*I)$, then φ is not in the underlying object language. Otherwise, the use of $(*I)$ creates a maximal formula, and there is a reduction of $\mathcal{D}$ that eliminates this maximal formula. By normalization, ultimately $\mathcal{D}$ is reduced to a derivation establishing $\vdash_N \Gamma : \varphi$. □

3.6.1.3 Structural intrinsic harmony

The above discussion of harmony was under the conception (along the I-view of PTS), that applying I-rules in the essential last step of a derivation is the *only* way to derive a conclusion canonically. Taking into account structural canonicity (cf. Definition 1.6.10), a additional need for a balance between rules emerges: a conclusion drawn via a structural rule (of the kind manipulating conclusions) should also be confronted with the I-rules. This leads to the following definitions.

Definition 3.6.22 (S-maximal formula). *A formula φ occurring as a node in an $\mathcal{N}$-derivation $\mathcal{D}$ is S-maximal iff it is either a conclusion of an application of an I-rule or of an S-rule, and at the same time the major premise of an application of an E-rule.*

I now recast the ("half of") intrinsic harmony, that of local-soundness, in terms of reducing $\mathcal{N}$-derivations containing an S-maximal formula.

Definition 3.6.23 (S-**local-soundness**). *An ND-system $\mathcal{N}$ is S-locally-sound iff every $\mathcal{N}$-derivation $\mathcal{D}$ containing an S-maximal formula can be reduced to an equivalent one (having the same conclusion and the same (or fewer) open assumptions) where that occurrence of the S-maximal formula does not appear.*

I will show S-local-soundness for a multi-conclusion ND-system in Section 4.2 and for a bilateral ND-system in Section 4.4.

3.6.1.4 Symmetric harmony

This notion[9] of harmony was first presented in Tennant [213], and subsequently refined in [214] and [215]. Its latest version is in Tennant [220], which summarizes the adventure of this definition, and discusses additional (mainly philosophical) aspects of PTS, such as its contribution to the evolutional possibility of logic; I consider those issues as being outside the scope of my presentation. In observing that the intrinsic harmony does not impose *uniqueness* on the E-rules that are harmonious w.r.t. given I-rules, and vice versa, it intends to base such uniqueness as a choice among the possible candidates based on *relative strength*.

Definition 3.6.24 (**symmetric harmony**). *The I-rules of a logical constant, say $(*I)$, are in harmony with its E-rules, denoted by $h(*I, *E)$, when*

1. *The major premise of any $(*E)$-rule, say $\varphi * \psi$, is the* weakest *major premise eliminable using $(*E)$, given the $(*I)$-rules. Moreover, to establish this property:*

   (a) All *premises of the $(*E)$-rule have to be used.*

   (b) All $(*I)$ *rules have to be applied, but*

   (c) No $(*E)$ *rule may be applied.*

2. *The conclusion of any $(*I)$-rule, say $\varphi * \psi$, is the* strongest *proposition introducible using $(*I)$, given $(*E)$-rules. To establish this property:*

   (a) All *premises of the $(*I)$-rule have to be used.*

   (b) All $(*E)$ *rules have to be applied, but*

   (c) No $(*I)$ *rule may be applied.*

Example 3.6.11. *Consider the intuitionistic rules for '$\vee$'. Suppose that $\vdash \Gamma, \varphi : \psi$ whenever $\vdash \Gamma, \alpha : \psi$ and $\vdash \Gamma, \beta : \psi$ (thus using all premises of $(\vee E)$, the single E-rule for '$\vee$'). By applying $(\vee I_1)$ and $(\vee I_2)$ (all the '$\vee$' I-rules), $\vdash \alpha : \alpha \vee \beta$ and $\vdash \beta : \alpha \vee \beta$, so that, by choosing ψ in the assumption as $\alpha \vee \beta$, $\vdash \Gamma, \varphi : \alpha \vee \beta$. Thus, $\alpha \vee \beta$ is indeed*

---

[9]The term is mine.

the weakest eliminable proposition using $(\vee E)$, given $(\vee I_1), (\vee I_2)$. Note that $(\vee E)$ was not applied in this proof.

Conversely, take $(\vee E)$ as given. Suppose that $\vdash\Gamma, \alpha : \varphi$ and $\vdash\Gamma, \beta : \varphi$ (using all the premises of all the $(\vee I)$-rules). By $(\vee E)$ (the only E-rule for '$\vee$'), also $\vdash\Gamma, \alpha\vee\beta : \varphi$. So, $\alpha\vee\beta$ is the strongest introducible proposition given $(\vee E)$.

In Section 3.7.1 I show the symmetric harmony of the other connectives.

Next, I establish that symmetric harmony disqualifies the *tonk*-like rules (cf. Section 3.2) from being considered meaning-conferring. The following variation of *tonk* (attributed by Tennant [220] to Crispin Wright, failing a previous formulation of symmetric harmony) has as its *I*-rules both $(\vee I)$-rules, and as its *E*-rules both $(\wedge E)$-rules. I omit contexts for simplicity.

$$\frac{\varphi}{\varphi \ tonk \ \psi} \ (tonk \ I_1) \qquad \frac{\psi}{\varphi \ tonk \ \psi} \ (tonk \ I_2)$$

$$\frac{\varphi \ tonk \ \psi}{\varphi} \ (tonk \ E_1) \qquad \frac{\varphi \ tonk \ \psi}{\psi} \ (tonk \ E_2) \qquad\qquad (3.6.21)$$

First, I need to show that if $(*) \vdash\chi : \varphi$ and $(**) \vdash\chi : \psi$, we can prove $\vdash\chi : \varphi \ tonk \ \psi$, using both assumptions $(*), (**)$ and applying both $(tonk \ I)$-rules. The "candidate proofs" might look as

$$\frac{\dfrac{\chi}{\varphi} \ (*)}{\varphi \ tonk \ \psi} \ (tonk \ I_1) \qquad \frac{\dfrac{\chi}{\psi} \ (**)}{\varphi \ tonk \ \psi} \ (tonk \ I_2)$$

but neither of those "proofs" uses both assumptions $(*), (**)$. The other direction fails for a similar reason.

The standard *tonk*-rules (3.2.3) also fail symmetric harmony, violating the prohibitions on applying $(tonk \ E)$ for the first task, and $(tonk \ I)$ for the second.

Uniqueness can be imposed as follows.

Definition 3.6.25 (uniquely harmonious rules).   • *Given $(*I)$, let $(*E)$ be the* strongest *E-rules for '$*$' s.t. $h(*I, *E)$*

   • *Given $(*E)$, let $(*I)$ be the* strongest *I-rules for '$*$' s.t. $h(*I, *E)$.*

3.6.1.5 Harmony in form

As already mentioned, this kind of harmony reflects, in accordance to Gentzen's remark (see p. 10), the "reading off" of the *E*-rules from the *I*-rules in a way differing

$$\frac{\Gamma : \varphi \wedge \psi \quad \Gamma, \varphi, \psi : \chi}{\Gamma : \chi} \ (\wedge GE) \qquad \frac{\Gamma : \varphi \to \psi \quad \Gamma : \varphi \quad \Gamma, \psi : \chi}{\Gamma : \chi} \ (\to GE)$$

Figure 3.1: General Elimination rules for conjunction and implication

from removing maximal formulas, by requiring the E-rules to have *a specific form*, given the I-rules. Another possibility, deviating from Gentzen's remark, is described below. The specific forms (in both cases) is induced by another principle, called "a stronger inversion principle" (see Negri [128]).

A stronger Inversion Principle: Whatever follows from the *direct grounds* for deriving a proposition must follow from that proposition.

Such an alternative form of E-rules was presented (e.g., Schroeder-Heister [183], Negri [128], von Plato [144], Tennant [217]) in the literature under the name *general elimination (GE)* rules, modelled after the $(\vee E)$ rule of NJ, where they are structured so as to draw an *arbitrary conclusion* from the major premise, in contrast to the specific conclusions drawn by NJ's E-rules. The main reason for introducing and studying those rules is their simpler connection to sequent-calculi and cut-elimination. Here, their special form plays an important role regarding their harmony with NJ's I-rules. The rules for conjunction and implication are presented in Figure 3.1.

To understand this specific form, that allows for the derivation of an *arbitrary* conclusion, consider again the E-rule for disjunction in NJ, repeated below.

$$\frac{\Gamma : (\varphi \vee \psi) \quad \Gamma, \varphi : \chi \quad \Gamma, \psi : \chi}{\Gamma : \chi} \ (\vee E)$$

We see that in order for an arbitrary conclusion χ to be drawn from a disjunction $\varphi \vee \psi$, this conclusion has to be derivable (with the aid of the auxiliary assumptions Γ), *from each* of the grounds for the introduction of $\varphi \vee \psi$ (serving as a discharged assumption), namely from φ (first minor premise) and from ψ (second minor premise). Mimicking this form of rule for other operators leads to what became to be known as *general elimination rules (GE-rules)*. A rule of this form was also proposed by von Plato [144] for independent reasons, related to the relationship between normal ND-derivations, and cut-free sequent-calculus derivations. For a more comprehensive presentation of the history of GE-rules, see von Plato [130].

Below (in Section 3.6.3), I present a precise construction of rules in this form, given the I-rules for an operator. The construction will rely on the classification of I-rules in Section 1.5. First, consider the resulting GE-rules for the conjunction and implication in minimal logic, repeated again below.

Conjunction:

$$\frac{\Gamma : \varphi \wedge \psi \quad \Gamma, \varphi, \psi : \chi}{\Gamma : \chi} \ (\wedge GE)$$

$$(3.6.22)$$

Here the arbitrary conclusion χ has to be derivable from the grounds for introducing $\varphi \wedge \psi$, namely, $\{\varphi, \ \psi\}$ (together). Note that the grounds used as assumptions are discharged by the application of the $(\wedge GE)$-rule.

Implication:

$$\frac{\Gamma : \varphi \to \psi \quad \Gamma : \varphi \quad \Gamma, \psi : \chi}{\Gamma : \chi} \ (\to GE)$$

$$(3.6.23)$$

This rule is of the same specific form as the $(\to I)$, in contrast to the $(\wedge I)$ rule, discharges an assumption. Thus, $(\to I)$ is hypothetical, while $(\wedge E)$ is categorical. We see that the classification of the I-rules affects the form of the GE-rules.

The standard NJ E-rules for conjunction and implication are special cases of their GE-rules, obtained by a specific choice of the arbitrary conclusion. This is the reason for the name 'general elimination rules'. For obtaining $(\wedge E_1)$, we choose χ, as φ, obtaining

$$\frac{\Gamma : \varphi \wedge \psi \quad \Gamma, \varphi, \psi : \varphi}{\Gamma : \varphi} \ (\wedge GE)$$

$$(3.6.24)$$

Since the minor premise is an axiom (with no need of discharging any assumptions), it can be omitted, resulting in the standard NJ rule. For obtaining $(\wedge E_2)$, we take χ in $(\wedge GE)$ as ψ and again omit the axiomatic premise. A similar choice of χ in $(\to GE)$ as ψ leads to the NJ standard rule for implication.

Conversely, the GE-rules are derivable from the standard rules E-rules. For example, for conjunction, suppose that $\mathcal{D}$ is the derivation in the minor premise of (3.6.22). Then, the usual $(\wedge GE)$ is derived from $(\wedge_1 E)$ and $(\wedge_2 E)$ as follows.

$$\begin{array}{c} \dfrac{\varphi \wedge \psi}{\varphi} \ (\wedge_1 E) \quad \dfrac{\varphi \wedge \psi}{\psi} \ (\wedge_2 E) \\ \mathcal{D} \\ \chi \end{array}$$

$$(3.6.25)$$

Similarly, for implication, with $\mathcal{D}$ the derivation in the minor premise of $(\to GE)$].

$$\begin{array}{c} \dfrac{\varphi \to \psi \quad \varphi}{\psi} \ (\to E) \\ \mathcal{D} \\ \xi \end{array}$$

$$(3.6.26)$$

Note that the equivalence of those GE-rules to the standard E-rules depends on the presence of the structural rules, in NJ incorporated into the operational rules.

As mentioned above, one may depart from Gentzen's view where the E-rules are to be read-off the I-rules, the latter having been obtained on independent grounds. For Gentzen, those ground consisted of capturing the general way mathematicians reason. If ND-systems are rather viewed as a tool of meaning-conferring rather than

$$\frac{[\varphi \wedge \psi]_i \quad\quad \vdots \quad\quad \chi \quad\quad \varphi \quad \psi}{\chi} \ (\wedge GI_i) \qquad \frac{[\varphi]_i \quad\quad \vdots \quad\quad \varphi \wedge \psi \quad \chi}{\chi} \ (\wedge GE_1^i) \qquad \frac{[\psi]_i \quad\quad \vdots \quad\quad \varphi \wedge \psi \quad \chi}{\chi} \ (\wedge GE_2^i) \qquad (3.6.29)$$

$$\frac{[\varphi \to \psi]_i \ \ [\varphi]_i \quad\quad \vdots \quad\quad \chi \quad\quad \chi}{\chi} \ (\to GI_1^i) \qquad \frac{[\varphi \to \psi]_i \quad\quad \vdots \quad\quad \chi \quad\quad \psi}{\chi} \ (\to GI_2^i) \qquad \frac{[\psi]_i \quad\quad \vdots \quad\quad \varphi \to \psi \quad \varphi \quad \chi}{\chi} \ (\to GE_i)$$

$$(3.6.30)$$

Figure 3.2: Milne's GI/GE-rules

capturing mathematical reasoning, one might as well read-off I-rules from E-rules, attributing a special form, referred to as *general introduction* (GI), to the I-rules. This approach is very suitable to the E-view of PTS. Such an approach was suggested recently by Milne (see [117, 118]). Milne's idea is based on the observation, that a formula $*(\varphi_1, \cdots, \varphi_n)$, with a generic main operator '$*$', can also appear in a derivation as an assumption, not necessarily having been deduced via an I-rule. one might ask in this case, under what conditions can an arbitrary χ be deduced from $*(\varphi_1, \cdots, \varphi_n)$ as a dischargeable assumption. Those conditions are that the premises of the suitable GI-rule for '$*$'. the general form of the rule '($*GI$)' is as follows,

$$\frac{[*(\varphi_1, \cdots, \varphi_n)]_i \ \ [\varphi_{i_1}]_i \quad\quad [\varphi_{i_k}]_i \quad\quad \vdots \quad\quad \vdots \quad\quad \chi \quad\quad \chi \ \cdots \ \chi \quad \varphi_{j_1} \ \cdots \ \varphi_{j_l}}{\chi} \ (*GI^i) \qquad (3.6.27)$$

where $k + l \leq n$ and $i_p \neq j_q$ for $1 \leq p \leq k$, $1 \leq q \leq l$.

The corresponding '($*GE$)' rule has the form

$$\frac{[*(\varphi_1, \cdots, \varphi_n)]_i \quad [\varphi_{r_1}]_i \quad\quad [\varphi_{r_t}]_i \quad\quad \vdots \quad\quad \vdots \quad\quad \chi \ \cdots \ \chi \quad \varphi_{s_1} \ \cdots \ \varphi_{s_u}}{\chi} \ (*GE^i) \qquad (3.6.28)$$

where $t + u \leq n$ and $r_p \neq s_q$ for $1 \leq p \leq t$, $1 \leq q \leq u$. It is not hard to realise that intrinsic harmony is also satisfied by those rules.

The (GI/GE)-rules proposed by Milne for conjunction and implication are presented in Figure 3.2. Together with the expected rules for disjunction and the following rules for negation

$$\frac{[\neg \varphi]_i \ \ [\varphi]_i \quad\quad \vdots \quad\quad \chi \quad\quad \chi}{\chi} \ (\neg GI^i) \qquad \frac{\varphi \quad \neg \varphi}{\chi} \ (\neg E) \qquad (3.6.31)$$

a harmonious ND-system for (propositional) Classical Logic is obtained (cf. the discussion in Chapter 4). Milne proposes also rules for the quantifiers, leading to first-order Classical Logic.

3.6.2   Stability

As stated above, Dummett [35] introduces this term for denoting a property of the I/E-rules, that of avoidance of undergeneration of warranted conclusions due to the E-rules being to weak w.r.t. the I-rules. Unfortunately, Dummett provides no precise definition of such a property, and focuses most of the considerations for emphasizing inferentialism-based PTS on harmony.

I start with a proposal of such a definition (in Davies and Pfenning [24], Pfenning and Davies [143]), an approximation to the informal notion of stability that suffices for many purposes, but will be shown not to suffice in all circumstances. The property *local-completeness* requires that an E-rule retains enough information *to reconstruct* the eliminated operator by the I-rules. Failing local-completeness means the E-rule is *too weak*, not allowing to conclude everything known. The definition is based on a transformation called *expansion* (generalizing the well-known η-expansion in the λ-calculus).

Definition 3.6.26 (expansion). *An* expansion *is a replacement of any derivation of* $\overset{\mathcal{D}}{\varphi} : \varphi$ *(for a compound φ) by an I-canonical derivation $\overset{\mathcal{D}'}{\varphi} : \varphi$ that includes an application of an E-rule to φ as a major premise (and due to its canonicity, also applications of the corresponding I-rules).*

The standard definition of expansion does not appeal to canonicity, just requires applications of E-rules and I-rules. In Francez [49] this issue is discussed in more detail.

Anticipating the discussion of the expansion property for NJ, below is an example of an expansion for '$\wedge$'.

$$\frac{\mathcal{D}}{(\varphi\wedge\psi)} \quad \leadsto^{\mathcal{E}} \quad \frac{\dfrac{\dfrac{\mathcal{D}}{(\varphi\wedge\psi)}}{\varphi}(\wedge E_1) \quad \dfrac{\dfrac{\mathcal{D}}{(\varphi\wedge\psi)}}{\psi}(\wedge E_2)}{(\varphi\wedge\psi)}(\wedge I) \tag{3.6.32}$$

Definition 3.6.27 (local-completeness). *The rules for an operator, say $*$, are locally-complete iff every derivation of the formula $\varphi * \psi$ can be expanded.*

Note the requirement of *existence* of an expansion, existentially quantifying on derivations, in contrast to the *universal* quantification on derivations exhibited in the definition of local-soundness.

Example 3.6.12. *Suppose one omits one of the $(\wedge E)$ rules, rendering the E-rules for '$\wedge$' too weak. Clearly, in that case the expansion (3.6.32) would not exist, violating local-completeness of the '$\wedge$'-rules.*

Additionally, *tunk* (cf. p 81) is easily seen not to expand, its E-rule being too weak w.r.t. its I-rule.

To see why local-completeness is only an approximation to the informal notion of stability, one can consider the *quantum disjunction* (see (3.6.36)), that does have an expansion (that of ordinary disjunction) in spite of being a weaker E-rule, prohibiting the use of an auxiliary context.

3.6.2.1   Comparing intrinsic harmony, symmetric harmony and conservative extension

In [208], Steinberger shows the pairwise incomparability of intrinsic harmony, symmetric harmony and conservative extension, refuting a conjecture about the equivalence of those three criteria for meaning-conferring attributed to Tennant [220], and, implicitly also present in Dummett ([35], p.250 and p. 291). I repeat below his arguments.

Conservative extension vs. intrinsic harmony: Mutual incomparability is shown as follows.

> **conservative extension does not imply intrinsic harmony:** Adding *tonk* to any *inconsistent* ND-system is obviously conservative. However, the *tonk*-rules have been shown to be disharmonious, not admitting reductions. If inconsistent systems are to be avoided, consider an ND-system with only rules for $\wedge$, $\vee$. This system is even strongly conservative. Adding '$\neg$' with the E-rule of DNE is still conservative, but no reductions seem to exist for '$\neg$'.
>
> Another example is from Nieuwendijk [131]. Consider the modal necessity operator '$\square$', governed by two sets of I/E-rules, where the syntax of φ determines[10] which version of the rules applies at a given stage of a derivation.
>
> $$\frac{\varphi}{\square\varphi}\,(\square I_1) \quad \frac{\square\varphi}{\varphi}\,(\square E_1) \quad \text{if '}\square\text{'does not occur in } \varphi \qquad (3.6.33)$$
>
> $$\frac{\varphi}{\square\square\varphi}\,(\square I_2) \quad \frac{\square\varphi}{\varphi}\,(\square E_2) \quad \text{if '}\square\text{'does occur in } \varphi \qquad (3.6.34)$$
>
> Suppose '$\square$' is added to some object language L not containing it. Note that in the extension both $\vdash\varphi : \square\varphi$ and $\vdash\square\varphi : \varphi$ hold. Consider any derivation $\mathcal{D}$ of φ where '$\square$' does not occur in φ. If '$\square$' occurs in this derivation, another derivation $\mathcal{D}'$ can be obtained, in which '$\square$' does not occur, by replacing in $\mathcal{D}$ every occurrence of a formula of the form $\square\psi$ by an occurrence of ψ. Thus, the extension is conservative.
>
> To see that the rules are disharmonious, let p be any atomic sentence, and consider the following derivation with a maximal formula.
>
> $$\frac{\dfrac{\square p}{\square\square\square p}\,(\square I_2)}{\square\square p}\,(\square E_2) \qquad (3.6.35)$$

---

[10]Notably, the rules in this example violate the formula generality condition (cf. Section 1.6.4)

Clearly, the maximal formula is irreducible!

intrinsic harmony does not imply conservative extension: As was already realized by Dummett, intrinsic harmony (local-soundness) does not imply conservative extension. Consider a logic that is a variant of intuitionistic logic, known as *Quantum Logic (QL)*, that has instead of the intuitionistic disjunction '∨' a connective '∨$_q$' (known as 'quantum disjunction'), the I-rule for which is the same as (∨I), but with the following weaker E-rule:

$$\frac{\Gamma : \varphi \vee_q \psi \quad \varphi : \chi \quad \psi : \chi}{\Gamma : \chi} \ (\vee_q E) \tag{3.6.36}$$

That is, (∨$_q$ does not allow the use of collateral assumptions Γ in deriving the arbitrary conclusion of $\varphi \vee_q \psi$; the conclusion should follow "in one step". Since both the reduction and expansion apply to ∨$_q$ in the same way as for '∨' (note that the ∨-expansion does not use collateral assumptions), the system is locally-sound (and also locally-complete). Note, however, that this rule violates the context generality property. In this weaker logic, the following distributivity is underivable.

$$\nvdash \varphi \wedge (\alpha \vee_q \beta) : (\varphi \wedge \alpha) \vee_q (\varphi \wedge \beta) \tag{3.6.37}$$

However, adding '∨' with its harmonious and stable rules does allow such a derivation of distributivity, thus non-conservatively extending the logic. Below, Equation 3.6.38, is the derivation (Dummett [35], p. 290, with slightly modified notation). What more can be learned from this phenomenon? A plausible conclusion is, that local-completeness is, as already mentioned, only an approximation to the informal notion of stability. A better approximation should be found, that will "catch" the weakness of ∨$_q$, a weakness not caught by still allowing expansion, as required by local-completeness. Another conclusion can be that the rules for '∨$_q$' do not qualify as meaning-conferring to start with, since the absence of a context variable in (∨$_q E$) violates the context-generality criterion, essential for qualification as part of a meaning-conferring ND-system.

intrinsic harmony vs. symmetric harmony: Mutual incomparability is shown as follows.

intrinsic harmony does not imply symmetric harmony: As we saw above, ∨$_q$ is locally-sound (admits reduction). However, its E-rule is not the strongest E-rule s.t. $h(\vee_q I, \vee_q E)$. The intuitionistic ∨I-rule also satisfies h but is stronger.

symmetric harmony does not imply intrinsic harmony: In [206], Steinberger considers a "rogue" quantifier '$\overline{\exists}$' (my notation), that has an E-rule like $\exists$, but without the side-condition on the freshness of the eigenvariable. The following derivation does not admit reduction, rendering the rule disharmonious.

$$\frac{\dfrac{\varphi(t_1)}{\overline{\exists} x . \varphi(x)} \ (\overline{\exists} I) \qquad [\varphi(t_2)]_1}{\varphi(t_2)} \ (\overline{\exists} E^1) \tag{3.6.39}$$

$$
\cfrac{
\cfrac{\varphi\wedge(\alpha\vee_q\beta)}{\alpha\vee_q\beta}\ (\wedge E)
\qquad
\cfrac{[\alpha]_1}{\alpha\vee\beta}\ (\vee_1 I)
\qquad
\cfrac{[\beta]_2}{\alpha\vee\beta}\ (\vee_2 I)
}{\alpha\vee\beta}\ (\vee_q E^{1,2})
$$

$$
\cfrac{
\cfrac{\dfrac{\varphi\wedge(\alpha\vee_q\beta)}{\varphi}\ (\wedge E)\qquad [\alpha]_3}{\varphi\wedge\alpha}\ (\wedge I)
}{(\varphi\wedge\alpha)\vee_q(\varphi\wedge\beta)}\ (\vee_q I)
$$

$$
\cfrac{
\cfrac{\dfrac{\varphi\wedge(\alpha\vee_q\beta)}{\varphi}\ (\wedge E)\qquad [\beta]_4}{\varphi\wedge\beta}\ (\wedge I)
}{(\varphi\wedge\alpha)\vee_q(\varphi\wedge\beta)}\ (\vee_q I)
$$

$$(\varphi\wedge\alpha)\vee_q(\varphi\wedge\beta)\quad (\vee E^{3,4})$$

$$(3.6.38)$$

symmetric harmony vs. conservative extension: Mutual incomparability is shown
as follows.

symmetric harmony does not imply conservative extension: The quantifier
'$\bar{\exists}$' considered above satisfies symmetric harmony, but clearly will be non-
conservative in any "reasonable" ND-system.

conservative extension does not imply symmetric harmony: Again, '$\vee_q$' considered above is conservative, say, over Intuitionistic Logic, but does not
satisfy symmetric harmony.

3.6.3   Relating harmony in form to intrinsic harmony

In this section, I present a general construction of GE-rules, *harmoniously induced* by
given I-rules. This construction realizes Gentzen's intention that E-rules can be "read
off" given I-rules. The correctness criterion of the construction is that the "read off"
GE-rules, which are harmonious by form by definition, are also provably intrinsically
harmonic w.r.t. the give I-rules. The Section is based on Francez and Dyckhoff [54].
I present the construction in stages, depending on the classification of the I-rules (see
Section 1.5).

Another formulation of the idea behind Prawitz's inversion-principle is, that any conclusion drawn from a formula with an introduced "main operator" δ, say $\varphi\delta\psi$, can be
drawn from ("is included in") *the grounds for introducing δ*.
What are these grounds?
For a categorical, unconditional I-rule, there is a natural answer: the premises of the
I-rule constitute the grounds for introduction (by that rule).

In Read [164], the following form of an E-rule is presented[11] as embodying harmony with given I-rules. This form will easily be seen as fitting categorical, and non-
parameterized I-rules. Suppose the I-rules of an operator δ, the main operator in φ,
can be schematically presented (omitting all shared contexts) as $\dfrac{\Pi_i}{\varphi}\,(\delta I)_i,\ i = 1, \cdots, n$.
Here Π_i is the collection of premises (formulas) of (δI_i), and constitutes the grounds
for introducing δ via (δI_i). Then, to be harmonious in form, the E-rule should combine all the grounds for introducing δ (by all of δ's I-rules) and use each of those
grounds as (discharged) assumptions for deriving an *arbitrary* conclusion ξ, thus having the following form:

$$\cfrac{\varphi \quad \cfrac{[\Pi_1]_1}{\begin{array}{c}\mathcal{D}'_1\\\xi\end{array}} \quad \cdots \quad \cfrac{[\Pi_n]_n}{\begin{array}{c}\mathcal{D}'_n\\\xi\end{array}}}{\xi}\,(\delta GE^{1,\cdots,n}) \qquad (3.6.40)$$

---

[11]A similar construction of E-rules from I-rules, fitting the categorical case, is presented in Prawitz
[151], with a more specific form of the assumptions discharged by the E-rule. The construction is used
there to define and establish functional completeness of a collection of connectives (without reference to
truth-tables).

Here $[\Pi]_i$ abbreviates $[\varphi_1]_i, \cdots, [\varphi_n]_i$, for $\Pi = \{\varphi_1, \cdots, \varphi_n\}$. Such a rule (and similar ones for the other connectives) are of the form of GE-rules.

A simple example is the GE-rule in (3.6.41) (Leblanc [103], Schroeder-Heister [183]) for conjunction, where $\Pi = \{\varphi, \psi\}$ are the grounds for introducing '$\wedge$' via its (only) I-rule ($\wedge I$), repeated below with contexts omitted:

$$\frac{\varphi \wedge \psi \qquad \begin{matrix} [\varphi]_1, [\psi]_2 \\ \vdots \\ \xi \end{matrix}}{\xi} \quad (\wedge GE^{1,2})$$

(3.6.41)

For a categorical, unconditional and non-parameterized rule, this form of a GE-rule indeed reflects the inversion idea: any arbitrary conclusion ξ that can be drawn from (the major premise) φ, can already be drawn from each of its grounds for introduction (all of them!) Π_i, $i = 1, \cdots, n$. Note that all those assumed grounds are discharged by the rule. The NJ disjunction-elimination rule is of this form to start with.

This leads directly to the well-known reduction, removing a maximal formula.

$$\frac{\begin{matrix} \hat{\mathcal{D}}_i \\ \Pi_i \\ \overline{\varphi} \end{matrix} (\delta I)_i \quad \begin{matrix} [\Pi_1]_1 \\ \mathcal{D}'_1 \\ \xi \end{matrix} \quad \cdots \quad \begin{matrix} [\Pi_n]_n \\ \mathcal{D}'_n \\ \xi \end{matrix}}{\xi} (\delta GE^{1,\cdots,n}) \quad \rightsquigarrow_r \quad \begin{matrix} \hat{\mathcal{D}}_i \\ \mathcal{D}'_i[\Pi_i := \hat{\Pi}_i] \\ \xi \end{matrix}$$

(3.6.42)

Recall that the availability of such a reduction constitutes part of the definition of intrinsic harmony. Note that the availability of this reduction rests on the admissibility of derivation composition, as noted also in Read [164] (referring to 'cut' instead of closure under derivation composition).

This leaves us with a question, what happens in the hypothetical, conditional and parameterized cases? As a first approximation, I first consider the effect of hypotheticalness on the harmoniously induced GE-rule. The effect of parameterization is deferred to the next stage. In the case of a hypothetical I-rule, that for implication, the generalized E-rule is (Dyckhoff [36], Read [164], von Plato [144]):

$$\frac{\varphi \rightarrow \psi \quad \varphi \quad \begin{matrix} [\psi]_i \\ \vdots \\ \xi \end{matrix}}{\xi} \quad (\rightarrow GE^i)$$

(3.6.43)

An analysis of the structure of this rule reveals, that one minor premise (φ) is the assumption of the corresponding (single) I-rule, while the "categorical part" of the I-rule (ψ) serves as an assumption of the GE-rule, discharged by it. I continue to refer to the categorical part as the '*grounds*', and refer to the assumptions on which the grounds depend (in the I-rule) as the ground's *support*. I now propose this structure as the *induced general harmonious form* of the E-rule, based on the following reformulation of (the shorter version of) the inversion-principle, that can serve as a semantic justification of the proposed GE-rule:

> Any conclusion drawn from a formula with an introduced main operator δ, say $\varphi\delta\psi$, can be drawn from ("is included in") the grounds for its introduction (all of them), *together with their respective supports (all of them)*.

Below is the proposal for a first approximation of the *harmoniously-induced GE-rule* (ignoring, for a while, conditionality and parameterization). Suppose the I-rules for an operator δ, the main operator in φ, are of the following form (contexts omitted).

$$
\cfrac{\Pi_i \quad \overset{\displaystyle [\Sigma_1^i]_1}{\underset{\vdots}{\alpha_1^i}} \quad \cdots \quad \overset{\displaystyle [\Sigma_{m_i}^i]_{m_i}}{\underset{\vdots}{\alpha_{m_i}^i}}}{\varphi} \ (\delta I^{1,\cdots,m_i})_i
\tag{3.6.44}
$$

for $1 \leq i \leq n$. The ith rule has $l_i = |\Pi_i|$ premises not discharging assumptions, and m_i discharging premises, each discharging a collection Σ of assumptions. These I-rules generate m_i GE-rules per I-rule, each GE-rule corresponding to one premise discharging discharged assumptions in the I-rule. The general form of (the first approximation of) the GE-rules is as follows.

$$
\cfrac{\varphi \quad \Sigma_k^i \quad \overset{\displaystyle [\Pi_1,\alpha_{k_1}^1]_1}{\underset{\vdots}{\xi}} \quad \cdots \quad \overset{\displaystyle [\Pi_n,\alpha_{k_n}^n]_n}{\underset{\vdots}{\xi}}}{\xi} \ (\delta GE_{i,k}^{1,\cdots,n})
\tag{3.6.45}
$$

$1 \leq i \leq n,\ 1 \leq k_i \leq m_i$.

Thus, the contribution of hypothetically in an I-rule is two-folded. Each of the supports becomes a premise (in the corresponding GE-rule), and all the grounds become dischargeable assumptions (in all GE-rules). This leads to the following reductions, where each $(\delta I)_i$ is confronted against each $(\delta E)_{i,k}$, for $1 \leq i \leq n$ and $1 \leq k \leq m_i$.

$$
\cfrac{\cfrac{\mathcal{D}_i \quad \overset{\mathcal{D}_1^i}{\underset{\alpha_1^i}{[\Sigma_1^i]_1}} \quad \cdots \quad \overset{\mathcal{D}_{m_i}^i}{\underset{\alpha_{m_i}^i}{[\Sigma_{m_i}^i]_{m_i}}}}{\varphi} \, (\delta I^{1,\cdots,m_i})_i \quad \overset{\mathcal{D}_{k_i}^i}{\underset{\Sigma_{k_i}^i}{}} \quad \overset{\mathcal{D}_1^*}{\underset{\xi}{[\Pi_1,\alpha_{k_1}^1]_1}} \quad \cdots \quad \overset{\mathcal{D}_n^*}{\underset{\xi}{[\Pi_n,\alpha_{k_n}^n]_n}}}{\xi} \, (\delta GE^{1,\cdots,n})_{i,k_i}
$$

$$
\rightsquigarrow_r
$$

$$
\cfrac{\hat{\mathcal{D}}_i \quad \overset{\mathcal{D}_{k_i}^i}{\underset{\alpha_{k_i}^i}{\mathcal{D}_{k_i}^i[\Sigma_{k_i}^i := \Sigma_{k_i}^i]}}}{\mathcal{D}_i^*[\Pi_i := \Pi_i, \alpha_{k_i}^i := \quad\quad\quad]}
$$

$$
\xi
\tag{3.6.46}
$$

Note the vectored notation for the categorical part in the (δI_i) application. The availability of this reduction depends on our assumption of closure of derivations under composition.

Example 3.6.13. *Consider the following example of a hypothetical and combining I-rule, for the ternary connective 'if-then-else' (abbreviated as ite)*

$$\frac{\Gamma, \phi : \psi \quad \Gamma, \neg\phi : \chi}{\Gamma : ite(\phi, \psi, \chi)} \ (iteI) \tag{3.6.47}$$

For this example, $n = 1$, and the single I-rule has no categorical premise, and has two premises discharging assumptions; thus, the two following GE-rules are induced.

$$\frac{ite(\varphi, \psi, \chi) \quad \varphi \quad \overset{[\psi]_i}{\underset{\xi}{\vdots}}}{\xi} \ (iteGE^i)_1 \qquad \frac{ite(\varphi, \psi, \chi) \quad \neg\varphi \quad \overset{[\chi]_i}{\underset{\xi}{\vdots}}}{\xi} \ (iteGE^i)_2 \tag{3.6.48}$$

The reductions are as follows

$$\frac{\dfrac{[\varphi]_1 \quad [\neg\varphi]_2}{\mathcal{D}_1 \quad \mathcal{D}_2}}{\dfrac{\psi \quad \chi}{ite(\varphi, \psi, \chi)} \ (iteI^{1,2}) \quad \mathcal{D}_1 \quad \dfrac{\overset{[\psi]_3}{\mathcal{D}_1^*}}{\xi}}{\xi} \ (iteGE^3)_1 \ \leadsto_r \quad \mathcal{D}_1^*[\psi := \mathcal{D}_1[\varphi := \overset{\mathcal{D}_1}{\varphi}]] \atop \xi \tag{3.6.49}$$

and

$$\frac{\dfrac{[\varphi]_1 \quad [\neg\varphi]_2}{\mathcal{D}_1 \quad \mathcal{D}_2}}{\dfrac{\psi \quad \chi}{ite(\varphi, \psi, \chi)} \ (iteI^{1,2}) \quad \mathcal{D}_2 \quad \dfrac{\overset{[\chi]_3}{\mathcal{D}_2^*}}{\xi}}{\xi} \ (iteGE^3)_2 \ \leadsto_r \quad \mathcal{D}_2^*[\chi := \mathcal{D}_2[\neg\varphi := \overset{\mathcal{D}_2}{\neg\varphi}]] \atop \xi \tag{3.6.50}$$

Clearly, the harmoniously-induced GE-rule for a categorical I-rule is the assumption-less special case, yielding the original formulation[12] of Read [164]. This construction explains the form of the generalized implication GE-rule (3.6.43), the I-rule of which is hypothetical: The one minor premise, φ, is the support of the I-rule, while the other minor premise is a derivation of ξ under the assumption of the ground ψ.

Example 3.6.14. *As a boundary case of a constant having an I-rule with* no *premises, consider* $\top$ *(verum) with the I-rule*

$$\frac{}{\top} \ (\top I) \tag{3.6.51}$$

Thus, in the construction of a harmoniously-induced GE-rule no rule is generated, and there is no possibility of having $\top$ as a maximum formula in need of reduction.

---

[12]The hypothetical case is treated there too, by appealing to heuristics drawn from Gentzen's sequent calculi, not fully formalized and justified.

I next refine the above approximation of the harmoniously induced GE-rule into its final form, by incorporating the effects of conditionality (freshness restriction on free variables) and parameterization. To simplify the notation, I assume one free variable and one parameter only. Passing to any finite number involves vectorizing the notation. Thus, suppose the i'th I-rule has the following form.

$$\frac{\Pi_i(x_i) \quad \alpha_1^i(X_1^i) \quad \cdots \quad \alpha_{m_i}^i(X_{m_i}^i)}{\varphi} \begin{array}{c} [\Sigma_1^i(X_1^i)]_1 \qquad [\Sigma_{m_i}^i(X_{m_i}^i)]_{m_i} \\ \vdots \qquad\qquad \vdots \end{array} (\delta I^{1,\cdots,m_i})_i$$

(3.6.52)

X_j^i fresh for $1 \le i \le n$, $1 \le j \le m_i$. Here x_i are the parameters, and X_i are the subjects of conditionalization (freshness requirement).

The harmoniously-induced GE-rule reflects the incorporation of variables by substitutions, that preserve validity by the freshness assumption.

$$\frac{\varphi \quad \Sigma_k^i[X_k^i := Y_k^i] \quad \begin{array}{c}[\Pi_1[x^i := y_1^i], \alpha_k^i[X_1^i] := Y_1^i]_1 \\ \vdots \\ \xi\end{array} \quad \cdots \quad \begin{array}{c}[\Pi_n[x_n^i := y_n^i], \alpha_k^i[X_1^i] := Y_1^i]_n \\ \vdots \\ \xi\end{array}}{\xi} (\delta GE_{i,k}^{1,\cdots,n})$$

(3.6.53)

$1 \le i \le n$, $1 \le k \le m_i$, and y_j^i are fresh, for $1 \le j \le k$. The resulting reductions are as in Equation 3.6.54. The availability of this reduction rests on closure under both substitution and composition. Clearly, for the categorical, unconditional, non-parameterized case, the reduction above yields the one in (3.6.42) observed in Read [164].

A hypothetical, parameterized and conditional I-rule with which I will be concerned in the second part of the book, on **PTS** for natural language, is that for *restricted universal quantification*, here in 1st-order logic syntax. Here 'e' stands for **every**.

$$\frac{\Gamma, \phi(y) : \psi(y)}{\Gamma : \forall x.\phi(x) \to \psi(x)} (eI) \quad y \text{ fresh for } \Gamma$$

(3.6.55)

For the restricted universal quantification I-rule (3.6.55), the generalized GE-rule harmoniously induced by this construction is

$$\frac{\forall x.\varphi(x) \to \psi(x) \quad \varphi(x := y) \quad \begin{array}{c}[\psi(x := y)]_i \\ \vdots \\ \xi\end{array}}{\xi} (eE^i), \quad y \text{ fresh}$$

(3.6.56)

Definition 3.6.28 (GE-harmony). *For an operator δ, its I-rules and E-rules are GE-harmonious iff the E-rules are the GE-rules harmoniously induced by the I-rules.*

Note that while the generation of harmoniously-induced GE-rules are functional, this function is not injective, and different sets of I-rules may lead to the same set of harmoniously-induced GE-rules; however, there is no reason to expect a one-one relationship.

$$\cfrac{\hat{\mathcal{D}}_i^i}{\Pi_i(x_i)} \quad \cfrac{[\Sigma_1^i(X_1^i)]_1 \cdots [\Sigma_{m_i}^i(X_{m_i}^i)]_{m_i}}{\cfrac{\mathcal{D}_1^i \qquad \mathcal{D}_{m_i}^i}{\cfrac{\alpha_1^i(X_1^i) \quad \cdots \quad \alpha_{m_i}^i(X_{m_i}^i)}{\varphi} \ (\delta I)_i}}$$

$$\cfrac{\Sigma_k^i(X_k^i := Y_k^i)}{\mathcal{D}_k^i}$$

$$\cfrac{[\Pi_1[x_1^i := y_1^i], \alpha_k^i[X_k^i := Y_k^i]]_1 \cdots [\Pi_n[x_n^i := y_n^i], \alpha_k^i[X_k^i := Y_k^i]]_n}{\cfrac{\mathcal{D}_1^* \qquad\qquad \mathcal{D}_n^*}{\xi \qquad\qquad \cdots \qquad\qquad \xi}} \ (\delta GE^{(1,\cdots,n)})_{i,k}$$

$$\rightsquigarrow_r$$

$$\xi$$

$$\cfrac{\hat{\mathcal{D}}_i^i[x_i := y_i] \quad \mathcal{D}_k^i[\Sigma_k^i(X_k^i) := \Sigma_k^i[X_k^i := Y_k^i]]}{\mathcal{D}_i^*[\Pi_i(x_i) := \Pi_i[x_i := y_i], \ \alpha_k^i(X_k^i) := \alpha_k^i[X_k^i := Y_k^i]\ ]}$$

$$\cfrac{}{\xi}$$

$$\mathcal{D}_k^i$$

(3.6.54)

I now establish the relationship between the two notions of harmony considered, and show that GE-harmony (under our extended GE-rule construction) is stronger than intrinsic harmony: the form of the GE-rules *guarantees* both local-soundness and local-completeness w.r.t. to the I-rules.

Theorem 3.6.3 (harmony implication). *In an ND-system closed under derivation composition, any operator δ, its GE-rules harmoniously-induced by its I-rules are intrinsically-harmonious.*

Proof. The reduction of a δ-maximal formula establishing local-soundness was already presented in (3.6.54). □

Note that the GE-rule for *tonk*, harmoniously-induced by its I-rule, is

$$\frac{\varphi \; tonk \; \psi \qquad \begin{matrix}[\psi]_i\\ \vdots\\ \chi\end{matrix}}{\chi} \; (tonkGE) \tag{3.6.57}$$

which is harmless. In particular, it does not produce the "regular" ($tonkE$) as a special case, and does not cause trivialization of '⊢'. As for *tunk* (p. 81), its harmoniously induced GE-rule is

$$\frac{\varphi \; tunk \; \psi \qquad \begin{matrix}[\varphi]_i, [\psi]_j\\ \vdots\\ \chi\end{matrix}}{\chi} \; (tunkGE^{i,j}) \tag{3.6.58}$$

actually identifying *tunk* with conjunction.

Note the strong dependence of the harmony implication theorem on the closure of the meaning-conferring ND-system under derivation composition. In Section 3.7.3, an example is presented of a logic the ND-system of which is *not* closed under derivation composition, and the impact of that non-closure on the harmony of that ND-system.

Unfortunately, the harmoniously-induced GE-rules construction is not strong enough[13] to yield local-soundness. For establishing local-completeness, I have to show a canonical open derivation, namely, *some* way to expand a derivation of φ (with main operator δ). The way I do it is to choose, for each $1 \le i \le n$, the "arbitrary conclusion" ξ in applications of the E-rule $\delta GE_{i,k}$ as $\alpha_{k,\,1\le k\le m_i}^i$ as well as Π_i itself, and take all of the supports, Σ_k^i, as assumptions (to be discharged by the δI-rule applications). We get the following expansion, see Equation 3.6.59. For simplicity of notation, I treat each Π and Σ as a single formula, and the potential parameterization is suppressed. where $\hat{\Pi}_i$ is $\Pi_i[x_i := y_i]$, $\hat{\Sigma}_j^i$ is $\Sigma_j^i[X_j^i := Y_j^i]$ and $\hat{\alpha}_j^i$ is $\alpha_j^i[X_j^i := Y_j^i]$, for fresh Ys and ys (in case the I-rule is conditional and/or parameterized).

As an example where the construction does work, consider the expansion for (ite), where $n = 1$ (and Π_1 is empty), the I-rules of which are given in (3.6.47), and E-rules

---

[13] In contrast to what was claimed in Francez and Dyckhoff [54].

$$
\cfrac{
\cfrac{
\begin{array}{c}\mathcal{D}\\ \varphi\\ \mathcal{D}\leadsto e\\ \varphi\\ [\hat{\Sigma}_1^i]\end{array}\quad [\hat{\Pi}_1,\hat{\alpha}_1^i]\ \cdots\ [\hat{\Pi}_n,\hat{\alpha}_1^i]
}{\hat{\Pi}_i^i}\ (\delta E_{i,1})
\quad
\cfrac{
\begin{array}{c}\mathcal{D}\\ \varphi\\ [\hat{\Sigma}_1^i]\end{array}\quad [\hat{\Pi}_1,\hat{\alpha}_1^i]\ \cdots\ [\hat{\Pi}_n,\hat{\alpha}_1^i]
}{\hat{\alpha}_1^i}\ (\delta E_{i,1})
\quad \cdots \quad
\cfrac{
\begin{array}{c}\mathcal{D}\\ \varphi\\ [\hat{\Sigma}_{m_i}^i]\end{array}\quad [\hat{\Pi}_1,\hat{\alpha}_{m_i}^i]\ \cdots\ [\hat{\Pi}_n,\hat{\alpha}_{m_i}^i]
}{\hat{\alpha}_{m_i}^i}\ (\delta E_{i,m_i})
}{\varphi}\ (\delta I_i^{1,\cdots,n})
$$

(3.6.59)

in (3.6.48).

$$\dfrac{\dfrac{ite(\varphi,\psi,\chi) \quad [\varphi]_1 \quad [\psi]_2}{\psi} \; (iteGE_1^2) \qquad \dfrac{ite(\varphi,\psi,\chi) \quad [\neg\varphi]_3 \quad [\chi]_4}{\chi} \; (iteGE_2^4)}{ite(\varphi,\psi,\chi)} \; (iteI^{1,3})$$

(3.6.60)

Next, consider cases where the construction of harmoniously-induced GE-rules fails to achieve local-completeness. The following I/E-rules for a ternary connective $*(\varphi,\psi,\chi)$ were presented in Olkhovikov and Schroeder-Heister [132]. The I-rules are[14] the following.

$$\dfrac{\begin{array}{c} [\varphi]_i \\ \vdots \\ \psi \end{array}}{*(\varphi,\psi,\chi)} \; (*I_1^i) \qquad \dfrac{\chi}{*(\varphi,\psi,\chi)} \; (*I_2)$$

(3.6.61)

Since there are two I-rules, each with one premise, the application of the procedure for generating the harmoniously-induced GE-rules in Francez and Dyckhoff [54] produces the following rule.

$$\dfrac{*(\varphi,\psi,\chi) \quad \varphi \quad \begin{array}{c} [\psi]_j \\ \vdots \\ \xi \end{array} \begin{array}{c} [\chi]_k \\ \vdots \\ \xi \end{array}}{\xi} \; (*GE^{j,k})$$

(3.6.62)

The rule is locally-sound. There are two reductions, one for each I-rule producing a maximal formula.

$$\dfrac{\dfrac{\begin{array}{c}[\varphi]_i\\ \mathcal{D}\\ \psi\end{array}}{*(\varphi,\psi,\chi)}(*I_1^i) \quad \mathcal{D}_1 \quad \begin{array}{c}[\psi]_j\\ \mathcal{D}_2\\ \varphi \quad \xi\end{array} \begin{array}{c}[\chi]_k\\ \mathcal{D}_3\\ \xi\end{array}}{\xi}(*GE^{j,k}) \quad \leadsto_r \quad \mathcal{D}_2[\psi := \begin{array}{c}\mathcal{D}[\varphi := \begin{array}{c}\mathcal{D}_1\\ \varphi\end{array}]\\ \psi \end{array}] \atop \xi$$

(3.6.63)

$$\dfrac{\dfrac{\begin{array}{c}\mathcal{D}\\ \chi\end{array}}{*(\varphi,\psi,\chi)}(*I_2) \quad \mathcal{D}_1 \quad \begin{array}{c}[\psi]_j\\ \mathcal{D}_2\\ \varphi \quad \xi\end{array} \begin{array}{c}[\chi]_k\\ \mathcal{D}_3\\ \xi\end{array}}{\xi}(*GE^{j,k}) \quad \leadsto_r \quad \mathcal{D}_3[\chi := \begin{array}{c}\mathcal{D}\\ \chi\end{array}] \atop \xi$$

(3.6.64)

The reductions are well-defined due to closure under composition of derivations. However, the rule is not locally-complete. An alleged expansion would be the following.

$$\dfrac{*(\varphi,\psi,\chi) \quad \dfrac{[\varphi]_i \quad [\psi]_j}{*(\varphi,\psi,\chi)}(*I_1^i) \quad \dfrac{[\chi]_k}{*(\varphi,\psi,\chi)}(*I_2)}{*(\varphi,\psi,\chi)} (*GE^{j,k})$$

(3.6.65)

---

[14]The original definition formulates the rule in terms of propositional variables only, in contrast to the current formulation using meta-variables over arbitrary formulas. This difference has no be bearing on the current discussion.

However, the premise $[\varphi]_i$, needed for the application of the GE-rule, is "snatched" by the previous application of $(*I_1^i)$, rendering the "deviation" above illegal.

Notably, the same harmoniously-induced GE-rule results from the "switched" I-rules, where the discharge of an assumption is delegated to $*I^2$, namely

$$
\frac{\psi}{*(\varphi, \psi, \chi)} \ (*I_1) \qquad \begin{array}{c} [\varphi]_i \\ \vdots \\ \chi \end{array} \frac{}{*(\varphi, \psi, \chi)} \ (*I_2^i) \tag{3.6.66}
$$

A similar problem arises in an example in Read [170]. Read considers a binary connective, '$\varphi \odot \psi$', with the following $\odot I$-rules.

$$
\begin{array}{c} [\varphi]_i \\ \vdots \\ \psi \end{array} \frac{}{\varphi \odot \psi} \ (\odot I_1^i) \qquad \begin{array}{c} [\psi]_j \\ \vdots \\ \varphi \end{array} \frac{}{\varphi \odot \psi} \ (\odot I_2^j) \tag{3.6.67}
$$

Here the construction of the harmoniously-induced GE-rule also does not yield a locally-complete rule.

The problem is related to the flatness of the resulting harmoniously GE-rule, and is further discussed in Section 6.6, under a different notion of harmony. A positive solution is obtained by the use of higher-level rules, discussed in Section 6.1.

3.7 Meeting the Criteria for Qualifying as meaning-conferring

In this section I inspect how do the rules of different logics meet the criteria for qualifying as meaning-conferring proposed at the beginning of the section, and mention some conclusions from the results of this inspection.

3.7.1 Intuitionistic Logic

Obviously, by its conception by Gentzen as "capturing" natural mathematical reasoning, Intuitionistic Logic answers positively the answerability criterion. Furthermore, all its rules are pure.

The next observation is, that NJ, the intuitionistic ND-system (when presented with GE-rules), meets the harmony and stability criteria.

harmony in form The harmony in form is obvious, as all the E-rules are GE-rules.

intrinsic harmony: I next turn to show that NJ is intrinsically harmonious, satisfying the first criterion qualifying it to serve as a meaning-conferring ND-system. This is shown by exhibiting reductions that invert every possible maximal formula in an NJ-derivation, in spite of this fact following from the general relationship between harmony in form and intrinsic harmony (cf. Section 3.6.3). Note that all the reductions are well-defined by the closure under derivation composition of NJ (proposition 2.1.1).

Inverting Conjunction:

$$\cfrac{\cfrac{\overset{\mathcal{D}_1}{\varphi} \quad \overset{\mathcal{D}_2}{\psi}}{(\varphi \wedge \psi)} (\wedge I)}{\varphi} (\wedge E_1) \quad \leadsto_r \quad \overset{\mathcal{D}_1}{\varphi} \tag{3.7.68}$$

Similarly for $(\wedge E_2)$. If the GE rule is used for elimination, the reduction is the following.

$$\cfrac{\cfrac{\overset{\mathcal{D}_1}{\varphi} \quad \overset{\mathcal{D}_2}{\psi}}{\varphi \wedge \psi} (\wedge I) \quad \cfrac{[\varphi]_1, [\psi]_2}{\overset{\mathcal{D}_3}{\chi}}}{\chi} (\wedge GE^{1,2}) \quad \leadsto_r \quad \mathcal{D}_3[\varphi := \overset{\mathcal{D}_1}{\varphi}, \psi := \overset{\mathcal{D}_2}{\psi}] \tag{3.7.69}$$

Inverting Disjunction:

$$\cfrac{\cfrac{\overset{\mathcal{D}}{\varphi}}{(\varphi \vee \psi)} (\vee I_1) \quad \cfrac{[\varphi]_i}{\overset{\mathcal{D}_1}{\chi}} \quad \cfrac{[\psi]_i}{\overset{\mathcal{D}_2}{\chi}}}{\chi} (\vee E^i) \quad \leadsto_r \quad \mathcal{D}_1[[\varphi]_i := \overset{\mathcal{D}}{\varphi}] \atop \chi \tag{3.7.70}$$

Note that $\overset{\mathcal{D}}{\varphi}$ is substituted for *every* use of $[\varphi]_i$ in $\mathcal{D}_1$. Similarly for the other case of $(\vee I_2)$.

Inverting Implication:

$$\cfrac{\cfrac{\overset{[\varphi]_i}{\overset{\mathcal{D}_1}{\psi}}}{(\varphi \to \psi)} (\to I^i) \quad \overset{\mathcal{D}_2}{\varphi}}{\psi} (\to E) \quad \leadsto_r \quad \mathcal{D}_1[[\varphi]_i := \overset{\mathcal{D}_2}{\varphi}] \atop \psi \tag{3.7.71}$$

If $(\to GE)$ is used, the reduction is

$$\cfrac{\cfrac{\overset{[\varphi]_i}{\overset{\mathcal{D}_1}{\psi}}}{\varphi \to \psi} (\to I^i) \quad \overset{\mathcal{D}_2}{\varphi} \quad \cfrac{[\psi]_j}{\overset{\mathcal{D}_3}{\chi}}}{\chi} \to GE^j) \quad \leadsto_r \quad \mathcal{D}_3[[\psi]_j := \cfrac{\mathcal{D}_1[[\varphi]_i := \overset{\mathcal{D}_2}{\varphi}]}{\psi}] \atop \chi \tag{3.7.72}$$

Note that since negation is a special case of implication (implying $\bot$), negation is invertible too via a special case of the implication reduction.

inverting ⊥: The condition is satisfied vacuously, as there are no I-rules for $\bot$, so it cannot occur as a maximal formula in an NJ-derivation.

Inverting universal quantification:

$$\frac{\dfrac{\mathcal{D}}{\varphi}\ (\forall I)}{\varphi[x := y]}\ (\forall E) \quad \leadsto_r \quad \frac{\mathcal{D}[x := y]}{\varphi[x := y]} \tag{3.7.73}$$

Inverting existential quantification:

$$\frac{\dfrac{\dfrac{\mathcal{D}}{\varphi[x := y]}}{\exists x.\varphi}\ (\exists I) \quad \dfrac{[\varphi]_i}{\dfrac{\mathcal{D}'}{\psi}}}{\psi}\ (\exists E_i) \quad \leadsto_r \quad \frac{\dfrac{\mathcal{D}}{\varphi[x := y]}}{\dfrac{\mathcal{D}'[x := y]}{\psi}} \tag{3.7.74}$$

Both reductions are well-defined in view of Proposition 2.1.2 (closure of NJ-derivations under term-substitution). I skip the reductions with GE-rules.

Immediate conclusions from the above reductions are the following propositions.

Proposition 3.7.5 (FA in IL). *IL satisfies* **FA**.

Proposition 3.7.6 (The disjunction property).

$$\vdash_{NJ} : \varphi \lor \psi \text{ iff } \vdash_{NJ} : \varphi \text{ or } \vdash_{NJ} : \psi \tag{3.7.75}$$

In particular,

$$\vdash_{NJ} : \varphi \lor \neg\varphi \text{ iff } \vdash_{NJ} : \varphi \text{ or } \vdash_{NJ} : \neg\varphi \tag{3.7.76}$$

The local-completeness of NJ: I next show below the expansions for NJ.

expanding conjunction: Shown in (3.6.32).

expanding disjunction:

$$\frac{\mathcal{D}}{(\varphi \lor \psi)} \quad \leadsto^{\mathcal{E}} \quad \frac{\dfrac{\mathcal{D}}{(\varphi \lor \psi)} \quad \dfrac{[\varphi]_i}{(\varphi \lor \psi)}\ (\lor_1 I) \quad \dfrac{[\psi]_i}{(\varphi \lor \psi)}\ (\lor_2 I)}{(\varphi \lor \psi)}\ (\lor E^{1,2}) \tag{3.7.77}$$

expanding implication:

$$\frac{\mathcal{D}}{(\varphi \to \psi)} \quad \leadsto^{\mathcal{E}} \quad \frac{\dfrac{\dfrac{\mathcal{D}}{(\varphi \to \psi)} \quad [\varphi]_i}{\psi}\ (\to E)}{(\varphi \to \psi)}\ (\to I^i) \tag{3.7.78}$$

Again, since negation is a special case of implication, it also expands.

expanding $\bot$*:* Here, one has to notice that there are no[15] *I*-rules for $\bot$ to re-introduce it after its elimination, so the following is a kind of "vacuous expansion" (see Prawitz [151]) .

$$\begin{array}{c} \mathcal{D} \\ \bot \end{array} \leadsto \mathcal{E} \quad \begin{array}{c} \mathcal{D} \\ \dfrac{\bot}{\bot} \ (\bot E) \end{array} \tag{3.7.80}$$

expanding universal quantification:

$$\begin{array}{c} \mathcal{D} \\ \forall x\varphi \end{array} \leadsto \mathcal{E} \quad \dfrac{\dfrac{\mathcal{D}}{\forall x.\varphi} \ (\forall E)}{\dfrac{\varphi[x := y]}{\forall x.\varphi}} \ (\forall I) \tag{3.7.81}$$

expanding existential quantification:

$$\begin{array}{c} \mathcal{D} \\ \exists x.\varphi \end{array} \leadsto \mathcal{E} \quad \dfrac{\dfrac{\mathcal{D}}{\exists x.\varphi} \quad \dfrac{[\varphi[x := y]]_i}{\exists x.\varphi} \ (\exists I)}{\exists x.\varphi} \ (\exists E^i) \tag{3.7.82}$$

3.7.2 Classical Logic

First, as noted by Hjortland [76], the effect of adding *tonk* to Classical Logic is even more devastating than its addition to Intuitionistic Logic. While in the latter it trivializes the consequence relation, for the former it leads, via (RAA), to actual (Post-) inconsistency, rendering every formula provable (under no assumptions)!

$$\dfrac{\dfrac{\dfrac{[\varphi]_1}{\varphi \ tonk \ \bot} \ (tonkI)}{\dfrac{\bot}{\neg\varphi} \ (RAA^1)} \quad \dfrac{\dfrac{[\neg\varphi]_2}{\varphi \ tonk \ \bot} \ (tonkI)}{\dfrac{\bot}{\neg\neg\varphi} \ (RAA^2)} (\neg E)}{\dfrac{\bot}{\varphi} \ (\bot E)} \tag{3.7.83}$$

An immediate observation (that was encountered before) is that (DNE) and (RAA) cause classical negation to extend non-conservatively the positive implicational fragment, by the derivations of Peirce's law in (2.2.16) and (2.2.20).

Obviously, both (DNE) and (RAA) fail to satisfy the complexity requirement, if complexity is taken as syntactic complexity. For another view on the complexity issue, see Kürbis [97].

---

[15]By Dummett's view, the infinitary *I*-rule (in 2.1.2) is also harmonic with $(\bot E)$. The expansion using this rule is

$$\begin{array}{c} \mathcal{D} \\ \bot \end{array} \leadsto \mathcal{E} \quad \dfrac{\dfrac{\mathcal{D}}{\dfrac{\bot}{p_1}} \ (\bot E) \quad \dots \quad \dfrac{\mathcal{D}}{\dfrac{\bot}{p_i}} \ (\bot E) \quad \dots}{\bot} \ (\bot I) \tag{3.7.79}$$

Next, observe that intrinsic harmony fails too for classical logic.

This can be seen easiest via the following irreducible derivation using (DNE), that has a maximal formula in the form of double negation.

$$
\begin{array}{c}
[\neg\varphi]_1 \\
\vdots \\
\dfrac{\bot}{\neg\neg\varphi} \; (\neg I^1) \\
\dfrac{}{\varphi} \; (DNE)
\end{array}
\tag{3.7.84}
$$

Another example, due to Prawitz, is the following irreducible derivation of LEM, also having an irremovable maximal formula.

$$
\dfrac{\dfrac{\dfrac{\dfrac{[\varphi]_1}{(\varphi\vee\neg\varphi)} \; (\vee I) \quad [\neg(\phi\vee\neg\varphi)]_2}{\neg\varphi} \; (\neg I)_1}{(\varphi\vee\neg\varphi)} \; (\vee I) \quad [\neg(\varphi\vee\neg\varphi)]_2}{\dfrac{\neg\neg(\varphi\vee\neg\varphi)}{\varphi\vee\neg\varphi} \; (DNE)} \; (\neg I)_2
\tag{3.7.85}
$$

Using (LEM) as an axiom raises the question of the relationship of this axiom to the determination of the meaning of '$\vee$'. Should it be viewed as another $(\vee I)$ rule? Consider the derivation below.

$$
\dfrac{\overline{\varphi\vee\neg\varphi} \quad \Gamma,\varphi:\psi \quad \Gamma,\neg\varphi:\psi}{\Gamma:\psi} \; (\vee E)
\tag{3.7.86}
$$

On the one hand, there is no reduction for this derivation. On the other hand, though, it is not clear whether $\varphi\vee\neg\varphi$ occurs in this derivation as a maximal formula.

Based on the above "disharmonious" behavior of classical logic, Dummett and Prawitz reject it as not amenable to proof-theoretic justification, hence unacceptable, and conclude that intuitionistic logic, that *is* proof-theoretically justifiable, is the "correct" logic.

Several responses were given, within PTS, to this claim, rebuking it from several angles.

- Accept the criteria for meaning-conferring, but maintain that their violations indicated above, attributed to CL itself, rely *on its presentation*, as described above. There are different presentation of CL that do satisfy harmony and conservative extension. Two such presentation are *multiple conclusion ND* (see Section 4.2) and *bilateralism* (see Section 4.4).

- Reject the proposed formulation of the criteria for qualifying as meaning-conferring and propose another formulation of proof-theoretic "well-behaved" criteria, met by CL. A revision of the notion of harmony is presented in Section 4.5.1. Revisions of the I/E-rules and their accompanying reductions is presented in Section 4.5.3.

- Revise the relevant notion of *complexity* so as not to be violated by Classical Logic. Such a revision is presented in Section 4.5.4.

- Accept the proposed criteria for meaning-conferring, but exempt negation from their scope. Kürbis, holding such a view ([97]) while adhering to the general PTS view, reaches the far-reaching conclusion that negation cannot (and should not) be defined proof-theoretically. This is also presented in Section 4.5.4.

It is interesting to note that while according to most views classical negation is to be "blamed" for the violations of the criteria for meaning-conferring (strongly stressed, e.g., in [99] and [97]), a different view is taken by Read, viewing intuitionistic implication as too weak, "blaming" it for CL being non-conservative over IL (see discussion on p. 143).

3.7.3 Relevant Logic

A consequence of Proposition (2.4.4) is the following theorem, ensuring the intrinsic harmony of $\mathbf{R}_\to$.

Theorem 3.7.4 (local-soundness of $\mathbf{R}_\to$). $\mathbf{R}_\to$ *is locally-sound.*

Proof. The required reduction is

$$
\cfrac{\cfrac{\begin{array}{c}[\varphi]_i\\ \mathcal{D}\\ \psi_\alpha \quad (i\epsilon\alpha)\end{array}}{(\varphi\to_{\mathbf{R}}\psi)_{\alpha-\{i\}}}\ (\to I^i_{\mathbf{R}}) \quad \begin{array}{c}\mathcal{D}'\\ \varphi_\beta\end{array}}{\psi_{\alpha\cup\beta-\{i\}}}\ (\to E_{\mathbf{R}}) \quad\rightsquigarrow_r\quad \cfrac{\mathcal{D}[[\varphi]_i := \begin{array}{c}\mathcal{D}'\\ \varphi_\beta\end{array}]^{\mathbf{R}}}{\psi_{\alpha\cup\beta-\{i\}}} \tag{3.7.87}
$$

which is well-defined by Proposition 2.4.4.   □

Example 3.7.15. *Consider the following* $\mathbf{R}_\to$ *derivation establishing*

$$
(\varphi\to_{\mathbf{R}}(\varphi\to_{\mathbf{R}}(\psi\to_{\mathbf{R}}\varphi)))_1, \psi_2\vdash_{\mathbf{R}_\to}(\varphi\to_{\mathbf{R}}(\varphi\to_{\mathbf{R}}\chi))_{1,2} \tag{3.7.88}
$$

$$
\cfrac{\cfrac{[\varphi]_5\quad (\varphi\to_{\mathbf{R}}(\varphi\to_{\mathbf{R}}(\psi\to_{\mathbf{R}}\varphi)))_1}{(\varphi\to_{\mathbf{R}}(\psi\to_{\mathbf{R}}\chi))_{1,5}}\ (\to_{\mathbf{R}}E)\quad \cfrac{\cfrac{\psi_2\quad \cfrac{\cfrac{[\varphi]_3\quad [(\varphi\to_{\mathbf{R}}(\psi\to_{\mathbf{R}}\chi))]_4}{(\psi\to_{\mathbf{R}}\chi)_{3,4}}\ (\to_{\mathbf{R}}E)}{\cfrac{\chi_{2,3,4}}{(\varphi\to_{\mathbf{R}}\chi)_{2,4}}\ (\to_{\mathbf{R}}I^3)}}{\cfrac{((\varphi\to_{\mathbf{R}}(\psi\to_{\mathbf{R}}\chi))\to_{\mathbf{R}}(\varphi\to_{\mathbf{R}}\chi))_2}{}}\ (\to_{\mathbf{R}}I^4)}{(\varphi\to_{\mathbf{R}}\chi)_{1,2,5}}\ (\to_{\mathbf{R}}E)}{(\varphi\to_{\mathbf{R}}(\varphi\to_{\mathbf{R}}\chi))_{1,2}}\ (\to_{\mathbf{R}}I^5)
$$

$$
\tag{3.7.89}
$$

The underlined $((\varphi \to_\mathbf{R}(\psi \to_\mathbf{R}\chi)) \to_\mathbf{R}(\varphi \to_\mathbf{R}\chi))_2$ is a maximal formula. The reduction removing it produces the following derivation.

$$
\cfrac{
  \cfrac{
    \cfrac{
      \cfrac{
        [\varphi]_3 \qquad
        \cfrac{[\varphi]_5 \quad (\varphi \to_\mathbf{R}(\varphi \to_\mathbf{R}(\psi \to_\mathbf{R}\chi)))_1}{(\psi \to_\mathbf{R}\chi)_{1,5}}\ (\varphi \to_\mathbf{R})
      }{(\psi \to \chi)_{1,3,5}}\ (\to_\mathbf{E})
    }{
      \cfrac{\psi_2 \qquad (\psi\to\chi)_{1,3,5}}{\chi_{1,2,3,5}}\ (\to_\mathbf{R}E)
    }
  }{(\varphi \to_\mathbf{R}\chi)_{1,2,3}}\ (\to_\mathbf{R}I^5)
}{(\varphi \to_\mathbf{R}(\varphi \to_\mathbf{R}\chi))_{1,2}}\ (\to_\mathbf{R}I^3)
$$

Incidentally, the derivation (3.7.89) is brought in Prawitz [146] as an example of an irreducible derivation in a variant R_p of relevant logic, in which a discharge always discharges all occurrences of the discharged formula (the occurrences need not be indexed at all), exhibiting an example of an essential non-closure under derivation composition.

As for the answer for the local-completeness of $\mathbf{R}_\to$, the positive answer here is much easier, as no derivation substitutions are involved. The expansion for '$\to_\mathbf{R}$' is a slight variation of that for the intuitionistic '$\to$' in (3.7.78).

$$
\cfrac{\mathcal{D}}{(\varphi \to_\mathbf{R}\psi)_\alpha} \quad \leadsto^\mathcal{E} \quad
\cfrac{
  \cfrac{
    \cfrac{\mathcal{D}}{(\varphi \to_\mathbf{R}\psi)_\alpha} \quad [\varphi]_i
  }{\psi_{\alpha \cup \{i\}}}\ (\to_\mathbf{R}E)
}{(\varphi \to_\mathbf{R}\psi)_\alpha}\ (\to_\mathbf{R}I^i)
\qquad (3.7.90)
$$

Note that the expansion restores the conclusion with the *same* set of dependency indices.

3.7.4 Modal Logic

I first consider harmony and stability.   The first observation is that the **S4** rules (in (2.5.40) and (2.5.41)) are locally-sound. The reductions are as follows.

$$
\cfrac{
  \cfrac{\cfrac{\mathcal{D}_1}{\varphi}}{\Box\varphi}\ (\Box I)
}{\varphi}\ (\Box E) \quad \leadsto_r \quad
\cfrac{\mathcal{D}_1}{\varphi}
\qquad (3.7.91)
$$

and

$$
\cfrac{
  \cfrac{\mathcal{D}_1}{\varphi}\ (\Diamond I) \qquad
  \cfrac{\cfrac{[\varphi]_1}{\mathcal{D}_2}}{\Diamond\psi}
}{\Diamond\psi}\ (\Diamond E^1) \quad \leadsto_r \quad
\mathcal{D}_2[\varphi := \cfrac{\mathcal{D}_1}{\varphi}]
\qquad (3.7.92)
$$

Similar reductions exist for **S5**.

However, some questions about the relationship between ML and harmony are raised by Read [167], based on the following two observations.

1. **S4** and **S5** differ in their I-rules but share a common E-rule.

2. **S4** and its weaker counterpart **K4** share their I-rules, but differ in their E-rules (recall that **K4** has *no* ($\square E$)-rule at all).

As an example of this puzzling situation, Read analyses the ($\lozenge I$)-rule, and asks which ($\lozenge E$)-rule would be licensed by it according to the Inversion Principle[16]. The answer Read gets is that it is ($\lozenge E$), *but without its side-condition*!

The puzzlement of Read is well-placed, but is not attributed to its true source. Harmony ascertains that the E-rules are not too strong w.r.t. the I-rules. However, the problem with the ML is that the E-rules are too weak w.r.t. the I-rules, namely the lack of *stability*. This is manifested by the lack of *expansions* realizing *local-completeness*. This is most clearly seen in the case of **T**. After eliminating $\Gamma : \square \varphi$ via ($\square E$), the result is $\Gamma : \varphi$. From this, $\Gamma : \square \varphi$ cannot be obtained via ($\square I$), because the side-condition requires the premise $\square \Gamma : \varphi$.

The lack of local-completeness in MLs was already noted by Pfenning and Davies [143]. Their amendment of the situation[17] is the formation of more complicated *hypothetical judgments*, embodied in a sequent with a partitioned antecedent (into a modal part and a "plain" part).

I like to pursue here Read's solution in [167] because he resorts to *labelled ND-systems*. General references on such systems are Gabbay [57] and Viganò [227]. The appeal of such system can be seen as serving two purposes.

1. Shed a light on one important role of labelling within the PTS programme, namely, imposing the balance between I/E-rules.

2. Put objects that are endowed an ontological status by MTS to a *formal* use as a mere artefacts of a proof-system, "possible worlds" in this specific case.

   I advance this view further in dealing with natural language in part **II** of the book.

Specifically, Read builds on a proposal by Simpson [201] of using "possible worlds" as formal label in rules and derivations. The use of those labels provides resources for finer distinction among otherwise "similar" rules. A similar approach (also for sequent calculi) is advocated by Negri (e.g., [129]). All this is explained in the next section.

For justification of adopting labelled ND as a corrective measure see Read [169].

---

[16]Read does not relate directly to the principle, but uses its contents.
[17]They are interested in the intuitionistic version of MLs, with the usual constructive flavor.

3.7.4.1   An harmonious and stable labelled ND-system for ML

Let **L** be a countable set of *labels*, ranged over by l, l_i, and let R be a relation over labels. In the proof-system, the objects over which the rules are defined (in the simple presentation mode) are of two sorts:

- Pairs of the for form $l : \varphi$, where $l \in \mathbf{L}$ and φ a formula in the object-language of the ML.

- Formulas $l_1 R l_2$, with $l_1, l_2 \in \mathbf{L}$.

The rules for the non-modal operators preserve the labels. I now can formulate the following I/E-rules

$$\dfrac{\begin{array}{c}[l_1 R l_2]_i \\ \vdots \\ l_2 : \varphi \end{array}}{l_1 : \Box\varphi}\ (\Box I^i), \ l_1 \neq l_2,\ l_2 \text{ fresh} \tag{3.7.93}$$

$$\dfrac{l_1 : \Box\varphi \quad l_1 R l_2}{l_2 : \varphi}\ (\Box E) \tag{3.7.94}$$

$$\dfrac{l_1 : \varphi \quad l_1 R l_2}{l_1 : \Diamond\varphi}\ (\Diamond I) \tag{3.7.95}$$

$$\dfrac{l_1 : \Diamond\varphi \qquad \begin{array}{c}[l_2 : \varphi]_i, [l_1 R l_2]_j \\ \vdots \\ l_3 : \chi \end{array}}{l_3 : \chi}\ (\Diamond E^{i,j}), \ l_2 \neq l_1, l_2 \neq l_3,\ l_2 \text{ fresh} \tag{3.7.96}$$

Clearly, the restrictions on labels in the rules ensure closure under derivation composition.

Below are shown the reduction and expansion establishing local-soundness and local-completeness.

Reduction for $\Box$:

$$\dfrac{\dfrac{\begin{array}{c}[l_1 R l_2]_i \\ \mathcal{D}_1 \\ l_2 : \varphi \end{array}}{l_1 : \Box\varphi}\ (\Box I^i) \quad \dfrac{\mathcal{D}_2}{l_1 R l_2}}{l_2 : \varphi}\ (\Box E) \quad \rightsquigarrow_r \quad \dfrac{\mathcal{D}_2}{\mathcal{D}_1[l_1 R l_2 := l_1 R l_2]}{l_2 : \varphi} \tag{3.7.97}$$

Expansion for $\Box$:

$$\dfrac{\dfrac{l_1 : \Box\varphi \quad [l_1 R l_2]_i}{l_2 : \varphi}\ (\Box E)}{l_1 : \Box\varphi}\ (\Box I^i) \tag{3.7.98}$$

Reduction for $\Diamond$:

$$\dfrac{\dfrac{\mathcal{D}_1 \quad \mathcal{D}_2}{\dfrac{l_2 : \varphi \quad l_1 R l_2}{l_1 : \Diamond \varphi}} (\Diamond I) \quad \dfrac{[l_2 : \varphi]_i, [l_1 R l_2]_j}{\dfrac{\mathcal{D}_3}{l_3 : \chi}} (\Diamond E^{i,j})}{l_3 : \chi}, \leadsto_r \quad \dfrac{\mathcal{D}_1 \qquad \mathcal{D}_2}{\dfrac{\mathcal{D}_3[l_2 : \varphi := l_2 : \varphi, l_1 R l_2 := l_1 R l_2]}{l_3 : \chi}}$$

$$(3.7.99)$$

Expansion for $\Diamond$:

$$\dfrac{l_1 : \varphi \qquad \dfrac{[l_1 : \varphi]_i, [l_1 R l_2]_j}{l_1 : \Diamond \varphi} (\Diamond I)}{l_1 : \varphi} (\Diamond E^{i,j})$$

$$(3.7.100)$$

So, local-completeness has been regained. The distinctions needed for distinguishing the rules for the different MLs, are imposed via the different properties of the relation R, parallel to the axioms assumed for the accessibility relation in the Kripke semantics, expressed here as additional rules that can be used in a derivation lRl'. For example,

$$\dfrac{l_1 R l_2 \quad l_2 R l_3 \quad \begin{matrix}[l_1 R l_3]_i \\ \vdots \\ l : \varphi\end{matrix}}{l : \varphi} (4^i) \qquad \dfrac{l_1 R l_2 \quad l_1 R l_3 \quad \begin{matrix}[l_2 R l_3]_i \\ \vdots \\ l : \varphi\end{matrix}}{l : \varphi} (5^i) \qquad (3.7.101)$$

Read views such rules as structural rules about labels.

3.7.5   Modal Logic and context generality

While the major criteria of harmony and stability can be successfully enforced via an appeal to labeled ND-systems, there is a major obstacle for obtaining a proof-theoretically defined meaning for ML. Consider again the most basic $(\Box I)$-rule, the I-rule for '$\Box$', in (2.5.40). Because of its side-condition on Γ, namely that its members have to be modal formulas, it obviously violates the context generality criterion (cf. Section 1.6.4)! Thus, this rule induces a circularity: '$\Box$' can only be added to an object language in which it is already present ... Note that the alternative rule $(\Box \hat{I})$ (in (2.5.47)) fares no better, as '$\Box$' already figures in its premises.

In [98] (Section 5), Kürbis provides, from a philosophical perspective, an "excuse" for the situation in which the meaning of modal operators (an in particular '$\Box$') cannot be fully determined by means of an ND-system qualifying as meaning-conferring. He sees an *implicit* modal aspect in the mere notion of using a rule as meaning-conferring, as such a rule expresses the conclusion of the rule *necessarily* following from the premises of the rule. He does not provide, however, any means of capturing the meaning of '$\Box$' *relative* to the implicit necessity involved in the notion of following a rule.

I find this argument unconvincing, and consider a PTS for modal operators as an issue still open.

3.8   Summary

In this chapter, the various criteria that an ND-system has to meet in order to qualify as meaning-conferring have been put to use. Some of the frequently encountered logics have been tested for meeting those criteria. The main observation is that classical logic (in its single-conclusion presentation) fails to meet harmony and stability, while intuitionistic logic does meet them. In some cases, like Relevant Logic, require some mild means in order to meet the criteria. Responses to the failure of classical logic to meet harmony are presented in Chapter 4.

Chapter 4

Alternative Presentations and the Justification of Classical Logic

4.1 Introduction

In this chapter, I consider two other presentations of Classical Logic, both in ND-form, but more general that the $SCND$ systems used so far, under the presentation in which Classical Logic ended up failing the criteria for qualifying as meaning-conferring.

The first approach, in Section 4.2, presents a *multiple-conclusion ND-system*. An early reference to multiple-conclusion logics is by Shoesmith and Smiley [200]. I will use a notation close to Boričić [16], where derivations maintain the tree-form. I mainly follow Read [172] in using this presentation as a justification of Classical Logic.

The second approach, in Section 4.4, presents a *bilateralism ND-system*, in which *denial* (or *rejection*) is primitive, on par with assertion. I mainly follow Rumfitt [179] in the argument for justifying Classical Logic; however, I view bilateralism as having a more general role in PTS, as explained in Section 4.4.1. In particular, I use it for conferring proof-theoretic meaning on negative determiners in natural language (see Section 10.4 in part II of the book).

After that, some other proof-theoretical justifications of Classical Logic are presented, either modifying the harmony requirement, or giving-up purity of the I-rules.

All those approaches offer a proof-theoretic justification of Classical Logic, thereby showing that any rivalry between logics (mainly intuitionistic vs. classical) needs to be settled on other grounds.

4.2 Multiple-Conclusion Natural-Deduction

4.2.1 Introduction

In this Section, I adapt the definition of all the $SCND$ notions to the $MCND$ setting. The presentation is based on Francez [44].

4.2.2 Object language, contexts and sequents

The object language L is defined exactly as for $SCND$-systems. Sequents for $MCND$ have the form $\Gamma : \Delta$, where Γ is the *left context*, a finite (possibly empty) sequence of L-formulas, and Δ is the *right context*, equally structured.

4.2.3 Rules

A rule in a $MCND$-system, say (R), again has (finitely many) *premises* and (finitely many) *conclusions*, again all objects depending on the presentation mode. In contrast to the $SCND$ case, where the distinction between additivity and multiplicativity of rules applied only to left contexts, here they apply both to left and to right contexts. I assume the following assumption, that will affect the construction of a harmoniously-induced $MCND$ GE-rule.

Assumption (structural rule-uniformity): In a $MCND$-system $\mathcal{N}$, a rule is multiplicative on both Γ and Δ, or is additive on both Γ and Δ.

This assumption excludes rules that are additive on the left context and multiplicative on the right context, or vice versa. Note that this assumption *does not* exclude that some rules in $\mathcal{N}$ are additive, while other rules are multiplicative – the assumption restricts the form of a single rule only.

Here too there are two modes of presentation, also referred to as 'simple' and 'logistic'.

Simple presentation: The objects of $\mathcal{N}$ in its simple presentation, serving as premises and conclusions of rules, are finite (possibly empty) sequences Δ of formulas of L.

Additive rule:

$$\frac{[\Sigma_1]_1 \quad\quad\quad [\Sigma_p]_p}{\varphi, \Delta} \quad (R_s^{\bar{i}})$$
$$\varphi_1, \Delta \quad \cdots \quad \varphi_p, \Delta$$

$$\frac{\varphi_1, \Delta \quad \cdots \quad \varphi_p, \Delta}{\varphi, \Delta} \quad (R_s^{\bar{i}}) \tag{4.2.1}$$

Here too premises may depend on additional, non-discharged, lateral assumptions, left implicit also, where there is a constraint that all premises depend on the same lateral assumptions.

How should the notation $[\Sigma]_k$ be read? If $\Sigma = \alpha_1, \cdots, \alpha_n$, then $[\Sigma]_k$ is to be read as $[\alpha_1]_k \cdots [\alpha_n]_k$. That is, a collection of assumptions each being a single formula, collectively discharged by any application of an instance of the rule. It *is not* to be read s a single assumption constituting a sequence of formulas, an assumption discharged by applications of instances of the rule. This convention allows for a certain compaction of the notation, already fairly complicated; once explained, it should not cause any confusion.

Multiplicative rule:

$$\frac{\varphi_1, \Delta_1 \quad \cdots \quad \varphi_p, \Delta_p}{\varphi, \Delta_1, \cdots, \Delta_p} \quad (R_s^{\bar{i}}) \tag{4.2.2}$$

Here the restriction on the equality of lateral assumptions is not imposed, each premise possibly depending on a different collection of lateral assumptions.

Logistic presentation: The objects of $\mathcal{N}$ in its logistic presentation, serving as premises and conclusions of rules, are *sequents* of the above form $\Gamma : \Delta$. In this, those objects resemble more the ones used for the Sequent Calculi presentation of logics. However, the central ingredients of $SCND$-systems, namely the use of I/E-rules and the discharge of assumptions, are preserved.

Additive:

$$\frac{\Gamma, \Sigma_1 : \varphi_1, \Delta \quad \cdots \quad \Gamma, \Sigma_p : \varphi_p, \Delta}{\Gamma : \varphi, \Delta} \quad (R_l) \tag{4.2.3}$$

Here both left context and right context are shared among the premises.

Multiplicative:

$$\frac{\Gamma_1, \Sigma_1 : \varphi_1, \Delta_1 \quad \cdots \quad \Gamma_p, \Sigma_p : \varphi_p, \Delta_p}{\Gamma_1, \cdots, \Gamma_p : \varphi, \Delta_1, \cdots, \Delta_p} \quad (R_l) \tag{4.2.4}$$

Here both left context and right context may vary with the premises.

The notions of rule-generality are inherited from the $SCND$-setting, and apply both to Γ and to Δ. The classification of the operational rules in Section 1.5.2.2 remains intact for $MCND$-rules too.

Anticipating the presentation of Classical Logic in Section 4.3, the following $(\wedge I)$ rule is an example of a logistically-presented additive categorical rule for conjunction in $MCND$-form.

$$\frac{\Gamma : \varphi, \Delta \quad \Gamma : \psi, \Delta}{\Gamma : \varphi \wedge \psi, \Delta} \ (\wedge I)$$

(4.2.5)

For an example of a logistically-presented additive hypothetical rule, the I-rule for implication, can be used.

$$\frac{\Gamma, \varphi : \psi, \Delta}{\Gamma : \varphi {\rightarrow} \psi, \Delta} \ (\rightarrow I)$$

(4.2.6)

A note about (abuse of) notation: Because sequences of formulas may have repetitions of the same formulas, a strict formulation of a rule discharging assumptions has to indicate which instances (if any) of an assumption is discharged. This might be done, for example, by using $\Gamma(\varphi)$ instead of Γ, φ as above. Under this strict notation, the $(\rightarrow I)$ rule would appear as follows:

$$\frac{\Gamma(\varphi) : \Delta(\psi)}{\Gamma : \Delta(\varphi {\rightarrow} \psi)} \ (\rightarrow I)$$

where the parenthetical occurrence is a distinguished one. Since the notation below becomes complicated anyway, I will relax this strictness. Furthermore, the notation will pretend as if the exchange structural rule is in force, and display the principal formulas as peripheral. Since those extra complications are orthogonal to the issue of harmony, no harm should be caused by this abuse of notation.

As for structural rules, they can be applied on *both* sides of ':', i.e., both on Γ and on Δ. Note that while both Γ and Δ use commas as formula separators, the meaning of the comma differs according to which context it is a part of.

In ([76], p. 141), Hjortland finds an apparent difficulty when explicit assumption recording (as an explicit left context) is desired. Because a wrong representation of this context, where Γ is just added to nodes in a derivation, that according to him look like Γ, Δ, the comma turns ambiguous. However, by strictly adhering to Gentzen's logistic-$MCSC$ form used in the sequent-calculi ([62], p. 150), where the nodes are *sequents* $\Gamma : \Delta$, there is no ambiguity, and, as mentioned above, commas in Γ are interpreted differently from commas in Δ – for Classical Logic (though not necessarily for arbitrary $MCND$-systems), the comma in Γ is read conjunctively, while the comma in Δ is read disjunctively.

4.2.4 $MCND$-Derivations

I assume also for $MCND$-derivations the usual definition of (tree-shaped) $\mathcal{N}$-derivations, ranged over by $\mathcal{D}$, again defined separately for the two presentation modes of $\mathcal{N}$.

Definition 4.2.29 (Derivations). *The definition splits into the simple and logistic presentations.*

Simple derivations:

$$\mathcal{D}_j$$

Additive: *If φ_j, Δ, for $1 \leq j \leq p$, are simply presented derivations with dependency sets $\mathbf{d}_{\mathcal{D}_1} = \cdots = \mathbf{d}_{\mathcal{D}_p} = \mathbf{d}$ with $\Sigma_j \subseteq \mathbf{d}$, and if*

$$
\frac{\begin{array}{ccc} [\Sigma_1]_1 & & [\Sigma_p]_p \\ \vdots & & \vdots \\ \varphi_1, \Delta & \cdots & \varphi_p, \Delta \end{array}}{\varphi, \Delta} \ (R_s^{\bar{i}}) \tag{4.2.7}
$$

is an instance of (simply presented) additive rule in $\mathcal{R}_\mathcal{N}$ with a fresh discharge label, then

$$
\frac{\mathcal{D}}{\varphi, \Delta} =^{\text{df.}} \frac{\begin{array}{ccc} [\Sigma_1]_1 & & [\Sigma_p]_p \\ \mathcal{D}_1 & & \mathcal{D}_p \\ \varphi_1, \Delta & \cdots & \varphi_p, \Delta \end{array}}{\varphi, \Delta} \ (R_s^{\bar{i}}) \tag{4.2.8}
$$

is a derivation with $\mathbf{d}_{\mathcal{D}} = \mathbf{d} - \cup_{1 \leq j \leq p} \hat{\Sigma}_j$

The derivations $\mathcal{D}_1, \cdots, \mathcal{D}_p$ are the direct sub-derivations of $\mathcal{D}$.

$$\mathcal{D}_j$$

Multiplicative: *If φ_j, Δ_j, for $1 \leq j \leq p$, are simply presented derivations with dependency sets $\mathbf{d}_{\mathcal{D}_j}$ with $\Sigma_j \subseteq \mathbf{d}_{\mathcal{D}_j}$, and if*

$$
\frac{\begin{array}{ccc} [\Sigma_1]_1 & & [\Sigma_p]_p \\ \vdots & & \vdots \\ \varphi_1, \Delta_1 & \cdots & \varphi_p, \Delta_p \end{array}}{\varphi, \Delta_1, \cdots, \Delta_p} \ (R_s^{\bar{i}}) \tag{4.2.9}
$$

is an instance of (simply presented) multiplicative rule in $\mathcal{R}_\mathcal{N}$ with a fresh discharge label, then

$$
\frac{\mathcal{D}}{\varphi, \Delta} =^{\text{df.}} \frac{\begin{array}{ccc} [\Sigma_1]_1 & & [\Sigma_p]_p \\ \mathcal{D}_1 & & \mathcal{D}_p \\ \varphi_1, \Delta_1 & \cdots & \varphi_p, \Delta_p \end{array}}{\varphi, \Delta_1, \cdots, \Delta_p} \ (R_s^{\bar{i}}) \tag{4.2.10}
$$

is a derivation with $\mathbf{d}_{\mathcal{D}} = \cup_{1 \leq i \leq p} \mathbf{d}_{\mathcal{D}_i} - \cup_{1 \leq j \leq p} \hat{\Sigma}_j$

The derivations $\mathcal{D}_1, \cdots, \mathcal{D}_p$ are the direct sub-derivations of $\mathcal{D}$.

Reminder: *Recall the convention as to how $[\Sigma]_k$ is to be read. Thus, the derivation $\mathcal{D}$ has $|\Sigma_1| + \cdots + |\Sigma_p|$ leaves (each being just a formula).*

$$\mathcal{D}$$

Logistic derivations: *Here a derivation $\Gamma : \Delta$ has explicit assumptions Γ, on which a (multiple) conclusion Δ depends.*

- *Every instance of an identity sequent $\varphi : \varphi$ is a derivation.*

- **Additive:** *If* $\Gamma, \Sigma_1 : \Delta, \cdots, \Gamma, \Sigma_p : \Delta$, *for some* $p \geq 1$, *are logistically presented derivations, and if*

$$\frac{\Gamma, \Sigma_1 : \Delta \quad \cdots \quad \Gamma, \Sigma_p : \Delta}{\Gamma : \Delta} \ (R_l) \tag{4.2.11}$$

is an instance of a (logistically presented) rule in $\mathcal{R}_{\mathcal{N}}$, then

$$\overset{\mathcal{D}}{\Gamma : \Delta} =^{\mathrm{df.}} \frac{\overset{\mathcal{D}_1}{\Gamma, \Sigma_1 : \Delta} \quad \cdots \quad \overset{\mathcal{D}_p}{\Gamma, \Sigma_p : \Delta}}{\Gamma : \Delta} \ (R_l) \tag{4.2.12}$$

is a logistically presented derivation.

Multiplicative: *If* $\overset{\mathcal{D}_1}{\Gamma_1, \Sigma_1 : \Delta_1}, \cdots, \overset{\mathcal{D}_p}{\Gamma_p, \Sigma_p : \Delta_p}$, *for some* $p \geq 1$, *are logistically presented derivations, and if*

$$\frac{\Gamma_1, \Sigma_1 : \Delta_1 \quad \cdots \quad \Gamma_p, \Sigma_p : \Delta_p}{\Gamma_1, \cdots, \Gamma_p : \Delta_1, \cdots, \Delta_p} \ (R_l) \tag{4.2.13}$$

is an instance of a (logistically presented) rule in $\mathcal{R}_{\mathcal{N}}$, then

$$\overset{\mathcal{D}}{\Gamma_1, \cdots, \Gamma_p : \Delta_1, \cdots, \Delta_p} =^{\mathrm{df.}} \frac{\overset{\mathcal{D}_1}{\Gamma_1, \Sigma_1 : \Delta_1} \quad \cdots \quad \overset{\mathcal{D}_p}{\Gamma_p, \Sigma_p : \Delta_p}}{\Gamma_1, \cdots, \Gamma_p : \Delta_1, \cdots, \Delta_p} \ (R_l) \tag{4.2.14}$$

is a logistically presented derivation.
The derivations $\mathcal{D}_1, \cdots, \mathcal{D}_p$ are the **direct sub-derivations** *of $\mathcal{D}$.*

Note that here too, assumptions are introduced into a derivation via identity axioms and remain as such as long as not discharged by some application of a rule. Note also that the definition adheres to the structural rule-uniformity assumption.

For $MCND$-derivations, there is also an easy mutual conversion between the the two presentation modes:

simple to logistic: Convert each node in $\overset{\mathcal{D}}{\psi, \Delta}$ to $\overset{\mathcal{D}}{\Gamma : \psi, \Delta}$, where $\Gamma = \mathbf{d}_{\mathcal{D}}$.

logistic to simple: Convert each node in $\overset{\mathcal{D}}{\Gamma : \psi, \Delta}$ to $\overset{\mathcal{D}}{\psi, \Delta}$, setting $\mathbf{d}_{\mathcal{D}} = \Gamma$.

Definition 4.2.30 (derivability). Δ *is* derivable *from Γ in $\mathcal{N}$, denoted by* $\vdash_{\mathcal{N}} \Gamma : \Delta$, *iff there exist a simple $\mathcal{N}$-derivation $\overset{\mathcal{D}}{\Delta}$ (respectively, a logistic derivation $\overset{\mathcal{D}}{\Gamma : \Delta}$) of Δ from Γ in $\mathcal{N}$.*

Thus, the logistic presentation keeps track *explicitly* of the assumptions Γ on which Δ depends. Note that $\vdash_{\mathcal{N}} : \Delta$ indicates derivability of Δ from an empty context, in which case Δ is referred to[1] as a *(formal) theorem* of $\mathcal{N}$.

---

[1] Usually, this notion is only used for $\Delta = \{\varphi\}$, a single conclusion.

Derivations in an $MCND$-system are also depicted as trees (of formulas, or of sequents, according to the presentation mode), where a node and its descendants are an instance (of an application of) one of the rules. In derivation-trees, the discharged occurrences of the assumptions are again enclosed in square brackets, and marked, $[...]_i$, for some (unique) index i, to match i on the applied rule-name, and are leaves in the tree depicting the derivation. Operational rules of an $MCND$-system also operate on the right context of a sequent *only*. I also use $\mathcal{D} : \Gamma : \Delta$ to indicate a specific derivation of that sequent. In such a derivation, Γ is also referred to as the *open assumptions* of $\mathcal{D}$. If Γ is empty, the derivation is *closed*. The definition of a canonical derivation remains unchanged, namely essentially ending with an application of an I-rule.

The property of *closure under composition* is very important also for $MCND$-derivations, again manifesting the *composability* of $MCND$-derivations. While derivation-composition in $SCND$-systems take place at the leaves only, for $MCND$-derivations a whole path (defined below) is modified by derivation-substitution.

Definition 4.2.31 (paths). *A* path *in a $MCND$-derivation $\mathcal{D}$ is a sequence d_i, $1 \leq i \leq m$ (for some natural number $m \geq 1$) of nodes (which depend on the presentation mode) in $\mathcal{D}$, s.t.:*

- d_1 *is a leaf, (an assumption formula φ for a simple $\mathcal{D}$, and a sequent $\varphi : \varphi$ for a logistic $\mathcal{D}$).*

- *for $1 \leq i < m$, d_i is a premise of some rule application in $\mathcal{D}$, the conclusion of which is d_{i+1}.*

- d_m *is the conclusion of $\mathcal{D}$.*

For a leaf d, let $\Pi_{\mathcal{D}}(d)$ be the path in $\mathcal{D}$ starting at d.

Definition 4.2.32 (closure under derivation composition). *Let $\mathcal{N}$ be an $MCND$-system.*

Simple derivations *: Let φ, Δ_1' and $\begin{matrix} \mathcal{D}' \\ \end{matrix}$ and $\begin{matrix} \varphi, \Delta \\ \mathcal{D} \\ \Delta_1 \end{matrix}$ be two simply presented $\mathcal{N}$ derivations. $\mathcal{N}$ is closed under derivation composition iff the result of prefixing $\mathcal{D}'$ to the result of adding Δ_1' to every node in $\Pi_{\mathcal{D}}(\varphi)$ (in $\mathcal{D}$) is a legal $\mathcal{N}$ derivation, denoted by $\mathcal{D}'' = \begin{matrix} \mathcal{D}' \\ \mathcal{D}[\Pi_{\mathcal{D}}(\varphi) := \varphi, \Delta_1'] \\ \Delta_1, \Delta_1' \end{matrix}$.*

Logistic derivations: *Let $\Gamma' : \varphi, \Delta'$ and $\begin{matrix} \mathcal{D}' \\ \end{matrix} \begin{matrix} \mathcal{D} \\ \end{matrix}$ and $\Gamma, \varphi : \Delta$ be two logistically presented $\mathcal{N}$-derivation. $\mathcal{N}$ is closed under derivation composition if the result of prefixing $\mathcal{D}'$ to the result of adding Γ' to every antecedent and Δ' to every succedent of every node in $\Pi_{\mathcal{D}}(\varphi)$ (in $\mathcal{D}$) is also a derivation, denoted $\mathcal{D}'' = \begin{matrix} \mathcal{D}' \\ \mathcal{D}[\Pi_{\mathcal{D}}(\varphi) := \Gamma' : \varphi, \Delta'] \\ \Gamma - \varphi, \Gamma' : \Delta, \Delta' \end{matrix}$.*

Thus, in $\mathcal{D}''$, any use in $\mathcal{D}$ of the assumption φ is replaced by re-deriving φ according to $\mathcal{D}'$ (from possibly additional assumptions, on which the conclusion of $\mathcal{D}''$ (which is the same as the conclusion of $\mathcal{D}$) now depends). This conclusion, however, depends no more on the assumption φ. Note that if the assumption φ is not actually used in $\mathcal{D}$, then $\mathcal{D}[\Pi(\varphi) := \overset{\mathcal{D}'}{\varphi}, \Delta] \equiv \mathcal{D}$.

The definition of derivation composition is naturally extended to multiple simulta-neous derivation compositions $\mathcal{D}[\Pi(\varphi_1) := \overset{\mathcal{D}'_1}{\varphi_1}, \Delta_1, \cdots, \Pi(\varphi_m) := \overset{\mathcal{D}'_m}{\varphi_m}, \Delta_m]$, for any $m \geq 1$. The satisfaction of the property needs to be shown for the specific $MCND$-systems used for qualifying as meaning-conferring.

Closure under derivation-composition is a necessary condition for reductions in $MCND$ derivations, and has to be established whenever needed.

Example 4.2.16. *Suppose we have the following two simple $MCND$-derivations $\mathcal{D}$, $\mathcal{D}'$, using some rules that are not further specified.*

$$\mathcal{D}:\quad \dfrac{\dfrac{\dfrac{\varphi, \Delta_1 \quad \Delta_2}{\Delta_3}\,(R_1) \quad \Delta_4}{\Delta_5}\,(R_2)}{} \qquad \mathcal{D}':\quad \dfrac{\dfrac{\dfrac{\Sigma_1 \quad \Sigma_2 \quad \Sigma_3}{\Sigma_4}\,(R_3) \quad \Sigma_5}{\varphi, \Sigma_6}\,(R_4)}{}$$

In $\mathcal{D}$, $\Pi_{\mathcal{D}}(\varphi)$ consists of the nodes (φ, Δ_1), Δ_3 and Δ_5. The resulting $\mathcal{D}[\Pi_{\mathcal{D}}(\varphi) := \overset{\mathcal{D}'}{\varphi, \Sigma_6}]$ is as follows.

$$\dfrac{\dfrac{\dfrac{\dfrac{\Sigma_1 \quad \Sigma_2 \quad \Sigma_3}{\Sigma_4}\,(R_3) \quad \Sigma_5}{\varphi, \Delta_1, \Sigma_6}\,(R_4) \quad \Delta_2}{\Delta_3, \Sigma_6}\,(R_1) \quad \Delta_4}{\Delta_5, \Sigma_6}\,(R_2)$$

The richness of structure in $MCND$ emphasizes even more strongly how structural rules participate in meaning conferring. As observed by Došen in [28], the very same operational rule in (4.2.6) can give rise to three different meanings of implication, by varying the structural assumptions.

classical: No structural restriction imposed.

intuitionistic: Abolishing weakening on the right, on Δ (rendering the system single-conclusion).

relevant: Abolishing weakening on the left, on Γ (with two variants, classical and intuitionistic) depending whether weakening on the right is retained or abolished.

4.2.5   Harmony and stability in $MCND$-systems

In view of the move from premises and conclusions as formulas to premises and conclusions as contexts, a reconsideration of local-soundness and local-completeness is due. The impact of the fact that in $MCND$-systems both premises and conclusions consist of (finite) collections of formulas is, as observed by Hjortland ([76], p. 139), that a maximal formula is *disjunctively situated* (D-situated) w.r.t. a right context Δ. This affects the form of the reductions needed to establish local-soundness. The situation is similar for the arbitrary conclusion to be drawn by a harmoniously-induced GE-rule, needing a finer analysis of the dependency on right and left contexts, affecting its construction.

This can be best understood in terms of the following generalization of Prawitz's inversion principle to the $MCND$-environment, to be called the D-inversion principle.

Definition 4.2.33 (D-saturation). *A collection of direct grounds $\varphi'_1, \Delta_1, \cdots, \varphi'_m, \Delta_m$ for the introduction of φ, Δ (premises of a suitable I-rule) is D-saturated iff $\cup_{1 \leq j \leq m} \Delta_j = \Delta$.*

The two relevant cases of D-saturation to emerge below are:

- Every single premise of an additive I-rule.

- The collection of *all* premises of a multiplicative I-rule.

The D-inversion Principle: Every conclusion D-situated w.r.t. a right context Δ drawn from φ, itself D-situated w.r.t. Δ, can already be drawn from any D-saturated grounds for introducing φ.

This principle will underly the harmoniously-induced GE-rules constructed below, allowing the reductions that establish harmony, and the expansions that establish stability.

4.2.5.1 $MCND$ harmoniously-induced GE-rules

In this section, I adapt and extend the procedure described in Francez and Dyckhoff [54] for generating the harmoniously-induced GE-rules (harmonious in form) from given I-rules, to an $MCND$-environment. Thereby, Gentzen's remark quoted above, about "reading off" the E-rules from the I-rules, is extended to cover also $MCND$-systems. Only propositional rules are handled here. Recall that while in $SCND$-systems a conclusion is a formula, for $MCND$-systems, a conclusion is a (finite) collection of formulas. So, deriving an *arbitrary* conclusion means deriving an *arbitrary such collection*, to be denoted Δ'. The examples anticipate the $MCND$ presentations of Classical Logic in Section 4.3.

As already mentioned, the key observation, already anticipated by Hjortland ([76], p. 139), is that the arbitrary conclusion Δ' inferred by applying a GE-rule has to be *disjunctively situated (D-situated)* w.r.t. the right context Δ of the major premise. In other words, right contexts are propagated from a D-situated formula to its arbitrary conclusions. The construction is done for both simple and logistic presentations, where the latter reflects possible structural impact by distinguishing multiplicative and additive I-rules. I consider separately the two modes of presentation (simple, logistic), and within each first the special case of categorical I-rules (more easily comprehended), to be followed by hypothetical I-rules. The resulting harmoniously-induced GE-rules reflect those distinctions.

4.2.5.2   Simple presentation

Additive I-rule: For additive rules, by definition, since all premises and the conclusion share the same right context Δ, each premise is, on its own, a D-exhaustive ground for introduction of the conclusion.

Categorical I-rule: Suppose the simply presented additive categorical I-rules of an operator δ, the main operator of φ, can be schematically presented

$$\frac{\varphi_i^1, \Delta \;\cdots\; \varphi_i^{m_i}, \Delta}{\varphi, \Delta} \;(\delta I)_i, \quad 0 \le i \le n \qquad (4.2.15)$$

The premises of (δI_i), each a finite collection of formulas, are denoted $\Delta_i^j = \varphi_i^j, \Delta$, $1 \le j \le m_i$. Note again that φ, as well as all φ_i^j, are D-situated w.r.t. the *same* right context Δ, whence the additivity of the I-rules.

Then, to be harmonious in form, a harmoniously-induced GE-rule should draw an arbitrary conclusion Δ', D-situated w.r.t. Δ, from every D-exhaustive ground for introducing φ (a single premise of the I-rule) using that ground as a discharged assumption. Thus, the harmoniously-induced GE-rules have the following form.

$$\frac{\varphi, \Delta \quad \begin{matrix} [\varphi_1^{j_1}]_{l_1} \\ \mathcal{D}'_{1,j_1} \\ \Delta' \end{matrix} \;\cdots\; \begin{matrix} [\varphi_n^{j_n}]_{l_n} \\ \mathcal{D}'_{n,j_n}, \\ \Delta' \end{matrix}}{\Delta', \Delta} \;(\delta GE_{j_1,\cdots,j_n}^{l_1,\cdots,l_n}), \quad 0 \le i \le n,\, 1 \le j_i \le m_i$$

$$(4.2.16)$$

This construction gives rise to the following reductions.

$$
\cfrac{
\cfrac{\hat{\mathcal{D}}_1 \quad\; \hat{\mathcal{D}}_{m_i}}{\varphi_i^1, \Delta \cdots \varphi_i^{m_i}, \Delta}(\delta I)_i
\qquad
\cfrac{[\varphi_1^{j_1}]_{l_1}}{\cfrac{\mathcal{D}'_{1,j_1}}{\Delta'}}
\quad \cdots \quad
\cfrac{[\varphi_n^{j_n}]_{l_n}}{\cfrac{\mathcal{D}'_{n,j_n}}{\Delta'}}
}{\Delta', \Delta}
\;(\delta GE_{j_1 \,\cdots\, j_n}^{l_1, \cdots, l_n})
\quad \leadsto_r
$$

$$
\cfrac{\hat{\mathcal{D}}_{j_i}}{\cfrac{\mathcal{D}'_{i,j_i}[\Pi_{\mathcal{D}'_{i,j_i}}(\varphi_i^{j_i}) := \varphi_i^{j_i}, \Delta]}{\Delta', \Delta}}
$$

$$(4.2.17)$$

Here, $\varphi_i^{j_i}$ is one of the premises of $(\delta I)_i$, so that the js span all those premises. The total number of harmoniously-induced GE-rules is $\Pi_{1 \leq i \leq n} m_i$.

For a simply presented categorical additive I-rule, this form of a harmoniously-induced GE-rule indeed reflects the D-inversion idea: any arbitrary conclusion Δ' that can be drawn from (the major premise) φ, Δ, can already be drawn (D-situated w.r.t. Δ) from each collection of its grounds of introduction, being singletons in this additive case. Every one of the generated harmoniously-induced GE-rules "prepares itself", so to speak, to "confront" every one of the (δI)-rules (via one of its premises), as reflected by the reductions.

Recall that the availability of such a reduction constitutes part of the definition of intrinsic harmony. Note also that the availability of this reduction rests on the closure under derivation composition. An instance of this rule (for additive conjunction) appears in Read [168]. In the examples to follow, I allow myself some relaxation of the strict indexing used everywhere whenever no confusion should arise.

Example 4.2.17. *Consider the simply presented additive categorical I-rules for disjunction.*

$$
\frac{\varphi, \Delta}{\varphi \vee \psi, \Delta}\,(\vee I)_1
\qquad
\frac{\psi, \Delta}{\varphi \vee \psi, \Delta}\,(\vee I)_2
\qquad\qquad (4.2.18)
$$

Here $n = 2$ (two $(\vee I)$ rules), $m_1 = m_2 = 1$ (one premise for each $(\vee I)$ rule), $\Delta_1^1 = \varphi, \Delta$ and $\Delta_2^1 = \psi, \Delta$. By applying the construction above, the resulting one $(\vee GE)$-rule is the following.

$$
\cfrac{\varphi \vee \psi, \Delta
\qquad
\cfrac{[\varphi]_{l_1}}{\cfrac{\mathcal{D}'_1}{\Delta'}}
\qquad
\cfrac{[\psi]_{l_2}}{\cfrac{\mathcal{D}'_2}{\Delta'}}
}{\Delta', \Delta}
\;(\vee E^{l_1, l_2})
\qquad\qquad (4.2.19)
$$

Example 4.2.18. *Consider the simply presented additive categorical I-rule for conjunction.*

$$
\frac{\varphi, \Delta \quad \psi, \Delta}{\varphi \wedge \psi, \Delta}\,(\wedge I)
\qquad\qquad (4.2.20)
$$

Here $n = 1$ (one $(\wedge I)$ rule), $m_1 = 2$ (two premises for this single rule), where $\Delta_1^1 = \varphi, \Delta$ and $\Delta_1^2 = \psi, \Delta$. By applying the construction above, the resulting two harmoniously-induced $(\wedge GE)$-rules are the following.

$$\cfrac{\begin{array}{cc} \begin{array}{c} [\varphi]_{l_1} \\ \mathcal{D}_1' \\ \varphi \wedge \psi, \Delta \end{array} & \Delta' \end{array}}{\Delta', \Delta} \; (\wedge GE^{l_1})_1 \qquad \cfrac{\begin{array}{cc} \begin{array}{c} [\varphi]_{l_2} \\ \mathcal{D}_2' \\ \varphi \wedge \psi, \Delta \end{array} & \Delta' \end{array}}{\Delta', \Delta} \; (\wedge GE^{l_2})_2 \qquad (4.2.21)$$

Continuing examples (4.2.17) and (4.2.18), the resulting reductions are as follows.

$$\cfrac{\begin{array}{ccc} \begin{array}{c} \hat{\mathcal{D}} \\ \varphi, \Delta \\ \overline{\varphi \vee \psi, \Delta} \; (\vee I)_1 \end{array} & \begin{array}{c} [\varphi]_{l_1} \\ \mathcal{D}_1' \\ \Delta' \end{array} & \begin{array}{c} [\psi]_{l_2} \\ \mathcal{D}_2' \\ \Delta' \end{array} \end{array}}{\Delta', \Delta} \; (\vee GE^{l_1, l_2}) \quad \leadsto_r \quad \cfrac{\begin{array}{c} \hat{\mathcal{D}} \\ \mathcal{D}_1'[\Pi_{\mathcal{D}_1'}(\varphi) := \varphi, \Delta] \end{array}}{\Delta', \Delta}$$

$$(4.2.22)$$

(and similarly for $(\vee I)_2$).

$$\cfrac{\begin{array}{ccc} \begin{array}{cc} \begin{array}{c} \hat{\mathcal{D}}_1 \\ \varphi, \Delta \end{array} & \begin{array}{c} \hat{\mathcal{D}}_2 \\ \psi, \Delta \end{array} \\ \hline \varphi \wedge \psi, \Delta \end{array} \; (\wedge I) & \begin{array}{c} [\varphi]_{l_1} \\ \mathcal{D}_1' \\ \Delta' \end{array} \end{array}}{\Delta', \Delta} \; (\wedge GE^{l_1})_1 \quad \leadsto_r \quad \cfrac{\begin{array}{c} \hat{\mathcal{D}}_1 \\ \mathcal{D}_1'[\Pi_{\mathcal{D}_1'}(\varphi) := \varphi, \Delta] \end{array}}{\Delta', \Delta}$$

$$(4.2.23)$$

(and similarly for $(GE\wedge)_2$).

hypothetical I-rule: Suppose the simply presented additive hypothetical I-rules for an operator δ, the main operator in φ, are of the following form.

$$\cfrac{\begin{array}{ccc} \begin{array}{c} [\Sigma_i^1]_1 \\ \mathcal{D}_1 \\ \psi_i^1, \Delta \end{array} & \cdots & \begin{array}{c} [\Sigma_i^{m_i}]_{m_i} \\ \mathcal{D}_{m_i} \\ \psi_i^{m_i}, \Delta \end{array} \end{array}}{\varphi, \Delta} \; (\delta I^{1, \cdots, m_i})_i, \qquad 1 \leq i \leq n \qquad (4.2.24)$$

The ith rule has m_i *possibly* discharging premises, each discharging a collection Σ_i^j of assumptions, the support of the ground. When $m_j = 0$ (for some j), the jth premise discharges no assumptions (a categorical premise). Once again, recall the convention as to how $[\Sigma]_k$ is read. If all premises are categorical, the hypothetical rule reduces to the special case of a categorical rule. These I-rules generate harmoniously-induced GE-rules based on the same D-exhaustive collections of grounds, each GE-rule corresponding to one premise discharging assumptions in the ith I-rule. Note that the arbitrary conclusion can be drawn (D-situated w.r.t. Δ) from the grounds *provided* that the corresponding support has been derived (as D-situated w.r.t. Δ). The total number of harmoniously-induced GE-rules is the same as in the categorical case. Thus, the contribution of hypotheticality in an I-rule is two-folded. Each of the supports becomes a premise (in the corresponding GE-rule), and all the grounds become dischargeable assumptions (in all GE-rules).

The general form of the harmoniously-induced GE-rule is as follows.

$$
\dfrac{\hat{\mathcal{D}} \quad \mathcal{D}^*_{1,j_1} \quad \mathcal{D}^*_{n,j_n} \quad \begin{array}{c} [\psi_1^{j_1}]_{l_1} \\ \mathcal{D}'_{1,j_1} \end{array} \quad \begin{array}{c} [\psi_n^{j_n}]_{l_n} \\ \mathcal{D}'_{n,j_n} \end{array}}{\Delta', \Delta} \; (\delta GE^{l_1,\cdots,l_n}_{j_1,\cdots,j_n})
$$

$$\varphi, \Delta \quad \Sigma_1^{j_1}, \Delta \;\cdots\; \Sigma_n^{j_n}, \Delta \quad \Delta' \;\cdots\; \Delta'$$

(4.2.25)

$1 \le i \le n$, $1 \le j_i \le m_i$. If $m_i = 0$, there is no derivation of the ith support.

This construction leads to the following reductions (4.2.26), where each (δGE) is again "confronted" against each (δI). Note that the notation $\Pi_{\mathcal{D}}(\varphi)$ is naturally extended to $\Pi_{\mathcal{D}}(\Sigma)$ pointwise. Note the nestedness of the reduction. The availability of this reduction depends on closure of derivations under derivation composition.

Example 4.2.19. *Consider implication, with the following $(\to I)$ simply presented hypothetical rule.*

$$
\dfrac{\begin{array}{c} [\varphi]_l \\ \mathcal{D} \\ \psi, \Delta \end{array}}{\varphi \to \psi, \Delta} \; (\to I^l)
$$

(4.2.27)

Here $n = 1$ (one rule, with no categorical premise), $m_1 = 1$ (one discharging premise), $\Sigma_1^1 = \varphi$. This simple additive hypothetical I-rule gives rise to one GE-rule, as follows.

$$
\dfrac{\hat{\mathcal{D}} \quad \mathcal{D}^* \quad \begin{array}{c} [\psi]_l \\ \mathcal{D}'_1 \end{array}}{\Delta', \Delta} \; (\to GE^l)
$$

$$\varphi \to \psi, \Delta \quad \varphi, \Delta \quad \Delta'$$

(4.2.28)

Continuing example (4.2.19), the reduction for implication is the following, see 4.2.29.

Example 4.2.20. *Consider another simply presented hypothetical additive I-rule for 'if ... then ... else' (ite), where $n = 1$, and the single I-rule has no categorical premise, and has two premises discharging assumptions;*

$$
\dfrac{\begin{array}{cc} [\varphi]_{l_1} & [\neg\varphi]_{l_2} \\ \mathcal{D}_1 & \mathcal{D}_2 \\ \psi, \Delta & \chi, \Delta \end{array}}{ite(\varphi, \psi, \chi), \Delta} \; (ite I^{l_1, l_2})
$$

(4.2.30)

thus, the two following GE-rules are harmoniously-induced.

$$
\dfrac{ite(\varphi, \psi, \chi), \Delta \quad \mathcal{D}^* \quad \begin{array}{c} [\psi]_1 \\ \mathcal{D}'_1 \end{array}}{\Delta', \Delta} \; (ite GE^1)_1
$$

$$\varphi, \Delta \quad \Delta'$$

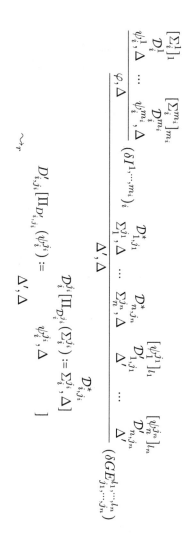

(4.2.26)

$$
\dfrac{
\begin{array}{c}
[\varphi]_1 \\
\mathcal{D} \\
\psi, \Delta
\end{array}
\ (\to I^1)
}{\varphi \to \psi, \Delta}
\qquad
\begin{array}{c}
\mathcal{D}^\star \\
\varphi, \Delta
\end{array}
\qquad
\begin{array}{c}
[\psi]_l \\
\mathcal{D}'_1 \\
\Delta'
\end{array}
\atop
\overline{\Delta', \Delta} \ (\to GE^l)
\qquad\leadsto_r\qquad
\begin{array}{c}
\mathcal{D}'_1[\Pi_{\mathcal{D}'_1}(\psi) :=
\begin{array}{c}
\mathcal{D}[\Pi_{\mathcal{D}}(\varphi) :=
\begin{array}{c}
\mathcal{D}^\star \\
\varphi, \Delta
\end{array}] \\
\psi, \Delta
\end{array}
] \\
\Delta', \Delta
\end{array}
$$

$$(4.2.29)$$

$$\frac{ite(\varphi,\psi,\chi),\Delta \quad \overset{\mathcal{D}^*}{\neg\varphi,\Delta} \quad \overset{[\chi]_1}{\underset{\mathcal{D}'_1}{\Delta'}}}{\Delta',\Delta} \; (iteGE^1)_2 \tag{4.2.31}$$

The reductions are as follows

$$\frac{\overset{[\varphi]_1 \quad [\neg\varphi]_2}{\underset{\psi,\Delta}{\mathcal{D}_1} \quad \underset{\chi,\Delta}{\mathcal{D}_2}} (iteI^{1,2}) \quad \overset{\mathcal{D}^*}{\varphi,\Delta} \quad \overset{[\psi]_3}{\underset{\mathcal{D}'_1}{\Delta'}}}{\Delta',\Delta} \; (iteGE^3)_1 \rightsquigarrow_r$$

$$\mathcal{D}'_1[\Pi_{\mathcal{D}'_1}(\psi) := \frac{\mathcal{D}_1[\Pi_{\mathcal{D}_1}(\varphi) := \overset{\mathcal{D}^*}{\varphi,\Delta}]}{\psi,\Delta}] \tag{4.2.32}$$

and

$$\frac{\overset{[\varphi]_1 \quad [\neg\varphi]_2}{\underset{\psi,\Delta}{\mathcal{D}_1} \quad \underset{\chi,\Delta}{\mathcal{D}_2}} (iteI^{1,2}) \quad \overset{\mathcal{D}^*}{\neg\varphi,\Delta} \quad \overset{[\chi]_3}{\underset{\mathcal{D}'_1}{\Delta'}}}{\Delta',\Delta} \; (iteGE^3)_2 \rightsquigarrow_r$$

$$\mathcal{D}'_1[\Pi_{\mathcal{D}'_1}(\psi) := \frac{\mathcal{D}_1[\Pi_{\mathcal{D}_1}(\neg\varphi) := \overset{\mathcal{D}^*}{\neg\varphi,\Delta}]}{\psi,\Delta}] \tag{4.2.33}$$

Clearly, the harmoniously-induced GE-rule for a categorical additive I-rule is the assumption-less special case, yielding the original formulation.

Multiplicative I-rule: Recall that here only the collection of all premises of an I-rule *together* are D-exhaustive.

Categorical I-rule: The simply presented multiplicative categorical I-rules of an operator δ, the main operator in φ, can be schematically presented as

$$\frac{\varphi_i^1,\Delta_i^1 \cdots \varphi_i^{m_i},\Delta_i^{m_i}}{\varphi,\Delta_i^1,\cdots,\Delta_i^{m_i}} \; (\delta I)_i, \quad 1 \le i \le n \tag{4.2.34}$$

Note that φ and all φ_i^j are D-situated w.r.t. possibly *different* right contexts Δ_i^j, whence the multiplicativity of the rule.

As mentioned above, no single premise can form a D-exhaustive ground, only all of the premises taken together. Intuitively, it is undeterminable which subset of $\Delta = \Delta_i^1,\cdots,\Delta_i^{m_i}$ is contributed by each separate premise. Therefore, the harmoniously-induced GE-rule should combine all the grounds for introducing δ (by all of δ's I-rules) and use each of those grounds as (discharged) assumptions for deriving an *arbitrary* conclusion Δ', thus having the following form:

$$\frac{\varphi, \Delta \qquad \begin{array}{c}[\varphi_1^1, \cdots, \varphi_1^{m_1}]_{l_1} \\ \mathcal{D}_1' \\ \Delta' \end{array} \quad \cdots \quad \begin{array}{c}[\varphi_n^1, \cdots, \varphi_n^{m_n}]_{l_n} \\ \mathcal{D}_n' \\ \Delta' \end{array}}{\Delta', \Delta} \; (\delta GE^{l_1, \cdots, l_n}) \tag{4.2.35}$$

Note that the decomposition of Δ into $\Delta_i^1, \cdots, \Delta_i^{m_i}$ is unavailable to the GE-rule, but becomes available during a reduction (see below), when the I-rule becomes available too.

For a simply presented multiplicative categorical rule, this form of the harmoniously-induced GE-rule indeed reflects the D-inversion idea: any arbitrary conclusion Δ' that can be drawn (D-situated w.r.t. Δ) from (the major premise, once δI) is determined) $\varphi, \Delta_i^1, \cdots, \Delta_i^{m_i}$, can already be drawn from the D-exhaustive grounds of introduction (all of them!) $[\varphi_i^j]$, $i = 1, \cdots, n$, $j = 1, \cdots, m_i$. Note again that all those assumed grounds are discharged by the rule.

The construction above leads directly to the following reduction, see Equation 4.2.36: Each such individual substitution adds *a possibly different* $\Delta_i^{m_i}$ to every node in every path $\Pi(\varphi_i^j)$.

Example 4.2.21. *Consider multiplicative conjunction $\otimes$ (known in Linear Logic as the tensor product, and in Relevant Logic as the intensional conjunction, or fusion), with the following I-rule.*

$$\frac{\varphi, \Delta_1 \quad \psi, \Delta_2}{\varphi \otimes \psi, \Delta_1, \Delta_2} \; (\otimes I) \tag{4.2.37}$$

Here $n = 1$ (one $(\otimes I)$ rule), $m_1 = 2$ (two premises for this single rule), where $\Delta_1^1 = \varphi, \Delta_1$ and $\Delta_1^2 = \psi, \Delta_2$. By applying the construction above, the resulting (single!) $(\wedge GE)$-rule is the following (formulated with the $(\otimes I)$ already in mind):

$$\frac{\varphi \otimes \psi, \Delta_1, \Delta_2 \qquad \begin{array}{c}[\varphi, \psi]_l \\ \mathcal{D}_1' \\ \Delta' \end{array}}{\Delta', \Delta_1, \Delta_2} \; (\otimes GE^l)_1 \tag{4.2.38}$$

Recall again that $[\varphi, \psi]_l$ represents two formula assumptions simultaneously discharged, not one assumption of a sequence (here of length 2). This is clearly seen by the reduction for multiplicative conjunction, presented below.

$$\frac{\dfrac{\begin{array}{cc}\mathcal{D}_1 & \mathcal{D}_2 \\ \varphi, \Delta_1 & \psi, \Delta_2\end{array}}{\varphi \otimes \psi, \Delta_1, \Delta_2} (\otimes I) \qquad \begin{array}{c}[\varphi, \psi]_l \\ \mathcal{D}_1' \\ \Delta'\end{array}}{\Delta', \Delta_1, \Delta_2} \; (\otimes GE^l)_1 \quad \leadsto_r$$

$$
\cfrac{
\cfrac{
\cfrac{\hat{\mathcal{D}}_1 \qquad \hat{\mathcal{D}}_{m_i}}{\varphi_i^1, \Delta_i^1 \cdots \varphi_i^{m_i}, \Delta_i^{m_i}}
}{
\cfrac{\varphi, \Delta_i^1, \ldots, \Delta_i^{m_i}}{\Delta', \Delta_i^1, \ldots, \Delta_i^{m_i}}\ (\delta I)_i
}
\qquad
\cfrac{[\varphi_1^1, \ldots, \varphi_1^{m_1}]_{l_1} \qquad [\varphi_n^1, \ldots, \varphi_n^{m_n}]_{l_n}}{
\cfrac{\mathcal{D}_1'}{\Delta'} \quad \cdots \quad \cfrac{\mathcal{D}_n'}{\Delta'}
}
}{\Delta'}\ (\delta GE^{l_1, \ldots, l_n})
$$

$$\rightsquigarrow_r$$

$$
\cfrac{
\mathcal{D}_i'\big[\Pi_{\mathcal{D}_i}(\varphi_i^1) := \varphi_i^1, \Delta_i^1, \ldots, \Pi(\varphi_i^{m_i}) := \varphi_i^{m_i}, \Delta_i^{m_i}\big]
}{
\cfrac{\hat{\mathcal{D}}_1 \qquad \hat{\mathcal{D}}_{m_i}}{\Delta', \Delta_i^1, \ldots, \Delta_i^{m_i}}
}
$$

$$(4.2.36)$$

$$\mathcal{D}'_1[\Pi_{\mathcal{D}'_1}(\varphi) := \overset{\mathcal{D}_1}{\varphi, \Delta_1}, \Pi_{\mathcal{D}'_1}(\psi) := \overset{\mathcal{D}_2}{\psi, \Delta_2}]$$
$$\Delta', \Delta_1, \Delta_2 \tag{4.2.39}$$

In comparing the multiplicative conjunction with the additive one in (4.2.18) in a sub-structural logic lacking contraction and weakening, like Linear Logic, we clearly get

$$\vdash \varphi \wedge \psi : \varphi, \quad \text{but} \quad \nvdash \varphi \otimes \psi : \varphi \tag{4.2.40}$$

The observation that the additive and multiplicative I-rules for conjunction yield different harmoniously-induced GE-rules (also for conjunction) appears already in Read [170] in a $SCND$-context, without resorting to additivity/multiplicativity of the right context Δ in general, and without any reference to D-exhaustiveness. I believe the latter provides a clearer insight to the source of the different results of the construction of harmoniously-induced GE-rules.

Hypothetical I-rule: Suppose a simply presented hypothetical multiplicative I-rule is the following:

$$
\begin{array}{ccc}
[\Sigma_i^1]_1 & & [\Sigma_i^{m_i}]_{m_i} \\
\mathcal{D}_1 & & \mathcal{D}_{m_i} \\
\psi_i^1, \Delta_i^1 & \cdots & \psi_i^{m_i}, \Delta_i^{m_i} \\
\hline
\varphi, \Delta_i^1, \cdots, \Delta_i^{m_i}
\end{array}
\; (\delta I^{1, \cdots, m_i})_i
\tag{4.2.41}
$$

The resulting harmoniously-induced simply presented multiplicative GE-rule is the following.

$$
\begin{array}{ccccc}
 & & [\psi_1^1, \cdots, \psi_1^{m^1}]_{l_1} & & [\psi_n^1, \cdots, \psi_n^{m^n}]_{l_n} \\
\mathcal{D}_1^* & \mathcal{D}_n^* & \mathcal{D}'_1 & & \mathcal{D}'_n \\
\varphi, \Delta \; \Sigma_i^1, \Delta & \cdots \; \Sigma_n^1, \Delta & \Delta' & \cdots & \Delta' \\
\hline
\multicolumn{5}{c}{\Delta', \Delta}
\end{array}
\; (\delta GE^{l_1, \cdots, l_n})
\tag{4.2.42}
$$

The resulting reduction is as follows (4.2.43).

Note that while the generation of harmoniously-induced GE-rules are functional, this function is not injective, and different sets of I-rules may lead to the same set of harmoniously-induced GE-rules; however, there is no reason to expect a one-one relationship, similarly to the $SCND$ case.

4.2.5.3 Logistic presentation

In this section, I turn to the construction of harmoniously-induced logistically-presented GE-rules, incorporating also the context Γ. In the multiplicative case, the context-uniformity assumption comes into place. I omit the specification of the resulting reductions, that become notationally cumbersome.

$$
\cfrac{
\begin{array}{ccc}
[\Sigma_i^1]_1 & & [\Sigma_i^{m_i}]_{m_i} \\
\mathcal{D}_1 & & \mathcal{D}_{m_i} \\
\psi_i^1, \Delta_i^1 & \cdots & \psi_i^{m_i}, \Delta_i^{m_i} \\
\hline
\varphi, \Delta_i^1, \ldots, \Delta_i^{m_i}
\end{array}
}{}
\ (\delta I^{1,\ldots,m_i})_i
$$

$$
\cfrac{
\begin{array}{ccc}
& \mathcal{D}_1^* & \\
\Sigma_1^1, \Delta_i & \cdots & \Sigma_n^1, \Delta_i \\
\end{array}
\qquad
\begin{array}{ccc}
[\psi_1^1,\ldots,\psi_1^{m_1}]_{l_1} & & [\psi_n^1,\ldots,\psi_n^{m_n}]_{l_n} \\
\mathcal{D}_1^* & \cdots & \mathcal{D}_n^* \\
\mathcal{D}_1' & & \mathcal{D}_n' \\
\Delta' & & \Delta'
\end{array}
}{
\Delta', \Delta_i^1, \ldots, \Delta_i^{m_i}
}
\ (\delta GE^{l_1,\ldots,l_n}) \ \leadsto_r
$$

$$
\begin{array}{c}
\mathcal{D}_1^* \\
\Sigma_1^1, \Delta_i \quad \cdots \quad \Sigma_n^1, \Delta_i
\end{array}
$$

$$
\mathcal{D}_i[\Pi_{\mathcal{D}_i}(\Sigma_i^1) := \Sigma_i^1, \Delta] \quad \cdots \quad \Pi_{\mathcal{D}_i}(\psi_i^{m_i}) :=
$$

$$
\begin{array}{c}
\mathcal{D}_1 \\
\psi_i^1, \Delta_i^1
\end{array}
\qquad\qquad
\begin{array}{c}
\mathcal{D}_{m_i} \\
\psi_i^{m_i}, \Delta_i^{m_i}
\end{array}
$$

$$
\mathcal{D}_i'[\Pi_{\mathcal{D}_i}(\psi_i^1) := \ \ \Delta', \Delta_i^1, \ldots, \Delta_i^{m_i} \ \ ]
$$

$$
\mathcal{D}_{m_i}[\Pi_{\mathcal{D}_{m_i}}(\Sigma_i^{m_i}) := \Sigma_i^{m_i}, \Delta]
$$

$$
(4.2.43)
$$

Additive I-rule: Once again, the discussion splits into the categorical case and hypothetical case.

Categorical I-rule: Suppose the ith logistically presented additive categorical I-rule for δ has the following form, where δ is the main operator of φ:

$$\frac{\Gamma:\varphi_i^1,\Delta \quad \cdots \quad \Gamma:\varphi_i^{m_i},\Delta}{\Gamma:\varphi,\Delta} \ (\delta I)_i \tag{4.2.44}$$

Then, the harmoniously-induced logistically presented GE-rules (again based on each premise forming a D-exhaustive ground) have the following form:

$$\frac{\Gamma:\varphi,\Delta \quad \Gamma,\varphi_1^{j_1}:\Delta' \quad \cdots \quad \Gamma,\varphi_n^{j_n}:\Delta'}{\Gamma:\Delta',\Delta} \ (\delta GE_{j_1,\cdots,j_n}) \tag{4.2.45}$$

Hypothetical I-rule: Suppose the ith logistically presented additive hypothetical I-rule for δ has the following form, where δ is the main operator of φ:

$$\frac{\Gamma,\Sigma_i^1:\varphi_i^1,\Delta \quad \cdots \quad \Gamma,\Sigma_i^{m_i}:\varphi_i^{m_i},\Delta}{\Gamma:\varphi,\Delta} \ (\delta I)_i \tag{4.2.46}$$

The general form of the harmoniously-induced logistically presented GE-rule is as follows:

$$\frac{\Gamma:\varphi,\Delta \quad \Gamma:\Sigma_1^{j_1},\Delta \quad \cdots \quad \Gamma:\Sigma_n^{j_n},\Delta \quad \Gamma,\psi_{j_1}^1:\Delta' \quad \cdots \quad \Gamma,\psi_{j_n}^n:d\Delta'}{\Gamma:\Delta',\Delta} \ (\delta GE_{j_1,\cdots,j_n}) \tag{4.2.47}$$

Multiplicative I-rule: Again, categorical and hypothetical I-rules are separately considered.

Categorical I-rule: Suppose the ith logistically presented multiplicative categorical I-rule for δ has the following form, where δ is the main operator of φ:

$$\frac{\Gamma_i^1:\varphi_i^1,\Delta_i^1 \quad \cdots \quad \Gamma_i^{m_i}:\varphi_i^{m_i},\Delta_i^{m_i}}{\Gamma_i^1,\cdots,\Gamma_i^{m_i}:\varphi,\Delta_i^1,\cdots,\Delta_i^{m_i}} \ (\delta I)_i \tag{4.2.48}$$

Then, the harmoniously-induced logistically presented GE-rule has the following form:

$$\frac{\Gamma:\varphi,\Delta \quad \Gamma_1,\varphi_1^1,\cdots,\varphi_1^{m_1}:\Delta' \quad \cdots \quad \Gamma_n,\varphi_n^1,\cdots,\varphi_n^{m_n}:\Delta'}{\Gamma_1,\cdots,\Gamma_n:\Delta',\Delta} \ (\delta GE) \tag{4.2.49}$$

Hypothetical I-rule: Suppose the ith logistically presented multiplicative hypothetical I-rule for δ has the following form, where δ is the main operator of φ:

$$\frac{\Gamma_i^1,\Sigma_i^1:\varphi_i^1,\Delta_i^1 \quad \cdots \quad \Gamma_i^{m_i},\Sigma_i^{m_i}:\varphi_i^{m_i},\Delta_i^{m_i}}{\Gamma_i^1,\cdots,\Gamma_i^{m_i}:\varphi,\Delta_i^1,\cdots,\Delta_i^{m_i}} \ (\delta I^{1,\cdots,m_i})_i \tag{4.2.50}$$

Then, the general form of the logistically presented harmoniously-induced GE-rule is as follows:

$$\frac{\Gamma:\varphi,\Delta \quad \Gamma_1:\Sigma_i^1 \quad \cdots \quad \Gamma_n:\Sigma_i^n \quad \Gamma_1,\psi_1^1,\cdots,\psi_1^{m_1}:\Delta' \quad \cdots \quad \Gamma_n,\psi_n^1,\cdots,\psi_n^{m_n}:\Delta'}{\Gamma_1,\cdots,\Gamma_n:\Delta',\Delta} \ (\delta GE) \tag{4.2.51}$$

4.2.6 GE-harmony in form implies Intrinsic Harmony

I now reestablish for $MCND$ the relationship between the two notions of harmony considered, and show that GE-harmony (under the extended harmoniously-induced GE-rule construction) is stronger than intrinsic harmony: the form of the GE-rules *guarantees* both local-soundness and local-completeness w.r.t. to the I-rules. I present only the case of simple presentation, logistic presentation being similar but notationally cumbersome.

Theorem 4.2.5 (harmony implication). *If $\mathcal{N}$ is closed under derivation composition, then for any operator δ, its GE-rules harmoniously-induced by its I-rules are intrinsically-harmonious.*

Proof: Assume the GE-rules for δ are harmoniously-induced by its I-rules.

Local-soundness: The reductions for a δ-maximal formula were already presented in (4.2.17), (4.2.36), (4.2.26) and (4.2.43).

Local-completeness: I have to show *some* way to expand a derivation of φ, Δ (with main operator δ). The way to do it differs for the additive and multiplicative cases, as the harmoniously-induced GE-rules generated differ. Only the simple presentation is shown.

  Additive rules: Again, the categorical case is separated, for convenience.

    Categorical rules: The exact specification of the expansion involves a lot of multiple indices, so instead of presenting it, I describe the idea of its construction, in stages, and then present an example. First, construct the following 1st-layer subderivation $\mathcal{D}_{j_1, \cdots, j_n}^{l_1, \cdots, l_n}$. The arbitrary conclusion Δ' is chosen as φ, Δ itself.

$$\cfrac{\cfrac{[\varphi_1^1]_{l_{j_1}} \quad \cdots \quad [\varphi_1^{m_1}]_{l_{j_1}}}{\varphi}(\delta I)_1 \quad \cdots \quad \cfrac{[\varphi_n^1]_{l_{j_n}} \quad \cdots \quad [\varphi_n^{m_n}]_{l_{j_n}}}{\varphi}(\delta I)_n}{\varphi, \Delta}(GE_{j_1, \cdots, j_n}^{l_{j_1}, \cdots, l_{j_n}})$$

    This 1st-layer subderivation uses one (δGE)-rule, and all the (δI)-rules. Note that it discharges *one* assumption[2] from each block of grounds for introduction via each $(\delta I)_i$. In the next layer, combine n 1st-layer subderivations into a 2nd-layer derivation, choosing the (GE)-rules applied (with a major premise as another copy of φ, Δ) so as again to discharge one (yet undischarged) assumption from each block of grounds. The second layer looks like

$$\cfrac{\overset{\varphi, \Delta}{\underset{\mathcal{D} \cdots}{}} \quad \overset{\varphi, \Delta}{\underset{\mathcal{D} \cdots}{}}}{\cfrac{\varphi, \Delta \quad \varphi, \Delta \quad \cdots \quad \varphi, \Delta}{\varphi, \Delta}}(\delta GE \cdots)$$

---

[2]This is not reflected in the notation; however, because of its complexity, I preferred not to complicate it more in order to enforce this, and just state it outside the rule itself.

At the nth (last layer), one appropriately chosen (GE)-rule combines the results of the previous layer, discharging the remaining assumptions in each block.

Hypothetical rules: Here the idea is similar, but the construction of the 1st-layer also requires establishing the supports $\Sigma_i^{j_i}$ in order to apply a (GE)-rule. The supports are assumed, and discharged by the applications of the (δI)-rules. The form of the 1st layer subderivation becomes the following.

$$\frac{\varphi, \Delta \ [\Sigma_{j_1}]\cdots \ \cdots \ [\Sigma_{j_n}]\cdots \quad \dfrac{[\varphi_1^1]_{l_{j_1}} \ \cdots \ [\varphi_1^{m1}]_{l_{j_1}}}{\varphi, \Delta} (\delta I^{\cdots})_1 \quad \cdots \quad \dfrac{[\varphi_n^1]_{l_{j_n}} \ \cdots \ [\varphi_n^{mn}]_{l_{j_n}}}{\varphi, \Delta} (\delta I^{\cdots})_n}{\varphi, \Delta} (GE^{l_{j_1},\cdots,l_{j_n}})_{j_1,\cdots,j_n}$$

The rest of the layered construction is like for the categorical case.

Multiplicative rules: Here the construction is simpler.

Categorical I-rule: Recall that only one harmoniously-induced GE-rule is generated, having as premises *all* the grounds for introduction, for all the (δI)-rules. The expansion is as follows.

$$\frac{\varphi, \Delta \quad \dfrac{[\Sigma_{j_1}^1]_1, [\varphi_1^1]_{l_{j_1}} \ \cdots \ [\varphi_1^{m1}]_{l_{j_1}}}{\varphi} (\delta I_1^1) \quad \cdots \quad \dfrac{[\Sigma_{j_n}^1]_n, [\varphi_n^1]_{l_{j_n}} \ \cdots \ [\varphi_n^{mn}]_{l_{j_n}}}{\varphi} (\delta I_n^n)}{\varphi, \Delta} (GE^{l_{j_1},\cdots,l_{j_n}}_{j_1,\cdots,j_n})$$

Hypothetical I-rule: Similar, with the Σ-supports assumed by the GE-rule and discharged by the I-rules. I omit the details.

Example 4.2.22. *Returning to conjunction, its expansions are as follows.*

Additive: *According to the construction, there are two layers only.*

$$\begin{array}{c} \mathcal{D} \\ \varphi \wedge \psi, \Delta \end{array} \rightsquigarrow_e$$

$$\frac{\begin{array}{c}\mathcal{D}\\\varphi\wedge\psi,\Delta\end{array} \quad \dfrac{\begin{array}{c}\mathcal{D}\\\varphi\wedge\psi,\Delta\end{array} \quad \dfrac{[\varphi]_{l_1} \quad [\psi]_{l_2}}{\varphi\wedge\psi}(\wedge I)}{\varphi\wedge\psi,\Delta}(\wedge GE_1^{l_1})}{\varphi\wedge\psi,\Delta}(\wedge GE_2^{l_2}) \tag{4.2.52}$$

Note, however, that there is in this case a simpler expansion.

$$\frac{\dfrac{\begin{array}{c}\mathcal{D}\\\varphi\wedge\psi,\Delta \quad [\varphi]_i\end{array}}{\varphi,\Delta}(\wedge GE^i) \quad \dfrac{\begin{array}{c}\mathcal{D}\\\varphi\wedge\psi,\Delta \quad [\psi]_j\end{array}}{\psi,\Delta}(\wedge GE^j)}{\varphi\wedge\psi,\Delta}(\wedge I) \tag{4.2.53}$$

Multiplicative:

$$\begin{array}{c}\mathcal{D}\\\varphi\otimes\psi,\Delta\end{array} \rightsquigarrow_e \quad \frac{\begin{array}{c}\mathcal{D}\\\varphi\otimes\psi,\Delta\end{array} \quad \dfrac{[\varphi,\psi]_i}{\varphi\otimes\psi}(\otimes I)}{\varphi\otimes\psi,\Delta}(\otimes GE^i) \tag{4.2.54}$$

To see an additive expansion[3] with more layers, consider 'exclusive or' (x).

Example 4.2.23. *The additive and multiplicative cases are separately considered.*

Additive: *Let the two simple additive (xI)-rules be the following.*

$$\dfrac{\varphi,\Delta \quad \neg\psi,\Delta}{\varphi x\psi,\Delta}\ (xI)_1 \qquad \dfrac{\neg\varphi,\Delta \quad \psi,\Delta}{\varphi x\psi,\Delta}\ (xI)_2 \qquad\qquad (4.2.55)$$

Four harmoniously-induced GE-rules are generated.

$$\dfrac{\varphi x\psi,\Delta \quad \begin{array}{c}[\varphi]_i\\ \mathcal{D}_1\\ \Delta'\end{array} \quad \begin{array}{c}[\psi]_j\\ \mathcal{D}_2\\ \Delta'\end{array}}{\Delta',\Delta}\ (xGE^{i,j})_1 \qquad \dfrac{\varphi x\psi,\Delta \quad \begin{array}{c}[\varphi]_i\\ \mathcal{D}_1\\ \Delta'\end{array} \quad \begin{array}{c}[\neg\varphi]_j\\ \mathcal{D}_2\\ \Delta'\end{array}}{\Delta',\Delta}\ (xGE^{i,j})_2$$

$$\dfrac{\varphi x\psi,\Delta \quad \begin{array}{c}[\neg\psi]_i\\ \mathcal{D}_1\\ \Delta'\end{array} \quad \begin{array}{c}[\psi]_j\\ \mathcal{D}_2\\ \Delta'\end{array}}{\Delta',\Delta}\ (xGE^{i,j})_3 \qquad \dfrac{\varphi x\psi,\Delta \quad \begin{array}{c}[\neg\psi]_i\\ \mathcal{D}_1\\ \Delta'\end{array} \quad \begin{array}{c}[\neg\varphi]_j\\ \mathcal{D}_2\\ \Delta'\end{array}}{\Delta',\Delta}\ (xGE^{i,j})_4$$

$$(4.2.56)$$

An expansion is as follows. For typographical reasons, it is displayed in parts, separating the layers. First, there are four 1st-level derivations. All applications of weakening are omitted.

$$\mathcal{D}_{1,1}: \quad \dfrac{\varphi x\psi,\Delta \quad \dfrac{[\varphi]_1 \quad [\neg\psi]_2}{\varphi x\psi,\Delta}\ (xI)_1 \quad \dfrac{\neg[\varphi]_3 \quad [\psi]_4}{\varphi x\psi,\Delta}\ (xI)_2}{\varphi x\psi,\Delta}\ (xGE^{1,4})_1$$

$$\mathcal{D}_{1,2}: \quad \dfrac{\varphi x\psi,\Delta \quad \dfrac{[\varphi]_5 \quad [\neg\psi]_6}{\varphi x\psi,\Delta}\ (xI)_1 \quad \dfrac{[\neg\varphi]_7 \quad [\psi]_2}{\varphi x\psi,\Delta}\ (xI)_2}{\varphi x\psi,\Delta}\ (xGE^{5,7})_2$$

$$\mathcal{D}_{1,3}: \quad \dfrac{\varphi x\psi,\Delta \quad \dfrac{[\varphi]_8 \quad [\neg\psi]_9}{\varphi x\psi,\Delta}\ (xI)_1 \quad \dfrac{\neg[\varphi]_{10} \quad [\psi]_{11}}{\varphi x\psi,\Delta}\ (xI)_2}{\varphi x\psi,\Delta}\ (xGE^{8,11})_1$$

$$\mathcal{D}_{1,4}: \quad \dfrac{\varphi x\psi,\Delta \quad \dfrac{[\varphi]_{12} \quad [\neg\psi]_{13}}{\varphi x\psi,\Delta}\ (xI)_1 \quad \dfrac{\neg[\varphi]_{10} \quad [\psi]_{14}}{\varphi x\psi,\Delta}\ (xI)_2}{\varphi x\psi,\Delta}\ (xGE^{13,14})_3$$

---

[3]The expansion for the additive case, in the single-conclusion case, was suggested to me by Stephen Read.

At the 2nd layer, there are two derivations, each one combining two 1st layer ones.

$$\mathcal{D}_{2,1}: \quad \dfrac{\overset{\mathcal{D}_{1,1} \quad \mathcal{D}_{1,2}}{\varphi x\psi, \Delta \quad \varphi x\psi, \Delta \quad \varphi x\psi, \Delta}}{\varphi x\psi, \Delta} \; (xGE^{2,6})_3$$

$$\mathcal{D}_{2,2}: \quad \dfrac{\overset{\mathcal{D}_{1,3} \quad \mathcal{D}_{1,4}}{\varphi x\psi, \Delta \quad \varphi x\psi, \Delta \quad \varphi x\psi, \Delta}}{\varphi x\psi, \Delta} \; (xGE^{10,13})_1$$

Finally, at the 3rd layer, there is one derivation, combining the two 2nd layer derivations, completing the discharge of the as yet undischarged assumptions.

$$\mathcal{D}_{3,1}: \quad \dfrac{\overset{\mathcal{D}_{2,1} \quad \mathcal{D}_{2,2}}{\varphi x\psi, \Delta \quad \varphi x\psi, \Delta \quad \varphi x\psi, \Delta}}{\varphi x\psi, \Delta} \; (xGE^{3,9})_4$$

Assumptions discharge: There are seven applications of GE-rules in the full derivations, and sixteen assumptions to discharge, where two pairs of assumptions are equi-labelled, discharged together.

- *The applications of GE-rules in 1st level subderivations discharge two (different) assumptions each.*
- *The applications of GE-rules in 2nd level subderivations discharge three assumptions each: xGE_3 one ψ and two $\neg\psi$s, and xGE_2 one φ and two $\neg\varphi$s.*
- *The application of the GE-rule on (the one) third level subderivation (ending the whole derivation) discharges two (different) assumptions.*

Multiplicative: *Here, only one harmoniously-induced GE-rule is generated.*

$$\dfrac{\varphi x\psi, \Delta \quad \overset{[\varphi, \neg\psi]_i}{\underset{\Delta'}{\overset{\mathcal{D}_1}{}}} \quad \overset{[\neg\varphi, \psi]_j}{\underset{\Delta'}{\overset{\mathcal{D}_2}{}}}}{\Delta', \Delta} \; (xGE^{i,j}) \tag{4.2.57}$$

The expansion establishing local completeness is as follows (with weakening applications suppressed).

$$\overset{\mathcal{D}}{\underset{\varphi x\psi, \Delta}{}} \; \leadsto_e \quad \dfrac{\overset{\mathcal{D}}{\varphi x\psi, \Delta} \quad \dfrac{[\varphi, \neg\psi]_i}{\varphi x\psi, \Delta}\,(xI)_1 \quad \dfrac{[\neg\varphi, \psi]_j}{\varphi x\psi, \Delta}\,(xI)_2}{\varphi x\psi, \Delta} \; (xGE)$$

Some more expansions
For implication, the resulting expansion is

$$\overset{\mathcal{D}}{\underset{\varphi \to \psi, \Delta}{}} \; \leadsto_e \quad \dfrac{\dfrac{\overset{\mathcal{D}}{\varphi \to \psi} \quad [\varphi]_1 \quad [\psi]_2}{\psi, \Delta}\,(\to GE^2)}{\varphi \to \psi, \Delta}\,(\to I^1) \tag{4.2.58}$$

$$\frac{\varphi,\Delta \quad \psi,\Delta}{\varphi \wedge \psi, \Delta} \ (\wedge I) \qquad \frac{\varphi \wedge \psi, \Delta}{\varphi, \Delta} \ (\wedge E_1) \qquad \frac{\varphi \wedge \psi, \Delta}{\psi, \Delta} \ (\wedge E_2)$$

$$\frac{\varphi,\Delta}{\varphi \vee \psi, \Delta} \ (\vee I_1) \qquad \frac{\psi,\Delta}{\varphi \vee \psi, \Delta} \ (\vee I_2) \qquad \frac{\varphi \vee \psi, \Delta}{\varphi, \psi, \Delta} \ (\vee E)$$

$$\frac{\begin{array}{c}[\varphi]_i\\ \mathcal{D}\\ \psi,\Delta\end{array}}{\varphi \to \psi, \Delta} \ (\to I^i) \qquad \frac{\varphi \to \psi, \Delta \quad \varphi,\Delta}{\psi, \Delta} \ (\to E)$$

$$\frac{\begin{array}{c}[\varphi]_i\\ \mathcal{D}\\ \Delta\end{array}}{\neg\varphi,\Delta} \ (\neg I^i) \qquad \frac{\neg\varphi,\Delta \quad \varphi,\Delta}{\Delta} \ (\neg E)$$

Figure 4.1: The simple MCND-presentation NC of propositional Classical Logic

Finally, consider the expansion for (ite), *where* $n = 1$, *the I-rules of which are given in (4.2.30), and GE-rules in (4.2.31).*

$$\frac{ite(\varphi,\psi,\chi) \quad [\varphi]_1 \quad [\psi]_2}{\psi, \Delta} \ (iteGE_1^2) \qquad \frac{ite(\varphi,\psi,\chi) \quad [\neg\varphi]_3 \quad [\chi]_4}{\chi, \Delta} \ (iteGE_2^4)$$
$$\frac{}{ite(\varphi,\psi,\chi), \Delta} \ (iteI^{1,3})$$

$$(4.2.59)$$

4.3   A Multiple-Conclusion presentation of Classical Logic

Following Boričić and Read [16, 164], the operatioinal rules for the connectives for a simple $MCND$-presentation NC of classical logic is presented in Figure 4.1. In addition, the previous structural rules are assumed too, operating on *both* sides of a sequent. The right rules are (RW) (right weakening), (RC) (right contraction) and (RE) (right exchange). Since only right rules are used in this section, the 'R' in their name is suppressed for brevity. In the presence of right rules the multiplicative and additive rules are equivalent, only the latter[4] are presented. I restrict the discussion to the propositional fragment, which suffices to make the point of proof-theoretical justification.

---

[4]Boričić (in [16]) and Cellucci (in [18]) present the former.

Remarks:

- All the rules are both pure and simple.

- In [18], Cellucci presents the following single $(\vee I)$ rule, that can be shown equivalent to the two rules of Boričić [16].

$$\frac{\varphi, \psi, \Delta}{\varphi \vee \psi, \Delta} \; (\vee I) \tag{4.3.60}$$

- Note that by substituting in all the rules of $MCND$ an empty right context Δ (except for $(\neg I)$), familiar single-conclusion $SCND$-rules are obtained.

- (LEM) is derivable as follows (contexts omitted).

$$\frac{\dfrac{\dfrac{[\varphi]_1}{\varphi \vee \neg \varphi} \; (\vee I_1)}{\dfrac{\varphi \vee \neg \varphi, \neg \varphi}{\varphi \vee \neg \varphi, \varphi \vee \neg \varphi} \; (\vee I_2)}}{\varphi \vee \neg \varphi} \; (C) \tag{4.3.61}$$

- As observed in Read [164], the definition of $\neg \varphi$ as $\varphi \to \perp$ (Definition 2.1.1) is not peculiar to intuitionistic logic, as has been generally believed. Whether this implication "is intuitionistic" or "is classical" depends on the rules for '$\to$'. The implication is classical here, yielding classical negation with the following E-rule for $\perp$.

$$\frac{\perp, \Delta}{\varphi, \Delta} \; (\perp E) \tag{4.3.62}$$

- In continuation to the previous comment, it was also observed by Read (in [164], p. 144) that the source of the non-conservativity of NK (Gentzen's classical ND-system) is that the theory does not specify completely the meaning of the classical implication; the current NC does! For example, below is an NC-derivation of Peirce's rule within the *positive* '$\to$'-fragment, without any appeal to negation.

$$\frac{[(\varphi \to \psi) \to \varphi]_2 \quad \dfrac{\dfrac{\dfrac{[\varphi]_1}{\varphi, \psi} \; (W)}{\varphi, \varphi \to \psi} \; (\to I^1)}{\varphi, \varphi} \; (\to E)}{\dfrac{\dfrac{\varphi, \varphi}{\varphi} \; (C)}{((\varphi \to \psi) \to \varphi) \to \varphi} \; (\to I^2)} \tag{4.3.63}$$

- The double-negation elimination rule (DNE) is derivable too, as shown below.

$$\frac{\neg \neg \varphi \quad \dfrac{\dfrac{\dfrac{[\varphi]_1}{\perp, \varphi} \; (W)}{\neg \varphi, \varphi} \; (\to I^1)}{\dfrac{\perp, \varphi}{\varphi, \varphi} \; (\perp E)}}{\dfrac{\dfrac{\perp, \varphi}{\varphi, \varphi}}{\varphi} \; (C)} \; (\to E) \tag{4.3.64}$$

$$\frac{\varphi\wedge\psi,\Delta \qquad \overset{\displaystyle [\varphi]_i,[\psi]_j}{\underset{\chi}{\overset{\mathcal{D}_2}{}}}}{\chi,\Delta} \ (\wedge GE^{i,j})$$

$$\frac{\varphi\vee\psi,\Delta \quad \overset{\displaystyle [\varphi]_i}{\underset{\chi}{\overset{\mathcal{D}_1}{}}} \quad \overset{\displaystyle [\psi]_j}{\underset{\chi}{\overset{\mathcal{D}_2}{}}}}{\chi,\Delta} \ (\vee GE^{i,j})$$

$$\frac{\varphi\rightarrow\psi,\Delta \quad \varphi,\Delta \quad \overset{\displaystyle [\psi]_i}{\underset{\chi}{\overset{\mathcal{D}}{}}}}{\chi,\Delta} \ (\rightarrow GE^i)$$

Figure 4.2: The harmoniously-induced GE-rules for NC

Theorem 4.3.6 (Closure of NC under derivation composition). *NC is closed under derivation composition.*

The proof is essentially the same as in the single conclusion case for Intuitionistic Logic, by induction on the derivation.

By applying the construction above, the following additive harmoniously-induced GE-rules are obtained (Figure 4.2).

The shared contexts (additive) formulation of NC with explicit left contexts is presented in Figure 4.3. Eliminations are via GE-rules. Formulating a multiplicative presentation is straightforward and omitted, as the two are equivalent in the presence of the standard structural rules.

Remarks:

- Since the object language of CL contains both '$\wedge$', '$\vee$' and '$\rightarrow$', it is possible to regard sequents $\Gamma\vdash\Delta$ here as expressing $\wedge_\Gamma\rightarrow\vee_\Delta$ (not necessarily possible for arbitrary object languages), with $\wedge_\varnothing = \bot$ and $\vee_\varnothing = \top$.

- As an example of a derivation from assumptions, below is a shared-context derivation for $\vdash_{NC}\neg(\varphi\wedge\psi) : \neg\varphi\vee\neg\psi$, which is not acceptable in Intuitionistic

$$\frac{}{\varphi:\varphi}\ (Ax)$$

$$\frac{\Gamma:\varphi,\Delta \quad \Gamma:\psi,\Delta}{\Gamma:\varphi\wedge\psi,\Delta}\ (\wedge I) \qquad \frac{\Gamma:\varphi\wedge\psi,\Delta \quad \Gamma,\varphi,\psi:\chi,\Delta}{\Gamma:\chi,\Delta}\ (\wedge E)$$

$$\frac{\Gamma:\varphi,\Delta}{\Gamma:\varphi\vee\psi,\Delta}\ (\vee I_1) \quad \frac{\Gamma:\psi,\Delta}{\Gamma:\varphi\vee\psi,\Delta}\ (\vee I_2) \quad \frac{\Gamma:\varphi\vee\psi,\Delta \quad \Gamma,\varphi:\chi,\Delta \quad \Gamma,\psi:\chi,\Delta}{\Gamma:\chi,\Delta}\ (\vee E)$$

$$\frac{\Gamma,\varphi:\psi,\Delta}{\Gamma:\varphi\to\psi,\Delta}\ (\to I) \qquad \frac{\Gamma:\varphi\to\psi,\Delta \quad \Gamma:\varphi,\Delta \quad \Gamma,\psi:\chi,\Delta}{\Gamma:\chi,\Delta}\ (\to E)$$

$$\frac{\Gamma,\varphi:\Delta}{\Gamma:\neg\varphi,\Delta}\ (\neg I) \qquad \frac{\Gamma:\neg\varphi,\Delta \quad \Gamma:\varphi,\Delta}{\Gamma:\Delta}\ (\neg E)$$

Figure 4.3: The additive logistic-presentation of NC

Logic.

$$\frac{\frac{\neg(\varphi\wedge\psi):\neg(\varphi\wedge\psi) \quad \frac{\dfrac{}{\varphi:\varphi}\ (Ax) \quad \dfrac{}{\psi:\psi}\ (Ax)}{\varphi,\psi:\varphi\wedge\psi}\ (\wedge I)}{\dfrac{\neg(\varphi\wedge\psi),\varphi,\psi:}{\neg(\varphi\wedge\psi):\neg\varphi,\neg\psi}\ (\neg I \times 2)}\ (\neg E)}{\dfrac{\neg(\varphi\wedge\psi):\neg\varphi\vee\neg\psi,\neg\psi}{\dfrac{\neg(\varphi\wedge\psi):\neg\varphi\vee\neg\psi,\neg\varphi\vee\neg\psi}{\neg(\varphi\wedge\psi):\neg\varphi\vee\neg\psi}\ (C)}\ (\vee I_2)}\ (\vee I_1) \qquad (4.3.65)$$

- NC restores the *symmetry* characteristic of Classical Logic. For example, parallel to $\vdash_{NC}\varphi,\psi:\varphi\wedge\psi$, we also have $\vdash_{NC}\varphi\vee\psi:\varphi,\psi$, in contrast to NK, in which no direct immediate conclusion can be drawn from a disjunction.

- As can be observed, all the NC-rules are pure. Furthermore, in [16], Boričić proves the normalizability (both weak and strong) of NC (indirectly, via cut-elimination of an equivalent sequent-calculus). As a consequence, NC is separable.

4.3.1 Reevaluating Classical Logic for meeting the meaning conferring criteria

First, as noted above, the addition of negation to the positive fragment *is* conservative. In addition, as was mentioned above, NC is both pure and separable.

Harmony in form obtains, by the GE form of the E-rules. The adaptation of the construction in Francez and Dyckhoff [54] shown above can "read of" the GE-rules from the I-rules. Note that, pace Hjortland ([76], p. 139), $(\neg E)$ *is* the GE-rule harmonically-induced by $(\neg I)$. This can be realized the easiest by comparison to the '$\rightarrow$'-rules. In '$(\rightarrow I)$, there are both a support (namely, φ), and a ground (categorical component, ψ). The former produces the premise $\Gamma : \varphi$ (of '$(\rightarrow E)$'), and the latter the premise $\Gamma, \psi : \chi, \Delta$. However, '$(\neg I)$' has only a support, again φ, producing the premise $\Gamma : \varphi$ for '$(\neg E)$' too, but *no explicit grounds*, hence no premise with an explicit arbitrary conclusion.

As for intrinsic harmony and stability, the reductions (which are not presented neither in Boričić [16] nor in Read [164], but are presented in Cellucci [18]) and expansions (not considered at all in Boričić [16], Cellucci [18] and Read [164]) are shown below.

Theorem 4.3.7 (local-soundness of NC). *NC is locally-sound.*

Proof. Bellow are the detour-eliminating reductions, with left contexts omitted. They are shown using the regular E-rules, that are derivable in NC from the GE-rules, and render the reductions easier to understand.

reducing conjunction:

$$\cfrac{\cfrac{\begin{matrix}\mathcal{D}_1 & \mathcal{D}_2 \\ \varphi, \Delta & \psi, \Delta\end{matrix}}{\varphi \wedge \psi, \Delta}\,(\wedge I)}{\varphi, \Delta}\,(\wedge E_1) \qquad \leadsto_r \qquad \begin{matrix}\mathcal{D}_1 \\ \varphi, \Delta\end{matrix} \qquad\qquad (4.3.66)$$

The case of $(\wedge \hat{E}_2)$ is similar.

reducing implication:

$$\cfrac{\cfrac{\begin{matrix}[\varphi]_i \\ \mathcal{D}_1 \\ \psi, \Delta\end{matrix}}{(\varphi \rightarrow \psi), \Delta}\,(\rightarrow I^i) \quad \begin{matrix}\mathcal{D}_2 \\ \varphi, \Delta\end{matrix}}{\psi, \Delta}\,(\rightarrow E) \qquad \leadsto_r \qquad \begin{matrix}\mathcal{D}_2 \\ \mathcal{D}_1[\Pi_{\mathcal{D}_1}(\varphi) := \varphi, \Delta] \\ \psi, \Delta\end{matrix} \qquad (4.3.67)$$

Note the dependence on the closure of NC under derivation composition, similarly to the single-conclusion case for Intuitionistic Logic.

reducing negation:

$$\cfrac{\cfrac{\begin{matrix}[\varphi]_i \\ \mathcal{D}_1 \\ \Delta\end{matrix}}{\neg\varphi, \Delta}\,(\neg I^i) \quad \begin{matrix}\mathcal{D}_2 \\ \varphi, \Delta\end{matrix}}{\Delta}\,(\neg E) \qquad \leadsto_r \qquad \begin{matrix}\mathcal{D}_2 \\ \mathcal{D}_1[\Pi_{\mathcal{D}_1}(\varphi) := \varphi, \Delta] \\ \Delta\end{matrix} \qquad (4.3.68)$$

There is an implicit use of right contraction in this derivation, as the substitution produces Δ, Δ in the r.h.s. of the conclusion.

$\square$

Theorem 4.3.8 (local-completeness of NC). *NC is locally-complete.*

Proof. Below are the required expansions.

expanding conjunction:

$$
\begin{array}{c}
\mathcal{D} \\
\varphi\wedge\psi,\Delta
\end{array}
\quad\leadsto^e\quad
\dfrac{\dfrac{\begin{array}{c}\mathcal{D}\\\varphi\wedge\psi,\Delta\end{array}}{\varphi,\Delta}(\wedge_1 E) \quad \dfrac{\begin{array}{c}\mathcal{D}\\\varphi\wedge\psi,\Delta\end{array}}{\psi,\Delta}(\wedge_2 E)}{\varphi\wedge\psi,\Delta}(\wedge I)
\tag{4.3.69}
$$

expanding implication:

$$
\begin{array}{c}
\mathcal{D} \\
(\varphi\to\psi),\Delta
\end{array}
\quad\leadsto^e\quad
\dfrac{\dfrac{\begin{array}{cc}\mathcal{D}\\(\varphi\to\psi),\Delta & [\varphi]_i,\Delta\end{array}}{\psi,\Delta}(\to E)}{(\varphi\to\psi),\Delta}(\to I^i)
\tag{4.3.70}
$$

expanding negation:

$$
\begin{array}{c}
\mathcal{D} \\
\neg\varphi,\Delta
\end{array}
\quad\leadsto^e\quad
\dfrac{\dfrac{\begin{array}{cc}\mathcal{D}\\\neg\varphi,\Delta & [\varphi]_i,\Delta\end{array}}{\Delta}(\neg E)}{\neg\varphi,\Delta}(\neg I^i)
\tag{4.3.71}
$$

$\square$

4.3.2 The effect of structural canonicity on multi-conclusion systems

In this section, I consider the effect of structural canonicity (Definition 1.6.10) in the context of $MCND$-systems. Recall that we now have structural rules that may modify right contexts, which requires the establishment of S-local-soundness to count as meaning-conferring. I show that NC is S-locally-sound. Note that the (RC)-rule does not produce new conclusions, but right weakening (RW) does. Thus, I have to show the reductions establishing S-local-soundness in reducing derivations in which a formula is concluded by (RW) and immediately eliminated.

The common structure of all the reductions is that right weakening followed by elimination is reduced to right weakening only (once or twice).

Implication:

$$\dfrac{\varphi,\Delta \quad \dfrac{\Delta}{\varphi\rightarrow\psi,\Delta}\,(RW)}{\psi,\Delta}\,(\rightarrow E) \qquad \leadsto_r \qquad \dfrac{\Delta}{\psi,\Delta}\,(RW) \tag{4.3.72}$$

Conjunction:

$$\dfrac{\dfrac{\Delta}{\varphi\wedge\psi,\Delta}\,(RW)}{\varphi,\Delta}\,(\wedge E_1) \qquad \leadsto_r \qquad \dfrac{\Delta}{\varphi,\Delta}\,(RW) \tag{4.3.73}$$

The reduction for $(\wedge E_2)$ is similar and ommitted.

Disjunction:

$$\dfrac{\dfrac{\Delta}{\varphi\vee\psi,\Delta}\,(RW)}{\varphi,\psi,\Delta}\,(\vee E) \qquad \leadsto_r \qquad \dfrac{\Delta}{\varphi,\psi,\Delta}\,(RW\otimes 2) \tag{4.3.74}$$

Negation:

$$\dfrac{\dfrac{\Delta}{\neg\varphi,\Delta}\,(RW) \qquad \varphi,\Delta}{\Delta}\,(\neg E) \qquad \leadsto_r \qquad \Delta \tag{4.3.75}$$

4.3.3 Criticizing multiple-conclusion natural-deduction as a way to confer meaning

While the appeal to multiple-conclusion natural-deduction restores harmony for Classical Logic, it raises the issue of adhering to the principle of answerability (Steinberger [207]) – does such a system of rules conform to our deductive inferential practice? In what sense does it reflect use? After all, harmony per se is a goal of PTS only as much as it serves as a qualification criterion for being meaning-conferring. A detailed negative answer to this question is presented by Steinberger in [207], where the final conclusion is that conclusions should remain single in order for the ND-system to be used to confer meaning. While in [173] Restall presents an argument in favor of multiple conclusions, it is based on one example – a certain way of proving by cases – claimed by Steinberger to be better handling by single-conclusion ND, appealing to disjunction elimination. A similar criticism is expressed by Dummett ([35], p. 187), as he inherently connects the conclusions set Δ with the disjunction of its elements, claiming that $MCND$ *enforces* the learning[5] of the meaning of '$\vee$'. Yet another similar opposition is expressed by Tennant ([215], p. 320).

However, such a criticism is based on a rather narrow view of what constitutes our inferential practices. Under a broader view, a positive answer to the above question might still be available. One attempt (admitted by the author to consist in a first

---

[5]One might wonder, though, why the notion of multiple premises does not raise an analogous criticism, forcing learning the meaning of '$\wedge$'.

approximation) is presented in Sandqvist [182]. It views a multiple conclusion sequent (in a slightly modified notation), for an arbitrary object language,

$$\Gamma = \{\gamma_1, \cdots, \gamma_m\} : \{\delta_1, \cdots, \delta_n\} = \Delta \tag{4.3.76}$$

as a *meta-rule*

$$\frac{\Theta : \gamma_1 \quad \cdots \quad \Theta : \gamma_m \quad \Theta, \delta_1 : \varphi \quad \cdots \quad \Theta, \delta_n : \varphi}{\Theta : \varphi} \tag{4.3.77}$$

For arbitrary Θ, φ (and for technical reasons, $n \geq 1$).

If one views a single-conclusion sequent $\Gamma : \varphi$ as a rationality requirement of a doxastic agent to accept φ whenever (s)he accepts every member of Γ, then the above rule expresses a more general rationality requirement of the agent. In case disjunction is present in the object language, as it is for Classical Logic, then the meta-rule is in line with the view of Δ as the disjunction of its elements, and the GE-rule for disjunction. For a detailed presentation of this view, with some technical results, the reader is referred to Sandqvist [182]. Another way of rebutting the offence of $MCND$ is presented by Hjortland (in [76], p. 89), relating it to *Bilateralism* (see next section).

While I find both of the arguments above as a convincing advance, I consider this issue, of relating multiple-conclusion natural-deduction to actual inferential practices, as still open, awaiting a fully satisfactory solution, which, I conjecture, will be found in due time.

4.4 Bilateralism

In this section, I present another approach to PTS, based on the central idea that *denial* should be taken as *primitive*, on par with assertion. While traditionally this approach is presented as a way to proof-theoretically justify Classical Logic, I view it as having merits of its own as a general approach to PTS. Therefore, in Section 4.4.1 (based on Francez [43]) I present this general view, fitting also for object languages lacking any kind of negation, deferring its use to justify Classical Logic to Section 4.4.3. In particular, this approach is used for defining proof-theoretic meanings of negative determiners in natural language (see Section 10.4).

4.4.1 Bilateralism in Proof-Theoretic Semantics

As mentioned above, *bilateralism* is an approach to meaning taking *denial* as a primitive attitude, on par with *assertion* (see, for example, Rumfitt [179], which also contains references to earlier work). Bilateralism puts forward a claim, that there are good reasons, pace Frege, *not* to regard a denial of φ to be adequately represented by an assertion of its (sentential) negation $\neg\varphi$; rather, it should be viewed primitive (this approach is called "rejectivism" in Humberstone [79]). For arguments to this end, see Restall [173].

The main features that distinguish a primitive denial from denying by asserting the negation, expressed by separating force (here, assertion and denial) from contents, are:

- In contrast to negation, primitive denial *cannot be iterated.*
  Thus, double-denial (the analog of double-negation) is not expressible, and hence the issue of its elimination does not arise.

- Primitive denial *cannot be embedded.*

Most often, bilateralism is brought in to "salvage" Classical Logic from shortcomings attributed to the latter by various proponents of PTS, as presented in Section 4.4.3.

On the other hand, bilateralism extends the scope of "use" by including both assertion and denial as exhibiting use.   In [179], Rumfitt identifies the use in English (and related languages) of rejection as the practice of answering by 'no' to a yes/no question. Answering by 'no' signals rejection. In a recent debate, Textor [222] argues against this specific rendering of 'use', while Incurvaty and Smith [82] refutes this argument. Be that as it may, it is hard to deny that denial (or rejection) is present in daily use of language. Thus, the answerability criterion (Section 1.6) is clearly met. A bilateral ND-systems extend the conventional ND-systems by having also I/E-rules for denial, in addition to the rules for assertion. In bilateral ND-systems denial can occur both as assumptions (possibly discharged ones) in arguments (formalized as derivations), and hence in premises of rules, as well as in conclusions of arguments and rules. I will mainly concentrate on the I-view of PTS, with occasional digressions to issues related to the E-view. Bilateralism can be seen as extending the scope of PTS. Just as on its inferentialism view truth is not *verification transcendent*, so is also falsity not *falsification transcendent*. On the pragmatism view, *commitment* is now encompassing both warranted conclusions as well as warranted "disconclusions" – derived denials.

I am interested mainly in the effect of bilateralism on proof-theoretic semantics, even for object languages lacking sentential negation altogether, but still in need to reflect some negative aspects in meaning. Some typical examples are listed below.

- Defining proof-theoretic semantics for fragments of natural language without sentential negation, but with negative determiners like no or at most n (see Section 10.4).

- In logic, an example is the addition of xor (exclusive Or) and ite (if-then-else) to a negation-less fragment (see examples 4.4.24 and 4.4.25 below).

- Defining *duals* of connectives like implication, that have no dual in a positive, negation-less logic (see example 4.4.32). Also, though not dealt with here, the definition of some *co-connectives*, like coimplication in the Heyting-Brouwer (bi-intuitionistic) logics.

In particular, I am interested in the revision bilateralism requires of the adequacy criteria for an ND-system to qualify as meaning conferring. I claim that for a bilateral ND-system to qualify as meaning conferring, a *two-dimensional* adequacy criterion is needed.

vertically: Relating I-rules to E-rules for balance. However, while traditional harmony and stability relate only I-rules for assertion to E-rules for assertion, in a bilateral system there is an additional need to relate I-rules for denial to E-rules for denial.

horizontally: This is a criterion having no counterpart in unilateral systems, relating for balance the I-rules for assertion to the I-rules for denial.

This issue is elaborated upon in Section 4.4.1.5.

In Restall [173], bilateralism is used to explain multiple-conclusion natural-deduction systems. In contrast, I am here interested in bilateralism in the usual, single conclusion, natural-deduction systems. Another use of bilateralism can be found in Restall [175], basing on it non-classical paraconsistent and paracomplete logics. To repeat, my main interest here is the impact of bilateralism on the qualification of ND-systems as constituting meaning-conferring devices, with an emphasis on language fragment lacking sentential negation altogether.

4.4.1.1 Preliminaries

I adopt the notation of Rumfitt [179] and add a *force marker*, '+' or '-', in front of each sentence of the object language. Thus, object language sentences are explicitly structured[6] into a *force indicator* and a *content* (cf. Incurvati and Smith [81]). I use σ (possibly subscripted) to range over force-marked sentences. Also, $\overline{\sigma}$, the *conjugate* of σ, indicates the *reversal* of the force marker of σ, where it is understood that $\overline{\overline{\sigma}} = \sigma$. A positively marked sentence of the form '$+\varphi$' is to be understood as indicating assertion of φ, while a negatively marked sentence '$-\varphi$' is to be understood as indicating denial of φ. In a bilateral ND-system, an operator, say '$*$', has two kinds of I-rules, marked $(*^+I)$ and $(*^-I)$, and two kinds of E-rules marked $(*^+E)$ and $(*^-E)$. The positive I-rules introduce $+(\varphi * \psi)$, while the negative I-rules introduce $-(\varphi * \psi)$. Similarly for elimination. In analogy of the I_*^+ providing grounds of assertion of $\varphi * \psi$, the I_*^- provide *grounds for denial* of $\varphi * \psi$. Note that this kind of bilateral ND-system is different from ND-systems for *refutation* (see, for example, Tranchini [224]), where the rules serve to draw a contradiction from premises.

---

[6]For Smiley [203], the force markers are part of the meta-language.

When constructing a bilateral extension of a given unilateral $\mathcal{N}D$-system, the positive rules for any introduced expression are what is called in Humberstone [79] "the positive sign analogues" of the unilateral rules for that expression, that is, the given I-rules, but with a '+' preceding each sentence in the premises and in the conclusion. An immediate conclusion of this convention is, that positive conclusions cannot have negative premises, while negative conclusions may have positive as well as negative premises. However, when the bilateral system is *not* an extension of a unilateral system, the above conclusion need not hold: a positive conclusion *may* have a negative premise, as shown by the following example.

Example 4.4.24. *Consider the familiar (binary) connective xor, having the following I-rules.*

$$\frac{\Gamma : +\varphi \quad \Gamma : -\psi}{\Gamma : +(\varphi \; xor \; \psi)} \; (xor^+I_1) \qquad \frac{\Gamma : -\varphi \quad \Gamma : +\psi}{\Gamma : +(\varphi \; xor \; \psi)} \; (xor^+I_2) \tag{4.4.78}$$

Example 4.4.25. *Let $ite(\varphi, \psi, \chi)$ denote the familiar ternary connective 'if φ then ψ else χ'. Its definition is given by the following I-rule, making use of a negative (discharged) assumption.*

$$\frac{\Gamma, +\varphi : +\psi \quad \Gamma, -\varphi : +\chi}{\Gamma : +ite(\varphi, \psi, \chi)} \; (ite^+ \; I) \tag{4.4.79}$$

The usual notion of a tree-shaped *derivation* is kept here too, again with $\mathcal{D}$ ranging over derivations.

The following definition of *(proof-theoretic) duality* of logical constants will be used below.

Definition 4.4.34 (proof-theoretic duality). *Let $*_1$, $*_2$ be two n-ary operators. They are (mutually) dual, i.e, $*_1^\circ = *_2$ and $*_2^\circ = *_1$, iff their respective I-rules satisfy the following condition. For brevity, all '+' markings are suppressed. Duality is denoted by '$\circ$'.*

additive rules: *The definition distinguishes between categorical and hypothetical I-rules.*

> **categorical rules:** *The operator $*_1$ has $m \geq 1$ categorical combining I-rules of the form*
>
> $$\frac{\Gamma : \sigma_1^k \quad \cdots \quad \Gamma : \sigma_{n_k}^k}{\Gamma : *_1(\varphi_1, \cdots, \varphi_n)} \; (*_1 I_k) \tag{4.4.80}$$
>
> *for $1 \leq k \leq m$, $n_k \geq 2$ and all the σs are signed occurrences of $\varphi_1, \cdots, \varphi_n$. Note that no assumptions are discharged by any $(*_1 I_k)$ rule.*
>
> *For each such rule, the operator $*_2$ has n_k non-combining categorical I-rules, with the jth rule of the form*
>
> $$\frac{\Gamma : \sigma_j^k}{\Gamma : *_2(\varphi_1, \cdots, \varphi_n)} \; (*_2 I_{k,j}) \qquad 1 \leq j \leq n_k \tag{4.4.81}$$

Again no assumption is discharged. An operator having such non-combining I-rules is referred to as splitting.

hypothetical rules: *The operator $*_1$ has $m \geq 1$ combining hypothetical I-rules of the form*

$$\frac{\Gamma, \sigma_1'^{\,k} : \sigma_1^k \quad \cdots \quad \Gamma, \sigma_{n_k}' : \sigma_{n_k}^k}{\Gamma : *_1(\varphi_1, \cdots, \varphi_n)} \ (*_1 I_k) \tag{4.4.82}$$

for $1 \leq k \leq m$, $n_k \geq 2$ and all the σs are signed occurrences of $\varphi_1, \cdots, \varphi_n$. Note that some of the premises need not have a discharged assumption.

For each such rule, the operator $*_2$ has n_k combining categorical I-rules, with the jth rule of the form

$$\frac{\Gamma : \overline{\sigma_j'^{\,k}} \quad \Gamma : \sigma_j^k}{\Gamma : *_2(\varphi_1, \cdots, \varphi_n)} \ (*_2 I_{k,j}) \tag{4.4.83}$$

multiplicative rules: *Again, there is a difference between categorical and hypothetical rules.*

categorical rules: *The operator $*_1$ has $m \geq 1$ combining categorical I-rules of the form*

$$\frac{\Gamma_1 : \sigma_1^k \quad \cdots \quad \Gamma_{n_k} : \sigma_{n_k}^k}{\Gamma_1 \cdots \Gamma_{n_k} : *_1(\varphi_1, \cdots, \varphi_n)} \ (*_1 I_k) \tag{4.4.84}$$

for $1 \leq k \leq m$, $n_k \geq 2$. Note again that no assumptions are discharged by the $(*_1 I)$ rules.

For each such rule, the operator $*_2$ has n_k non-combining categorical I-rules, with the jth rule of the form

$$\frac{\Gamma : \sigma_j^k}{\Gamma : *_2(\varphi_1, \cdots, \varphi_n)} \ (*_2 I_{k,j}) \quad 1 \leq j \leq n \tag{4.4.85}$$

Again no assumption is discharged. An operator having such I-rules is also referred to as splitting.

hypothetical rules: *The operator $*_1$ has m combining hypothetical I-rules of the form*

$$\frac{\Gamma_1, \sigma_1' : \sigma_1^k \quad \cdots \quad \Gamma_{n_k}, \sigma_{n_k}' : \sigma_{n_k}^k}{\Gamma_1 \cdots \Gamma_{n_k} : *_1(\varphi_1, \cdots, \varphi_{n_k})} \ (*_1 I_k) \tag{4.4.86}$$

for $1 \leq k \leq m$.

For each such rule, the operator $*_2$ has n_k combining categorical I-rules, with the jth rule of the form

$$\frac{\Gamma : \overline{\sigma_j'^{\,k}} \quad \Gamma : \sigma_j^k}{\Gamma : *_2(\varphi_1, \cdots, \varphi_n)} \ (*_2 I_{k,j}) \tag{4.4.87}$$

Remarks:

1. By this definition, in contrast to positive unilateral ND-systems, in bilateral ND-systems every operator has a dual (of the same arity).

2. The usual '$\vee$' (disjunction) and '$\wedge$' (conjunction), with additive I-rules involving no negative premises, are mutually dual, where '$\wedge$' is combining and '$\vee$' is splitting. Here $n = 2$ and $m = 1$.

3. A typical operator the I-rule of which discharges an assumption is implication, with $n = 2$ and $m = 1$. For its dual, see example 4.4.26 below.

4. The intuitionistic linear logic, additive '$\oplus$' and multiplicative '$\otimes$', with I-rules below (cf. Negri [127]), are mutually dual, where '$\otimes$' is combining and '$\oplus$' is splitting.

$$\frac{\Gamma_1 : \varphi \quad \Gamma_2 : \psi}{\Gamma_1\Gamma_2 : \varphi \otimes \psi} \ (\otimes I) \qquad \frac{\Gamma : \varphi}{\Gamma : \varphi \oplus \psi} \ (\oplus I_1) \quad \frac{\Gamma : \psi}{\Gamma : \varphi \oplus \psi} \ (\oplus I_2) \qquad (4.4.88)$$

5. The condition on the I-rules of dual operators impose a restriction on their *harmoniously induced GE-rules*, when the latter are constructed as in subsection 3.6.3.

Example 4.4.26. *The rules for the intuitionistic implication are given in (4.4.89) below.*

$$\frac{\Gamma, +\varphi : +\psi}{\Gamma : +(\varphi \to \psi)} \ (\to^+ I) \quad \frac{\Gamma : +(\varphi \to \psi) \quad \Gamma : +\varphi \quad \Gamma, +\psi : +\xi}{\Gamma : +\xi} \ (\to^+ E) \qquad (4.4.89)$$

The dual of the implication[7] '$\to^\circ$' is defined by the I-rule below.

$$\frac{\Gamma : -\varphi \quad \Gamma : +\psi}{\Gamma : +(\varphi \to^\circ \psi)} \ (\to^\circ I) \qquad (4.4.90)$$

Example 4.4.27. *The dual of xor, which has two combining categorical I-rules given in example 4.4.24, is given by the hypothetical I-rule below,*

$$\frac{\Gamma, +\varphi : +\psi \quad \Gamma, +\psi : +\varphi}{\Gamma : +(\varphi \ xor^\circ \ \psi)} \ (xor^\circ I) \qquad (4.4.91)$$

which is the familiar (material) equivalence connective. Note that xor plays the role of $*_2$ in the definition of duality, while xor° plays the role of $*_1$.

The following corollary follows directly from the definition of duality above.

Corollary 1 (idempotence of duality). *For every connective '$*$': $*^{\circ\circ} = *$.*

By the construction of the harmoniously induced GE-rules in Section 3.6.3, a support of an I-rule turns into a premise of the induced GE-rule. Supports have an important role in inducing negative I-rules from positive ones, as described in the horizontal inversion principle below (Section 4.4.1.7).

---

[7]Note that this is a different connective than the one known as *co-implication* (or *subtraction*), studied in the Brouwer-Heyting logic.

4.4.1.2 Sub-structurality and PTS

In passing, I would like to elaborate a little more[8] on the effect of *sub-structurality* (the distinction between additive and multiplicative rules), which is not common in the PTS literature, on the construction of harmoniously-induced GE-rules (see the discussion on 'harmony in form' below). Recall that a more detailed exposition of this matter was presented in Section 4.2.5, in a more suitable setting of *multi-conclusion natural-deduction systems*, that have also right-contexts. In particular, that section includes a *justification* of the effect described below. Here, I will only exemplify this effect, to clarify the issue.

Example 4.4.28. *Consider the additive and multiplicative conjunctions, '$\wedge$' and '$\otimes$', respectively. Their I-rules are the following.*

$$\frac{\Gamma:\varphi \quad \Gamma:\psi}{\Gamma:\varphi\wedge\psi} \ (\wedge I) \qquad \frac{\Gamma_1:\varphi \quad \Gamma_2:\psi}{\Gamma_1\Gamma_2:\varphi\otimes\psi} \ (\otimes I) \qquad\qquad (4.4.92)$$

When constructing harmoniously-induced GE-rules, the result for the multiplicative conjunction '$\otimes$' (the one actually given in Francez and Dyckhoff [54]) is (with context omitted)

$$\frac{\varphi\otimes\psi \quad \overset{\displaystyle [\varphi,\psi]_i}{\overset{\vdots}{\chi}}}{\chi} \ (\otimes GE^i) \qquad\qquad (4.4.93)$$

On the other hand, an application of the construction to the additive conjunction '$\wedge$' (as explained in detail in Section 4.2.5.1) yields the following two GE-rules (simply presented).

$$\frac{\varphi\wedge\psi \quad \overset{\displaystyle [\varphi]_i}{\overset{\vdots}{\chi}}}{\chi} \ (\wedge_1 GE^i) \qquad \frac{\varphi\wedge\psi \quad \overset{\displaystyle [\psi]_j}{\overset{\vdots}{\chi}}}{\chi} \ (\wedge_2 GE^j) \qquad\qquad (4.4.94)$$

A similar bifurcation would result for implication, had both additive and multiplicative implications been considered. I will not pursue this distinction further here, as only additive bilateral ND-systems are considered in the rest of the chapter.

4.4.1.3 Coherence (of contexts)

The objects over which the ND-rules are defined are sequents of one of the forms $\Gamma:+\varphi$ and $\Gamma:-\varphi$. A context Γ is here taken to contain both positively and negatively marked sentences. An important property of contexts is that of *coherence*, formally defined below. It is the counterpart of *consistency* in a unilateral logic with explicit sentential negation. It relies on the two attitudes of assertion and denial being mutually exclusive.

---

[8]I am indebted to Stephen Read for lengthy discussions on the topic discussed in this Section.

Definition 4.4.35 (Coherent contexts). *A context* Γ *is coherent, denoted by* $\mathcal{C}(\Gamma)$, *iff for every sentence* φ *(in the object language), at most one of* $\Gamma : +\varphi$ *and* $\Gamma : -\varphi$ *is derivable. Otherwise,* Γ *is incoherent.*

In particular, in the presence of the identity axioms $\Gamma, \sigma : \sigma$ (for both positively and negatively signed σ), any context $\Gamma, +\varphi, -\varphi$, containing an assertion and a denial of the *same contents*, is incoherent. The incoherence of such contexts reflects the role of the signs as indicating forces that cannot be held together as forces of the same content, an important attribute of bilateralism.

However, the distinction between coherent and incoherent contexts is not sufficient for endowing an ND-system with the expression of bilateralism, based on contrariety of the two forces of assertion and denial. The latter effect is established by means of appropriate *structural rules*, extending the meaning of the underlying derivability relation (cf. the remark on p. 42).

$$(a) \;\; \Gamma, +\varphi, -\varphi : \sigma \; (INC) \quad (b) \quad \frac{\Gamma, +\varphi : +\psi \quad \Gamma, +\varphi : -\psi}{\Gamma : -\varphi} \; (CO) \qquad (4.4.95)$$

The (INC) rule establishes the incoherence of an incoherent context, that cannot be coherently used in deductions, by letting *every* sentence being derivable from it (a property known also as *explosion*). The second rule, (CO), a *coordination* rule in the terminology of Rumfitt [179], allows denial of an assumption the assertion of which leads to derivability of the same content but with reversed forces. To obtain classical strength, the rule (CO) needs to be strengthened, allowing also the reversal of force from assuming a denied sentence $-\varphi$ leading to incoherence (see Section 4.4.3).

As for "positive languages", the required effect of bilateralism is obtained via the structural rules in (4.4.95) and the vertical and horizontal balance as specified in (4.4.1.5).

4.4.1.4   Canonicity

As mentioned above, meaning is not based on arbitrary derivations, but on *canonical ones*, depending of the PTS view chosen. Note that in the definition canonicity for a bilateral ND-system the I/E-rules can be either positive or negative. Thus, we have both canonical proofs (in the positive case) and canonical denials (in the negative case).

4.4.1.5   Bilateral harmony

I now return the criteria to be imposed on an ND-system to qualify as meaning-conferring, and principles guaranteeing those criteria are met. In the unilateral setting (as in Section 3.6.1), these criteria are "vertical", in relating E-rules to I-rules. In a bilateral setting, the vertical criteria split into two groups, for the respective positive and negative groups of rules. However, there is in the bilateral case a need to relate positive rules to negative rules, to guarantee those rules justify being rules for assertion and denial; clearly, a certain dependency among those two groups is required, so that $+\varphi$ and $-\varphi$ satisfy (mutual) *contrariety*.

4.4.1.6　Vertical balance

Recall that harmony requires a certain balance between the I-rules and the E-rules, none of which overpowering the other in strength.

Traditionally, in a unilateral logic, harmony is associated with the *inversion principle* Prawitz [146] (see p. 82). Recall that it states that every conclusion derivable from (a compound) φ is also a conclusion derivable from any grounds for asserting φ. To distinguish this principle from a second one, not present in the unilateral case, I will refer to it as the *vertical inversion* principle.

As we saw in Chapter 3, this principle can be "implemented" in two ways, leading to two ways of capturing the informal notion of harmony: *harmony in form* and *intrinsic harmony*. In a bilateral logic, if unilateral rules are equated with the corresponding positive rules, the above principle guarantees that the *positive E*-rules are not too strong w.r.t. the *positive I*-rules. Clearly, the negative E-rules also need to be shown not to be too strong w.r.t. the *negative I*-rules. Such a need was noticed in Hjortland [76] (p. 78). To see this need, consider the following example.

Example 4.4.29. *Suppose the following two negative rules constitute the denial rules for 'tonk'.*

$$\frac{\Gamma : -\varphi}{\Gamma : -(\varphi\ tonk\ \psi)}\ (tonk^-I) \qquad \frac{\Gamma : -(\varphi\ tonk\ \psi)}{\Gamma : -\psi}\ (tonk^-E) \qquad\qquad (4.4.96)$$

Clearly, similarly to the unilateral/positive case, those rules trivialize the derivability relation in letting $\vdash - \varphi : -\psi$ for any φ, ψ. This observation leads to an enhancement of the vertical inversion principle to bilateralism, encompassing negative rules too.

The bilateral vertical inversion principle:

1. Any conclusion of an assertion of (a compound) φ is also a conclusion of any grounds for assertion of φ.

2. Any conclusion of a denial of (a compound) φ is also a conclusion of any grounds for denial of φ.

For harmony in form, the bilateral vertical inversion principle requires that negative E-rules are also of the GE-form. A simple argument can establish that the construction in Section 3.6.3 can be applied in the negative case too.

For intrinsic harmony, the bilateral vertical inversion principle requires the existence of both positive and negative detour reductions. A negative reduction for disjunction is mentioned in Hjortland [76].

Example 4.4.30. *As an example of a negative reduction, anticipating the $PNJB$-rules in Figures 4.4and 4.5, consider a negative reduction for disjunction.*

$$
\cfrac{\cfrac{\mathcal{D}_1 \quad \mathcal{D}_2}{-\varphi \quad -\psi}}{\cfrac{-(\varphi \vee \psi)}{\xi}} (\vee^- I) \quad \cfrac{[-\varphi]_i, [-\psi]_j}{\cfrac{\mathcal{D}_3}{\xi}} (\vee^- E^{i,j}) \quad \rightsquigarrow_r \quad \mathcal{D}_3[-\varphi := \cfrac{\mathcal{D}_1}{-\varphi}, -\psi := \cfrac{\mathcal{D}_2}{-\psi}] \quad (4.4.97)
$$

In addition, in the bilateral setting, we need to require under local-completeness the existence of both positive and negative expansions, balancing also the negative E-rules w.r.t. the negative I-rules.

4.4.1.7   Horizontal balance

Horizontal balance should relate positive and negative rules, to render assertion and denial enjoy contrariety[9] forces. It also accounts for a certain symmetry found amongst those rules. In continuation to Gentzen's remark to the effect that I-rules are self-justifying, and E-rules are justified by the I-rules by the vertical inversion principle, one can see the positive I-rules as a definition that should justify also the negative I-rules. Also, in the same way that the role of the I/E-rules are variably classifiable as to which is self-justifying and which group justifies the other group, in the bilateral case one can view the *negative I*-rules as being self-justifying, and the positive I-rules being justified by the negative ones. I will not pursue here this direction further.

Clearly, grounds for denial should be related to denial of grounds for assertion. However, a finer distinction is needed to get this relationship right. The approach proposed is more similar to harmony in form than to intrinsic harmony.

Consider first a simple example.

Example 4.4.31. *Suppose the usual '$\wedge$'-rules are given, to constitute the positive rules in a bilateral ND-system:*

$$
\cfrac{\Gamma : +\varphi \quad \Gamma : +\psi}{\Gamma : +(\varphi \wedge \psi)} (\wedge^+ I) \quad \cfrac{\Gamma : +(\varphi \wedge \psi) \quad \Gamma, +\varphi, +\psi : \xi}{\Gamma : \xi} (\wedge^+ E) \quad (4.4.98)
$$

Now suppose, that the following single negative I-rule is added, supposedly capturing (on its own) the introduction of the denial of conjunction.

$$
\cfrac{\Gamma : -\varphi}{\Gamma : -(\varphi \wedge \psi)} (\wedge^- I) \quad (4.4.99)
$$

---

[9]For the classical case, contradictory forces are needed.

Pretheoretically, this seems wrong! This negative I-rule is too weak w.r.t. the positive I-rules of conjunction. Similarly,

$$\frac{\Gamma : -\varphi \quad \Gamma : -\psi}{\Gamma : -(\varphi \wedge \psi)} \ (\wedge^- I)$$

                                                                                    (4.4.100)

seems, pretheoretically, too strong w.r.t. the positive I-rules of conjunction.

A conjunction should be deniable upon the denial of each conjunct (separately). But which principle would bring this about?

The horizontal inversion principle:

constants with categorical I-rules: Let φ have one of those constants as its main operator.

1. Any denial of a ground for assertion of a categorical combining (compound) φ is a ground for denial of φ.

2. The collection of denials of a ground for assertion of a categorical splitting (compound) φ is a ground for denial of φ.

3. Any conclusion of a denial of a categorical combining (compound) φ is also a conclusion of each of the grounds of assertion of φ.

4. Any conclusion of a denial of a categorical splitting (compound) φ is also a joint conclusion of the grounds for assertion of φ.

The first principle indicates that the negative I-rules (of the main operator of a categorical combining φ) are neither too weak nor too strong relative to the positive I-rules (of the same operator). The second principle assures the same for a categorical splitting logical constant. This balance can be seen as a generalization of (classical) De-Morgan's laws, specifically relating negation and the dual pair of conjunction/disjunction (see below). The other two principles relate the positive and negative E-rules to each other.

The fact that the required balance between positive and negative rules is a generalization of De-Morgan's laws comes as no surprise. This fact just reflects the connection between the force of denial and top-level negation. The same holds for the ability to *define* negation in terms of denial, as done by Rumfitt in [179]. When defining negation this way, De-Morgan's rule are imposed by the horizontal inversion principle merely due to the form of the '$\wedge$'-rules and '$\vee$'-rules, a form also responsible to their duality as discussed above.

constants with a hypothetical I-rule: The positive I-rule of such a constant discharges assumptions, its support. The support is "temporarily" added to form parts of the grounds for asserting the conclusions of the premises of the rule. To deny the conclusion of the rule (with the constant as its main operator), one has, therefore, to accept the support while denying the conclusions of the premises.

Consider a constant with a hypothetical I-rule. Let φ have the constant as its main operator.

1. Any assertion of the support together with the denial of the conclusion of the premise is a ground for denial of φ.

2. Any conclusion of the denial of φ is also a joint conclusion of the assertion of the support and denial of the conclusion of the premise.

Returning to Example 4.4.31, since the grounds for assertion of $+(\varphi \wedge \psi)$ are $\Gamma = \varphi, \psi$ (a categorical combining constant), having only the negative I-rule in (4.4.99) is too weak, as both the denials of the separate conjuncts have to be, by the first horizontal inversion principle, grounds for denial of the conjunction. What is needed is a second negative I-rule for conjunction, namely

$$\frac{\Gamma : -\psi}{\Gamma : -(\varphi \wedge \psi)} \ (\wedge^- I)$$

(4.4.101)

Similarly, the rule in (4.4.100) is too strong.

Example 4.4.32. *For an example of denial I/E-rules for an operator with an hypothetical rule, consider the familiar example of intuitionistic implication, the I/E-rules of which were presented in Example (4.4.26). Here the (single) support is φ, and the conclusion of the (single) premise is ψ. By the horizontal inversion principle, the following negative I/E-rules for implication are justified.*

$$\frac{\Gamma : +\varphi \quad \Gamma : -\psi}{\Gamma : -(\varphi \rightarrow \psi)} \ (\rightarrow^- I) \qquad \frac{\Gamma : -(\varphi \rightarrow \psi) \quad \Gamma, +\varphi, -\psi : \xi}{\Gamma : \xi} \ (\rightarrow^- E)$$

(4.4.102)

Definition 4.4.36 (bilateral harmony). *A bilateral ND-system is bilaterally harmonious iff it satisfies both the vertical and the horizontal inversion principles.*

The overall relationship induced by bilateral harmony on the rules of a bilateral ND-system can be summarised by the following *commuting diagram* for an generic operator '$*$'. I use hh and vh to abbreviate 'horizontal harmony' and 'vertical harmony', respectively. The arrows marked 'i' indicate 'inducing', thus the relationship holds by construction. The relationships shown by the other arrows need to be verified to establish the commuting of the diagram.

$$
\begin{array}{ccc}
(*^+ I) & \xrightarrow[i]{hh} & (*^- I) \\
{\scriptstyle vh} \downarrow {\scriptstyle i} & & \downarrow {\scriptstyle vh} \\
(*^+ E) & \xrightarrow{hh} & (*^- E)
\end{array}
$$

(4.4.103)

This diagram reflects the view that the positive I-rules are those determining meaning (according to the PTS inferentialism approach). Similar diagrams, differing in the identity and direction of the 'i'-marked arrows, can summarise the pragmatism approach of PTS, where the E-rules are given as meaning determining. In addition, similar diagrams fit the less common view, the point of departure of which are the negative rules. I'll not pursue further those extra possibilities here.

The following proposition, following directly from the horizontal inversion principle, relates denial and duality.

Proposition 4.4.7. *The premises of a negative I-rule for a connective, say '$*$', are the premises of the positive I-rule of $*°$, but with a reversed force indicator.*

For conjunction and disjunction, this amounts to the usual (classical) De-Morgan laws. For implication, the premises of the I-rule of its dual (4.4.90) are $\Gamma : -\varphi$ and $\Gamma : +\psi$. Hence, its negative I-rule is

$$\frac{\Gamma : +\varphi \quad \Gamma : -\psi}{\Gamma : -(\varphi \to \psi)} \ (\to^- I) \tag{4.4.104}$$

The main theorem that characterizes a bilaterally-harmonious ND-system is the following.

Theorem 4.4.9 (coordination admissibility). *The bilateral structural rule CO (see (4.4.95 (b))) is admissible in any bilaterally-harmonious ND-system.*

Proof: The proof is by induction on φ. W.l.o.g, we consider additive rules.

atomic conclusion: Let p be any atomic sentence. Suppose

$$(a) \ \Gamma, +\varphi : +p \ \text{ and } \ (b) \ \Gamma, +\varphi : -p \tag{4.4.105}$$

Since p is atomic, (4.4.105) is only possible if both (a) and (b) are axioms, hence $+p, -p \in \Gamma$. This renders Γ incoherent, and by (INC) (4.4.95 (a)) $\Gamma : -\varphi$.

compound conclusion: Assume, as an inductive assumption, that (CO) holds for $\varphi_1, \cdots, \varphi_n$. Consider an arbitrary n-ary connective '$*$'. Suppose that

$$(1) \ \Gamma, +\varphi : + * (\sigma_1, \cdots, \sigma_n) \quad (2) \ \Gamma, +\varphi : - * (\sigma_1, \cdots, \sigma_n) \tag{4.4.106}$$

Recall that each σ_i is a signed occurrence of some φ_j. We have to show that $\vdash \Gamma : -\varphi$. We distinguish cases according to the form of $(*I)$.

categorical splitting rules: Consider the $(*I_{k,j})$-rules, given by (4.4.81). By vertical harmony, $(*E)$ can be assumed to be a GE-rule of the form

$$\frac{\Gamma : + * (\varphi_1, \cdots, \varphi_n) \quad \Gamma, \sigma_1^k : \xi \quad \cdots \quad \Gamma, \sigma_{n_k}^k : \xi}{\Gamma : \xi} \ (*^+ E) \tag{4.4.107}$$

By Proposition 4.4.7 and vertical harmony, the negative $(*E)$-rule has the form

$$\frac{\Gamma : - * (\varphi_1, \cdots, \varphi_n) \quad \Gamma, \sigma_1^k, \cdots, \sigma_{n_k}^k : \xi}{\Gamma : \xi} \ (*^- E) \tag{4.4.108}$$

Hence, by assumption (4.4.106, (1)), we get, for some $1 \le \hat{j} \le n_k$ and $\xi = \sigma_{\hat{j}}$

$$\Gamma, +\varphi : \sigma_{\hat{j}} \tag{4.4.109}$$

W.l.o.g, assume $\sigma_{\hat{j}} = +\varphi_{\hat{j}}$. Thus,

$$\Gamma, +\varphi : +\varphi_{\hat{j}} \tag{4.4.110}$$

Similarly, by assumption (4.4.106, (2)), we get for every $1 \leq j \leq n_k$ and $\xi = \overline{\sigma_j}$

$$\Gamma, +\varphi : \overline{\sigma_j} \tag{4.4.111}$$

In particular, for $\hat{j}$

$$\Gamma, +\varphi : \overline{\sigma_j} = -\varphi_{\hat{j}} \tag{4.4.112}$$

By the inductive assumption on $\varphi_{\hat{j}}$, $\Gamma : -\varphi$, as required.

categorical combining rules: The argument is dual and omitted.

hypothetical rules: Consider the kth positive I-rule as given by (4.4.82). By vertical harmony, the jth ($*E$)-rule can be assumed to have the form

$$\frac{\Gamma : + * (\varphi_1, \cdots, \varphi_n) \quad \Gamma : +\sigma_j'^k \quad \Gamma, \sigma_j^k : \xi}{\Gamma : \xi} \; (*^+E) \tag{4.4.113}$$

By proposition 4.4.7 and vertical harmony, the jth negative ($*E$)-rule has the form

$$\frac{\Gamma : - * (\varphi_1, \cdots, \varphi_n) \quad \Gamma, +\sigma_j'^k, -\sigma_j^k : \xi}{\Gamma : \xi} \; (*^-E) \tag{4.4.114}$$

Hence, by assumption (4.4.106, (1)), we get, for some $1 \leq \hat{j} \leq n_k$ and $\xi = \sigma_{\hat{j}}^k$

$$(i) \; \Gamma, +\varphi, +\sigma_{\hat{j}}'^k : +\sigma_{\hat{j}}^k \tag{4.4.115}$$

W.l.o.g, assume $\sigma_{\hat{j}} = +\varphi_{\hat{j}}$. Thus,

$$(i) \; \Gamma, +\varphi, +\varphi_{\hat{j}}'^k : +\varphi_{\hat{j}}^k \tag{4.4.116}$$

Similarly, by assumption (4.4.106, (2)), we get for $\xi = \overline{\sigma_{\hat{j}}^k}$

$$(ii) \; \Gamma, +\varphi : \sigma_{\hat{j}}'^k, \quad (iii) \; \Gamma, +\varphi : \overline{\sigma_{\hat{j}}^k} \tag{4.4.117}$$

Namely,

$$(ii) \; \Gamma, +\varphi : +\varphi_{\hat{j}}'^k, \quad (iii) \; \Gamma, +\varphi : -\varphi_{\hat{j}}^k \tag{4.4.118}$$

By weakening applied to (iii), we get

$$(iv) \; \Gamma, +\varphi, [+\varphi_{\hat{j}}'^k] : -\varphi_{\hat{j}}^k, \tag{4.4.119}$$

By the induction hypothesis on (i) and (iv), we get

$$(v) \; \Gamma, +\varphi : -\varphi_{\hat{j}}'^k \tag{4.4.120}$$

Finally, by the induction hypothesis on (ii) and (v), we get $\Gamma : -\varphi$, as required.

To get a better grasp on the idea of this proof, I repeat it for concrete connectives in the next section.

$$\overline{\Gamma, +\varphi : +\varphi}\ (Ax^+)$$

$$\frac{\Gamma : +\varphi \quad \Gamma : +\psi}{\Gamma : +(\varphi \wedge \psi)}\ (\wedge^+ I) \qquad \frac{\Gamma : +(\varphi \wedge \psi) \quad \Gamma, +\varphi, [+\psi]_j : +\xi}{\Gamma : +\xi}\ (\wedge^+ E)$$

$$\frac{\Gamma : +\varphi}{\Gamma : +(\varphi \vee \psi)}\ (\vee^+ I_1) \qquad \frac{\Gamma : +\psi}{\Gamma : +(\varphi \vee \psi)}\ (\vee^+ I_2)$$

$$\frac{\Gamma : +(\varphi \vee \psi) \quad \Gamma, +\varphi : +\xi \quad \Gamma, +\psi : +\xi}{\Gamma : +\xi}\ (\vee^+ E)$$

$$\frac{\Gamma, +\varphi : +\psi}{\Gamma : +(\varphi \rightarrow \psi)}\ (\rightarrow^+ I) \qquad \frac{\Gamma : +(\varphi \rightarrow \psi) \quad \Gamma : +\varphi \quad \Gamma, +\psi : +\xi}{\Gamma : +\xi}\ (\rightarrow^+ E)$$

Figure 4.4: The PNJB rules for assertion

$$\overline{\Gamma, -\varphi : -\varphi}\ (Ax^-)$$

$$\frac{\Gamma : -\varphi}{\Gamma : -(\varphi \wedge \psi)}\ (\wedge^- I_1) \qquad \frac{\Gamma : -\psi}{\Gamma : -(\varphi \wedge \psi)}\ (\wedge^- I_2) \qquad \frac{\Gamma : -(\varphi \wedge \psi) \quad \Gamma, -\varphi : \xi \quad \Gamma, -\psi : \xi}{\Gamma : \xi}\ (\wedge^- E)$$

$$\frac{\Gamma : -\varphi \quad \Gamma : -\psi}{\Gamma : -(\varphi \vee \psi)}\ (\vee^- I) \qquad \frac{\Gamma : -(\varphi \vee \psi) \quad \Gamma, -\varphi, -\psi : \xi}{\Gamma : \xi}\ (\vee^- E)$$

$$\frac{\Gamma : +\varphi \quad \Gamma : -\psi}{\Gamma : -(\varphi \rightarrow \psi)}\ (\rightarrow^- I) \qquad \frac{\Gamma : -(\varphi \rightarrow \psi) \quad \Gamma, +\varphi, -\psi : \xi}{\Gamma : \xi}\ (\rightarrow^- E)$$

Figure 4.5: The PNJB rules for denial

4.4.2 A Bilaterally Harmonious Positive Logic

The above considerations lead to the following *positive* bilateral logic (in Figures 4.4 and 4.5), satisfying all the above stated criteria for qualifying as meaning-conferring. The logic can be viewed as an extension of the (unilateral) positive fragment of Gentzen's NJ, but with E-rules in GE form, referred to as $PNJB$. The following two (simply presented) regular (negative) E-rules are clearly derivable.

$$\frac{-(\varphi \rightarrow \psi)}{+\varphi}\ (\rightarrow^- E_1) \qquad \frac{-(\varphi \rightarrow \psi)}{-\psi}\ (\rightarrow^- E_2) \tag{4.4.121}$$

As usual, discharged assumptions are enclosed in square brackets and indexed, where an application of a rule discharging an assumption indicates the index of the discharged assumption.

Note again that all the E-rules (both positive and negative) are in GE form, allowing to draw an *arbitrary conclusion* from the major premise, whether positive or negative. This rule format exhibits vertical harmony for both the positive and the negative rules. Below are two examples of (simply presented) derivations in $PNJB$.

Example 4.4.33. *A derivation establishing*

$$\vdash_{PNJB} + (\varphi{\rightarrow}\psi), -(\varphi{\rightarrow}\chi) : -(\psi{\rightarrow}\chi)$$

$$\cfrac{\cfrac{\cfrac{-(\varphi{\rightarrow}\chi)}{+\varphi}\ (\rightarrow^- E) \quad +(\varphi{\rightarrow}\psi)}{+\psi}\ (\rightarrow^+ E) \quad \cfrac{-(\varphi{\rightarrow}\chi)}{-\chi}\ (\rightarrow^- E)}{-(\psi{\rightarrow}\chi)}\ (\rightarrow^- I) \tag{4.4.122}$$

Example 4.4.34. *A derivation establishing*

$$\vdash_{PNJB} - ((\varphi{\rightarrow}\psi){\wedge}(\psi{\rightarrow}\chi)) : -(\varphi{\rightarrow}\chi)$$

$$\cfrac{-((\varphi{\rightarrow}\psi){\wedge}(\psi{\rightarrow}\chi)) \qquad \cfrac{\cfrac{\cfrac{[-(\varphi{\rightarrow}\psi)]_1}{+\varphi}\ (\rightarrow^- E_1) \quad \cfrac{[-(\psi{\rightarrow}\chi)]_2}{-\chi}\ (\rightarrow^- E_2)}{-(\varphi{\rightarrow}\chi)}\ (\rightarrow^- I)}{}}{-(\varphi{\rightarrow}\chi)}\ (\wedge^- E^{1,2}) \tag{4.4.123}$$

Theorem 4.4.10 (bilateral harmony of $PNJB$). *The logic $PNJB$ is bilaterally-harmonious.*

Proof. Both vertical and horizontal harmony need to be shown.

vertical harmony: **harmony in form:** This is evident by inspection of the E-rules (both positive and negative), all in the GE form.

    intrinsic harmony: The reductions for establishing positive local-soundness are the well-known reductions from Prawitz [146]. The positive expansions for establishing local-completeness are also well-known, see Pfenning and Davies [143, 24]. The negative reductions and expansions are simple analogues of the positive ones. For example, the negative reduction for conjunction is

$$\cfrac{\cfrac{\mathcal{D}}{-\varphi} \atop \cfrac{-\varphi}{-(\varphi{\wedge}\psi)}\ (\wedge^- I_1) \qquad \cfrac{[-\varphi]_i \quad [-\psi]_j}{\mathcal{D}_1 \quad \mathcal{D}_2 \atop \xi \qquad \xi}}{\xi}\ (\wedge^- E^{i,j}) \quad\leadsto_r\quad \cfrac{\mathcal{D}}{\mathcal{D}_1[-\varphi := -\varphi] \atop \xi} \tag{4.4.124}$$

The reduction for $(\wedge^- I_2)$ is similar. I omit the other negative reductions. The negative expansion for conjunction is

$$\cfrac{\mathcal{D}}{-(\varphi{\wedge}\psi)}\ \leadsto_e\quad \cfrac{\cfrac{\mathcal{D}}{-(\varphi{\wedge}\psi)} \quad \cfrac{[-\varphi]_i}{-(\varphi{\wedge}\psi)}\ (\wedge^- I_1) \quad \cfrac{[-\psi]_j}{-(\varphi{\wedge}\psi)}\ (\wedge^- I_2)}{-(\varphi{\wedge}\psi)}\ (\wedge^- E^{i,j}) \tag{4.4.125}$$

Again, the other negative expansions are similar and omitted.

horizontal harmony: This is again established by a simple inspection. For example, the dual '∧' and '∨', combining and splitting, respectively, conform to the requirement. So does '→', having an hypothetical I-rule.

$\square$

Proposition 4.4.8 (admissibility of (CO)**).** *The coordination rule* (CO) *is admissible for* $PNJB$.

Proof. Though this is just a special case of Theorem 4.4.9, that proof is instantiated for the concrete connectives of $PNJB$, for better comprehension of the argument involved.

The proof is by induction on σ.

atomic conclusion: Identical to the atomic case of the general theorem.

compound conclusion: Assume, as an inductive assumption, that (CO) holds for φ_1 and φ_2.

    disjunction: Suppose

$$(1)\ \Gamma, +\varphi : +(\varphi_1 \vee \varphi_2)\ \text{ and }\ (2)\ \Gamma, +\varphi : -(\varphi_1 \vee \varphi_2) \qquad (4.4.126)$$

We need to show $\Gamma : -\varphi$. From (4.4.126, (1)), we get that either

$$\Gamma, +\varphi : +\varphi_1 \qquad (4.4.127)$$

or

$$\Gamma, +\varphi : +\varphi_2 \qquad (4.4.128)$$

In addition, from (4.4.126, (2)), we get that both

$$\Gamma, +\varphi : -\varphi_1 \qquad (4.4.129)$$

and

$$\Gamma, +\varphi : -\varphi_2 \qquad (4.4.130)$$

If (4.4.127) is the case, than by applying the induction hypothesis on (4.4.127) and (4.4.129), we get $\Gamma : -\varphi$, as required. In the other case, applying the induction hypothesis on (4.4.128) and (4.4.130) yields the result.

conjunction: A dual argument – omitted.

implication: Suppose

$$(1)\ \Gamma, +\varphi : +(\varphi_1 \rightarrow \varphi_2)\ \text{ and }\ (2)\ \Gamma, +\varphi : -(\varphi_1 \rightarrow \varphi_2) \qquad (4.4.131)$$

We need to show $\Gamma : -\varphi$. From (4.4.131, (1)) we get

$$(i)\ \Gamma, +\varphi, +\varphi_1 : +\varphi_2 \qquad\qquad (4.4.132)$$

From (4.4.131, (2)) we get

$$(ii)\ \Gamma, +\varphi : +\varphi_1 \quad \text{and} \quad (iii)\ \Gamma, +\varphi : -\varphi_2 \qquad\qquad (4.4.133)$$

By applying weakening to (4.4.131, iii), we get

$$\Gamma, +\varphi, +\varphi_1 : -\varphi_2 \qquad\qquad (4.4.134)$$

Applying the induction hypothesis to (4.4.132,(i)) and (4.4.134), we get

$$\Gamma, +\varphi : -\varphi_1 \qquad\qquad (4.4.135)$$

And, finally, applying again the induction hypothesis on (4.4.133,(ii)) and (4.4.135) yields $\Gamma : -\varphi$, as required.

□

4.4.3   A bilateral justification of Classical Logic

In this subsection, I present, following Rumfitt [179], a bilateral presentation of Classical Logic, using a bilateral ND-system extending $PNJB$, to be called $BICL$. This presentation is a proof-theoretical justification of Classical Logic. Early references for bilateral justification of Classical Logic are Price [154, 155]. See also Smiley [203] and Bendal [12]. I follow here Rumfitt [179].

Two issues are involved in this extension.

1. Defining negation.

2. Strengthening the structural rules.

4.4.3.1   Defining negation bilaterally

The I/E-rules for negation are presented in Figure 4.6 (cf. also Price [154], p. 167). I deviate from Rumfitt's original presentation in choosing GE-rules for elimination. The regular E-rules for negation are the following special case.

$$\frac{\Gamma : +\neg\varphi}{\Gamma : -\varphi}\ (\neg^+ E) \qquad \frac{\Gamma : -\neg\varphi}{\Gamma : +\varphi}\ (\neg^- E) \qquad\qquad (4.4.136)$$

$$\frac{\Gamma : -\varphi}{\Gamma : +\neg\varphi} \ (\neg^+ I) \qquad \frac{\Gamma : +\neg\varphi \quad \Gamma, -\varphi : \sigma}{\Gamma : \sigma} \ (\neg^+ E)$$

$$\frac{\Gamma : +\varphi}{\Gamma : -\neg\varphi} \ (\neg^- I) \qquad \frac{\Gamma : -\neg\varphi \quad \Gamma, +\varphi : \sigma}{\Gamma : \sigma} \ (\neg^- E)$$

Figure 4.6: The bilateral negation I/E-rules

Thus, the definition of negation is based on the approach that asserting $\neg\varphi$ is warranted by denying φ, and denying $\neg\varphi$ is warranted by denying $\neg\varphi$, with the harmoniously induced GE-rules for eliminating negation.

Obviously, the rules for '$\neg$' are pure.

The rules for negation *are* also harmonious, both in form (GE-rules) and vertically intrinsically, as evident from the following reductions establishing local-soundness.

$$\dfrac{\dfrac{\dfrac{\mathcal{D}}{-\varphi}}{+\neg\varphi} \ (\neg^+ I)}{-\varphi} \ (\neg^+ E) \quad \leadsto_r \quad \dfrac{\mathcal{D}}{-\varphi} \tag{4.4.137}$$

and

$$\dfrac{\dfrac{\dfrac{\mathcal{D}}{+\varphi}}{-\neg\varphi} \ (\neg^- I)}{+\varphi} \ (\neg^- E) \quad \leadsto_r \quad \dfrac{\mathcal{D}}{+\varphi} \tag{4.4.138}$$

Horizontal intrinsic harmony (not considered by Rumfitt) holds too, by the structure of the I-rules (both categorical). Furthermore, local-completeness (not considered by Rumfitt) holds too, as evident from the following expansions.

$$\dfrac{\mathcal{D}}{+\neg\varphi} \quad \leadsto_e \quad \dfrac{\dfrac{\dfrac{\mathcal{D}}{+\neg\varphi}}{-\varphi} \ (\neg^+ E)}{+\neg\varphi} \ (\neg^+ I) \tag{4.4.139}$$

and

$$\dfrac{\mathcal{D}}{-\neg\varphi} \quad \leadsto_e \quad \dfrac{\dfrac{\dfrac{\mathcal{D}}{-\neg\varphi}}{+\varphi} \ (\neg^- E)}{-\neg\varphi} \ (\neg^- I) \tag{4.4.140}$$

4.4.3.2 Structural bilateral rules for reaching classical strength

The bilateral negation rules suffice for deriving (the bilateral presentation of) (DNE).

$$\dfrac{\dfrac{+\neg\neg\varphi}{-\neg\varphi} \ (\neg^+ E)}{+\varphi} \ (\neg^- E) \tag{4.4.141}$$

However, the rules for '$\neg$' do not suffice for reaching full classical strength. For example, (LEM) is not yet derivable (it has an empty Γ!), since the I/E-rules for '$\neg$', as already mentioned, are categorical, not discharging assumptions.

In Rumfitt [179], dealing explicitly with Classical Logic, two bilateral structural rules are added (under the name of *coordination rules*, cf. (4.4.95)), rendering the negative I-rules for conjunction to be determined by the positive ones. The first one is a strengthening of (4.4.95 b), which was restricted to positive assumptions only, formulated using $\perp$ instead of an explicit contradiction. For the expression of those rules, $\perp$ is used. In keeping with Tennant [216] and Rumfitt [179], $\perp$ is viewed here not as a propositional constant but as a logical punctuation sign, belonging to the language in which rules are formulated (like a comma within a sequent). Consequently it does not occur with a force marker attached.

The new rule allows for a *reversal* of the judgement (in both directions), once a contradiction is reached. The similarity to (DNE) is not surprising.

$$\frac{\Gamma, \sigma : \perp}{\Gamma : \overline{\sigma}} \; (RED) \qquad \frac{\Gamma : \sigma \quad \Gamma : \overline{\sigma}}{\Gamma : \perp} \; (LNC) \qquad\qquad (4.4.142)$$

Rumfitt shows (in [179]) that it suffices to have (LNC) for atomic sentences only, implying its admissibility to arbitrary sentences.

What justifies the inclusion of those coordination rules? This question goes back to the observation (in Section 1.6.2) about the role of the structural rules take, as constituting the background of the meaning definition, in the form of characterizing the "bare" derivability relation. So, the bilateral structural rules are part of the inferential practice involved with the incorporation of denial as primitive, and, as observed by Murzi [125], part of the "classicism" of the resulting logic. As admitted in Rumfitt [179], (RED^i) does not originate from bilateralism per se; rather, its role is to enable reaching classical strength in a bilateral formalization of Classical Logic. To see that classical strength has been established, below is a derivation of (LEM).

Example 4.4.35.

$$\frac{\frac{[-(\varphi \vee \neg\varphi)]_1 \quad \dfrac{[+\varphi]_2}{+(\varphi \vee \neg\varphi)} \; (+\vee I_1)}{\dfrac{\perp}{-\varphi} \; (RED^2)} \qquad \frac{[-(\varphi \vee \neg\varphi)]_1 \quad \dfrac{\dfrac{[+\neg\varphi]_3}{+(\varphi \vee \neg\varphi)} \; (+\vee I_2)}{\quad} (LNC)}{\dfrac{\dfrac{\dfrac{\perp}{-\neg\varphi} \; (RED^3)}{+\neg\neg\varphi} \; (\neg^+ I)}{\dfrac{+\varphi}{} \; (DNE)} (LNC)}}{\dfrac{\perp}{+(\varphi \vee \neg\varphi)} \; (RED^1)} \qquad\qquad (4.4.143)$$

As another example of the classical strength of $BICL$, consider the following derivation of Peirce's law.

Example 4.4.36.

$$\cfrac{[-\varphi]_2 \quad \cfrac{\cfrac{[+((\varphi\to\psi)\to\varphi)]_1 \quad \cfrac{\cfrac{[-\varphi]_2 \quad [+\varphi]_3}{+\psi}\,(INC)}{+(\varphi\to\psi)}\,(\to^+I^3)}{+\varphi}\,(\to^+E)}{\cfrac{\bot}{+\varphi}\,(RED^1)}\,(LNC^2)}{+(((\varphi\to\psi)\to\varphi)\to\varphi)}\,(\to^+I^1)$$
$$(4.4.144)$$

This presence of additional structural rules appeals to, as described in Section 1.6.2, the revised notion of canonicity, namely st-canonicity. Thus, the derivation of (LEM) in (4.4.143), ending in an application of a structural rule, *is* according to the meaning (of disjunction, in this case).

Notably, the rule (RED) is impure in that it explicitly mentions $\bot$. An alternative, pure form of a bilateral Reduction structural rule was proposed by Smiley [203]:

$$\cfrac{\begin{array}{cc}[\sigma]_i & [\sigma]_j \\ \vdots & \vdots \\ \alpha & \overline{\alpha}\end{array}}{\overline{\sigma}}\,(SR^{i,j})$$
$$(4.4.145)$$

Its classical strength can be seen from the following derivation of (LEM).

$$\cfrac{\cfrac{[-(\varphi\vee\neg\varphi)]_1}{-\varphi}\,(\vee^-E_1) \qquad \cfrac{\cfrac{[-(\varphi\vee\neg\varphi)]_2}{-(\neg\varphi)}\,(\vee^-E_2)}{+\varphi}\,(\neg^-E)}{+(\varphi\vee\neg\varphi)}\,(SR^{1,2})$$
$$(4.4.146)$$

A useful derived structural rule, called '*reversal*' in Rumfitt [179], is a kind of a bilateral contra-position rule (BCP).

$$\cfrac{\Gamma,\alpha:\beta}{\Gamma,\overline{\beta}:\overline{\alpha}}\,(BCP)$$
$$(4.4.147)$$

In a sense, this rule embodies in NC the horizontal harmony discussed above. Its derivation is as follows.

$$\cfrac{\cfrac{\Gamma,\overline{\beta},\alpha:\overline{\beta} \quad \cfrac{\Gamma,\alpha:\beta}{\Gamma,\overline{\beta},\alpha:\beta}\,(W)}{\cfrac{\Gamma,\overline{\beta},\alpha:\bot}{\Gamma,\overline{\beta}:\overline{\alpha}}\,(RED)}\,(LNC)}{}$$
$$(4.4.148)$$

This rule yields yet another bilateral structural rule, useful for case analysis.

$$\cfrac{\Gamma,+\varphi:\xi \quad \Gamma,-\varphi:\xi}{\Gamma:\xi}\,(Cases)$$
$$(4.4.149)$$

Its derivation is:

$$\frac{\dfrac{\Gamma, +\varphi : \xi}{\Gamma, \overline{\xi} : -\varphi} \; (BCP) \quad \dfrac{\Gamma, -\varphi : \xi}{\Gamma, \overline{\overline{\xi}} : +\varphi} \; (BCP)}{\dfrac{\Gamma, \overline{\xi} : \perp}{\Gamma : \xi[= \overline{\overline{\xi}}]} \; (RED)} \; (LNC)$$

$$(4.4.150)$$

As observed by Rumfitt ([179], p. 808-9) $BICL$ is separable, in contrast to the standard unilateral presentation of CL. The result is obtained by adapting a proof of Bendall [11] of the separability of a close variant of the bilateral presentation of CL.

4.4.4   The effect of structural canonicity on bilateral systems

In this section, I consider the effect of structural canonicity on bilateral ND-systems that also has structural rules that allow drawing conclusions (the coordination rules). I will show the S-local-soundness of $BICL$. For each connective, there are separate reductions for the two force markers. For two connectives (implication and conjunction) I also present the reductions arising from standard E-rules, in addition to the GE-rules.

Implication:

    Assertion:

$$\frac{[-(\varphi\to\psi)]_i \atop {\mathcal{D}_1 \atop \dfrac{\perp}{+(\varphi\to\psi)} \; (RED^i)} \quad \mathcal{D}_2 \quad {[+\psi]_j \atop {\mathcal{D}_3 \atop \xi}}}{\xi} \; (\to^- E^j) \qquad \leadsto_r \qquad \frac{\dfrac{[-\psi]_i \quad {\mathcal{D}_2 \atop +\varphi}}{-(\varphi\to\psi)} \; (\to^- I) \atop {\mathcal{D}_1 \atop {\dfrac{\perp}{+\psi} \; (RED^i) \atop {\mathcal{D}_3 \atop \xi}}}}{}$$

$$(4.4.151)$$

Had we used the standard $(\to E)$, namely Modus Ponens, the reduction would be

$$\frac{{\mathcal{D}_2 \atop +\varphi} \quad \dfrac{[-(\varphi\to\psi)]_i \atop {\mathcal{D}_1 \atop \dfrac{\perp}{+(\varphi\to\psi)} \; (RED^i)}}{+\psi} \; (\to^+ E)}{} \qquad \leadsto_r \qquad \frac{\dfrac{{\mathcal{D}_2 \atop +\varphi} \quad [-\psi]_j}{-(\varphi\to\psi)} \; (\to^- I) \atop {\mathcal{D}_1 \atop \dfrac{\perp}{+\psi} \; (RED^j)}}{}$$

$$(4.4.152)$$

Denial: Since there are two negative E-rules, there are two reductions.

-

$$\frac{[+(\varphi\to\psi)]_i \atop {\mathcal{D}_1 \atop \dfrac{\perp}{-(\varphi\to\psi)} \; (RED_1)}}{+\varphi} \; (\to^- E_i) \qquad \leadsto_r \qquad \frac{\dfrac{[-\varphi]_i}{+(\varphi\to\psi)} \; (*) \atop {\mathcal{D}_1 \atop \dfrac{\perp}{+\varphi} \; (RED_i)}}{}$$

$$(4.4.153)$$

where the sub-derivation indicated by $(*)$ is the following.

$$\frac{\dfrac{[\varphi]_i \quad [+\varphi]_j \quad [-\psi]_k}{\dfrac{\bot}{+\psi}\ (RED^k)}\ (LNC)}{+(\varphi\to\psi)}\ (\to^+ I^j)$$

•

$$\frac{\dfrac{[+(\varphi\to\psi)]_i}{\mathcal{D}_1}}{\dfrac{\bot}{-(\varphi\to\psi)}\ (RED^i)}\ (\to^- E_2)}{-\psi}\ (\to^- E_2)
\qquad \leadsto_r \qquad
\frac{\dfrac{[+\psi]_i}{+(\varphi\to\psi)}\ (\to^+ I^l)}{\dfrac{\mathcal{D}_1}{\dfrac{\bot}{-\psi}\ (RED^i)}}
\qquad (4.4.154)$$

Note the vacuous discharge in the first step of the derivation.

Conjunction:

Assertion:

$$\frac{\dfrac{\dfrac{[-(\varphi\wedge\psi)]_i}{\mathcal{D}_1}}{\dfrac{\bot}{+(\varphi\wedge\psi)}\ (RED^i)}{\xi} \qquad \dfrac{[+\varphi]_j, [+\psi]_k}{\dfrac{\mathcal{D}_2}{\xi}}}{\xi}\ (\wedge^+ E^{j,k}) \qquad \leadsto_r$$

$$\frac{\dfrac{[-\varphi]_i}{-(\varphi\wedge\psi)}\ (\wedge_1^- I) \qquad \dfrac{[-\psi]_j}{-(\varphi\wedge\psi)}\ (\wedge_2^- I)}{\dfrac{\mathcal{D}_1}{\dfrac{\bot}{+\varphi}\ (RED^i)} \qquad \dfrac{\mathcal{D}_1}{\dfrac{\bot}{+\psi}\ (RED^j)}} \\ \dfrac{\mathcal{D}_2}{\xi} \qquad (4.4.155)$$

Using the standard $(\wedge E)$-rule, the reduction would be

$$\frac{\dfrac{[-(\varphi\wedge\psi)]_i}{\mathcal{D}_1}}{\dfrac{\bot}{+(\varphi\wedge\psi)}\ (RED^i)}{+\varphi}\ (\wedge^+ E_1)
\qquad \leadsto_r \qquad
\frac{\dfrac{[-\varphi]_j}{-(\varphi\wedge\psi)}\ (\wedge^- I_1)}{\dfrac{\mathcal{D}_1}{\dfrac{\bot}{+\varphi}\ (RED^j)}}
\qquad (4.4.156)$$

Denial:

$$\frac{\dfrac{\dfrac{[+(\varphi\wedge\psi)]_i}{\mathcal{D}_1}}{\dfrac{\bot}{-(\varphi\wedge\psi)}\ (RED^i)}{\xi} \qquad \dfrac{[-\varphi]_j}{\dfrac{\mathcal{D}_2}{\xi}} \qquad \dfrac{[-\psi]_j}{\dfrac{\mathcal{D}_3}{\xi}}}{\xi}\ (\wedge^- E^{j,k}) \qquad \leadsto_r$$

$$\frac{[+\varphi]_i \quad [+\psi]_j}{+(\varphi \wedge \psi)} \ (\wedge^+ I)$$
$$\mathcal{D}_1$$
$$\frac{\bot}{-\varphi} \ (RED^i) \qquad [-\psi]_k$$
$$\mathcal{D}_2 \qquad\qquad \mathcal{D}_3$$
$$\frac{\xi \qquad\qquad\qquad \xi}{\xi} \ (Cases^{j,k}) \qquad\qquad (4.4.157)$$

Disjunction:

    Assertion:

$$[-(\varphi \vee \psi)]_i$$
$$\mathcal{D}_1 \qquad\qquad\qquad [+\varphi]_j \quad [+\psi_k$$
$$\frac{\bot}{+(\varphi \vee \psi)} \ (RED^i) \qquad \mathcal{D}_2 \qquad \mathcal{D}_3$$
$$\frac{\qquad\qquad\qquad\qquad \xi \qquad\quad \xi}{\xi} \ (\vee^+ E^{j,k}) \qquad \leadsto_r$$

$$\frac{[-\varphi]_i \quad [-\psi]_j}{-(\varphi \vee \psi)} \ (\vee^- I)$$
$$\mathcal{D}_1$$
$$\frac{\bot}{+\varphi} \ (RED^i) \qquad [+\psi]_k$$
$$\mathcal{D}_2 \qquad\qquad \mathcal{D}_3$$
$$\frac{\xi \qquad\qquad\qquad \xi}{\xi} \ (Cases^{j,k}) \qquad\qquad (4.4.158)$$

    Denial:

$$[+(\varphi \vee \psi)]_i$$
$$\mathcal{D}_1 \qquad\qquad\qquad [-\varphi]_j, [-\psi]_k$$
$$\frac{\bot}{-(\varphi \vee \psi)} \ (RED^i) \qquad \mathcal{D}_2$$
$$\frac{\qquad\qquad\qquad\qquad \xi}{\xi} \ (\vee^- E^{j,k}) \qquad \leadsto_r$$

$$\frac{[+\varphi]_i}{+(\varphi \vee \psi)} \ (\vee_1^- I) \qquad \frac{[+\psi]_j}{+(\varphi \vee \psi)} \ (\vee_2^- I)$$
$$\mathcal{D}_1 \qquad\qquad\qquad\qquad \mathcal{D}_1$$
$$\frac{\bot}{-\varphi} \ (RED^i) \qquad\quad \frac{\bot}{-\psi} \ (RED^j)$$
$$\mathcal{D}_2$$
$$\xi \qquad\qquad\qquad\qquad\qquad\qquad (4.4.159)$$

Negation:

    Assertion:

$$\frac{[+\varphi]_i}{-(\neg\varphi)} \ (\neg^- I)$$
$$[-(\neg\varphi)]_i \qquad\qquad\qquad \mathcal{D}_1$$
$$\mathcal{D}_1 \qquad\qquad [-\varphi]_j$$
$$\frac{\bot}{+(\neg\varphi)} \ (RED^i) \quad \mathcal{D}_2 \qquad\qquad \frac{\bot}{-\varphi} \ (RED^i)$$
$$\frac{\qquad\qquad\qquad\quad \xi}{\xi} \ (\neg^+ E^j) \quad \leadsto_r \qquad \mathcal{D}_2$$
$$\xi \qquad\qquad (4.4.160)$$

Denial:

$$\frac{[-\varphi]_i}{+(\neg\varphi)} \,(\neg^+I)$$

$$\begin{array}{c} [+(\neg\varphi)]_i \\ \mathcal{D}_1 \\ \hline \frac{\bot}{-(\neg\varphi)} \,(RED^i) \end{array} \quad \begin{array}{c} [+\varphi]_j \\ \mathcal{D}_2 \\ \xi \end{array}$$

$$\frac{}{\xi} \,(\neg^-E^j) \qquad \rightsquigarrow_r$$

$$\begin{array}{c} [-\varphi]_i \\ +(\neg\varphi) \,(\neg^+I) \\ \mathcal{D}_1 \\ \frac{\bot}{+\varphi} \,(RED^i) \\ \mathcal{D}_2 \\ \xi \end{array} \qquad (4.4.161)$$

4.4.5 Relating multiple conclusions to bilateralism

In addition to both kinds of ND-systems having a proof-theoretic "well-behaving" presentation of Classical Logic, the two ND-systems are more closely related. As observed by Hjortland [76] (p. 89), the following relationship holds.

Proposition 4.4.9.   *1. If*

$$\vdash_{MCND} \Gamma = \{\varphi_1, \cdots, \varphi_n\} : \{\psi_1, \cdots, \psi_m\} = \Delta$$

then

$$\vdash_{BICL} +\varphi_1, \cdots, +\varphi_n, -\psi_1, \cdots, -\psi_{m-1} : +\psi_m$$

2. If

$$\vdash_{BICL} +\varphi_1, \cdots, +\varphi_n, -\psi_1, \cdots, -\psi_m : \sigma$$

then, if $\sigma = +\chi$ (for some χ)

$$\vdash_{MCND} \{\varphi_1, \cdots, \varphi_n\} : \{\psi_1, \cdots, \psi_m, \chi\}$$

else, if $\sigma = -\chi$ (for some χ), then

$$\vdash_{MCND} \{\varphi_1, \cdots, \varphi_n, \chi\} : \{\psi_1, \cdots, \psi_m\}$$

So, by this proposition, there is a *structural* faithful translation between both frameworks. Thus, instead of a purely assertional understanding of $\vdash_{MCND} \Gamma : \Delta$, according to which understanding asserting (simultaneously) all members of Γ implies asserting at least one member of Δ, we have another understanding of $\vdash_{MCND} \Gamma : \Delta$ (according to its faithful translation), as asserting (simultaneously) all members of Γ implies the incoherence of denying (simultaneously) all members of Δ.

This understanding escapes Dummett's and Tennant's "enforcement of disjunction" criticism. A detailed discussion of this issue can be found in Restall's works ([173, 174, 175] and [176].

4.5 Alternative conceptions of harmony justifying Classical Logic

In this section, I review approaches to PTS that accept its ideas, but evade the revisionist approach to logic by modifying the criterion of harmony.

4.5.1 Harmony as functional inversion

In [233], Weir starts from an observation of a difference in the effect of the reductions showing intrinsic harmony of '∧' and '∨'. By a judicious choice of the arbitrary conclusion of $(\land E)$ as one of the conjuncts, we can obtain a reduction *restoring* the stage in a derivation just before the application of $(\land I)$; moreover, the same holds in a derivation that first applies $(\land E)$ and immediately afterwards reintroduces by an application of $(\land I)$.

$$\dfrac{\dfrac{\varphi \quad \psi}{\varphi \land \psi}\ (\land I)}{\varphi}\ (\land E) \qquad \dfrac{\varphi \land \psi}{\varphi}\ (\land E) \qquad \dfrac{\dfrac{\varphi \land \psi}{\psi}\ (\land E)}{\varphi \land \psi}\ (\land I)$$

On the other hand, for '∨':

- There is no way to choose as the arbitrary conclusion of the local-soundness reduction to restore one of the disjuncts, and

- The is no local-completeness expansion that applies $(\lor E)$ to $\varphi \lor \psi$ immediately after its introduction.

The rules for '→', once the discharged assumption in $(\to I)$ is taken into account, behave like those of the conjunction.

This observation is generalized by Weir to a general requirement, still going by the name of harmony, of *all* constants having I/E-rules the composition of which restores the premises. Weir refers to this requirement also as the (stronger) *inversion principle*. This inversion principle takes care of *both* aspect of the I/E-equilibrium, in contrast to Prawitz's principle in Section 3.6.1.1, the latter still needing stability for a full equilibrium.

What about negation and disjunction?

Negation can be defined, as seen before, as a special case of implication, leading for the satisfaction of the I/E-invertibility.

$$\dfrac{\dfrac{[\varphi]_1}{\vdots}}{\dfrac{\bot}{\neg\varphi}\,(\neg I^1)}\quad\dfrac{}{\bot}\,(\neg E)\qquad\qquad \dfrac{\dfrac{\neg\varphi\quad [\varphi]_1}{\bot}\,(\neg E)}{\neg\varphi}\,(\neg I^1)$$

As for disjunction, Weir proposes the following rules, where the classical disjunction is denoted by '$\vee_c$', for distinction from the rules for intuitionistic disjunction.

$$\dfrac{\dfrac{\neg[\psi]_i}{\vdots}{\varphi}}{\varphi\vee_c\psi}\,(\vee_{c,1}I^i)\qquad \dfrac{\dfrac{\neg[\varphi]_i}{\vdots}{\psi}}{\varphi\vee_c\psi}\,(\vee_{c,2}I^i)\qquad \dfrac{\varphi\vee_c\psi\quad \neg\varphi}{\psi}\,(\vee_{c,1}E)\qquad \dfrac{\varphi\vee_c\psi\quad \neg\psi}{\varphi}\,(\vee_{c,2}E)$$

$$(4.5.162)$$

Milne has a similar proposal ([114], p. 83). Clearly, corresponding GE-rules can be formulated. The following reduction and expansion establish their I/E-invertibility.

$$\dfrac{\dfrac{[\neg\psi]_1}{\vdots}{\varphi}}{\dfrac{\varphi\vee_c\psi}{\varphi}}\,(\vee_{c,1}I^1)\quad\dfrac{}{\varphi}\quad \neg\psi}{}\,(\neg\vee_{c,1}E)\qquad\qquad \dfrac{\dfrac{\varphi\vee_c\psi\quad [\neg\varphi]_1}{\psi}\,(\vee_{c,1}E)}{\varphi\vee_c\psi}\,(\vee_{c,1}I^1)$$

$$(4.5.163)$$

(and similarly for the other pair of rules). Classical strength is indicated, for example, by the following derivation of (LEM).

$$\dfrac{[\neg\varphi]_1}{\varphi\vee_c\neg\varphi}\,(\vee_c I^1)$$

In Weir [233], rules for the quantifiers are given too.

The trouble with the proposed rules for disjunction is that they are not separable, since they mention negation explicitly. This generates a strong linkage between (classical) disjunction and negation. A much weaker sub-formula property is induced by this rules, allowing also negations of sub-formulas of the premises and the conclusion (see Weir [233] for more details).

Classical harmony comes for a high price!

4.5.2   Harmony with I/E-rules re-conceived

In [202], Slater regains harmony for Classical Logic by revising the view of what I/E-rules are. His first claim is that Gentzen (and everyone else following him) was mistaken in conceiving negation as a (unary) *sentential operator*. Whatever rules that were considered I/E-rules for negation should really be conceived as I/E-rules for *contradictions*! By this view, while mostly the role of I/E-rules is to introduce/eliminate a relation between proposition expressed via a sentence (having a main operator), it is *not always* the case. Contradiction is exactly such an exceptional case. Indeed, in Intuitionistic Logic the relationship of φ and $\neg\varphi$ is *not* that of contradiction; rather, it is one of *contrariety*, the difference between the two being exactly the disputed LEM, namely $\varphi \vee \neg\varphi$. Indeed, there is no *unique* contrariety proposition of φ that could be expressed by a sentential connective, say '$\neg$'. The contradiction is indeed a *symmetric* relation among propositions, and $\neg\varphi$ should be viewed as the contradictory proposition to φ inasmuch as φ should be viewed as the contradictory proposition to $\neg\varphi$, whence the equivalence between φ and $\neg\neg\varphi$. Contradiction is signalled by '$\neg$' through *its presence and absence*!

Turning now to the I/E-rules for contradiction (not for negation as a connective), there are three of them.

$$\begin{array}{ccc} [\varphi]_i & [\neg\varphi]_i & \\ \vdots & \vdots & \\ \dfrac{\bot}{\neg\varphi}\ (contra_1 I^i) & \dfrac{\bot}{\varphi}\ (contra_2 I^i) & \dfrac{\varphi \quad \neg\varphi}{\bot}\ (contraE) \end{array} \qquad (4.5.164)$$

Thus, (RAA) $(contra_2 I$ here) is viewed as a second I-rule for contradiction, in contrast to its traditional view as an E-rule for negation. There is a similar proposal by Milne in [114].

Those rules give rise to the following reductions, establishing intrinsic harmony (local-soundness).

$$\dfrac{\dfrac{\begin{array}{c}[\varphi]_1\\ \mathcal{D}::\\ \dfrac{\bot}{\neg\varphi}\ (contra_1 I^1)\end{array} \quad \varphi}{\bot}\ (contraE) \quad \leadsto_r \quad \begin{array}{c}\varphi\\ \mathcal{D}\\ \bot\end{array}} \qquad (4.5.165)$$

and

$$\dfrac{\dfrac{\begin{array}{c}[\neg\varphi]_1\\ \mathcal{D}::\\ \dfrac{\bot}{\varphi}\ (contra_2 I^1)\end{array} \quad \varphi}{\bot}\ (contraE) \quad \leadsto_r \quad \begin{array}{c}\neg\varphi\\ \mathcal{D}\\ \bot\end{array}} \qquad (4.5.166)$$

A noticeable effect of those reductions is the absence of the sub-formula property for normal derivations. It is due to the premise $\neg\varphi$ of (RAA) (i.e., of $(contra_2 I)$) not being a sub-formula of the conclusion φ; but, it *is* a sub-formula of the equivalent $\neg\neg\varphi$! According to Slater ([202], p. 190)

> So the sub-formula principle obviously only arises in forms of 'logics' that concern themselves with signs and not with what those signs signify.

Once again, Classical harmony comes for a high price!

Now comes Slater's second claim. The tripartite view of I/E-rules is not specific to contradiction; rather, *all* of Classical Logic is endowed with such a view. As for

$$\frac{\Gamma:\varphi\vee\chi \quad \Gamma:\psi\vee\xi}{\Gamma:(\varphi\wedge\psi)\vee(\chi\vee\xi)} \ (\wedge I_m) \quad \frac{\Gamma:\varphi\vee\chi}{\Gamma:(\varphi\vee\psi)\vee\chi} \ (\vee_1 I_m) \quad \frac{\Gamma:\psi\vee\chi}{\Gamma:(\varphi\vee\psi)\vee\chi} \ (\vee_2 I_m)$$

$$\frac{\Gamma,\varphi:\psi\vee\chi}{\Gamma:(\varphi\to\psi)\vee\chi} \ (\to I_m) \quad \frac{\Gamma,\varphi:\chi}{\Gamma:\neg\varphi\vee\chi} \ (\neg I_m)$$

Figure 4.7: Milne's impure ND-rules for Classical Logic

conjunction and disjunction, each of them has three rules already in Intuitionistic Logic, rules transferred to Classical Logic. But what about implication? Slater see's the second $(\to I)$ rule as being implicitly present in Gentzen's presentation, by the possibility of a vacuous discharge of a formula in $(\to I)$, giving rise to $\psi\vdash\varphi\to\psi$. So, Slater proposes to add as an explicit $(\to I)$ rule

$$\frac{\psi}{\varphi\to\psi} \ (\to_2 I) \tag{4.5.167}$$

He motivates this rule by the presence, in natural language, of uses of 'if' which do not convey any implication. Two such examples are:

Biscuit Conditionals: There is jam in the cupboard, if you want some
>   Clearly, such a sentence does not pretend to claim that your desires imply the presence of jam in the cupboard.

Counterfactuals: If there is jam in the cupboard, I am a dutchman
>   Again, there is no claim made by asserting this sentence that the presence of jam in the cupboard implies anything about the nationality of the speaker.

There is vast literature (in linguistics) on the truth-conditions of such sentences.

Finally, the tripartite view is extended to quantifiers' rules too, by an intriguing appeal to Hilbert's ϵ-calculus. See Slater [202] for further details.

4.5.3 Impure I-rules

In [115], Milne proposes a justification of Classical Logic, inspired by multiple-conclusion proof-systems, by which I-rules for one connective may be impure, namely, refer to other connectives, disjunction in this case. The proposed I-rules are presented in Figure 4.7. The rule names are subscripted with m. I present an additive version, instead of Milne's original multiplicative version. The E-rules remain the intuitionistic ones. To see classical strength, below is a derivation of (LEM), by choosing in

$(\neg I_m)$ the auxiliary χ as φ itself.

$$\frac{\Gamma, \varphi : \varphi}{\Gamma : \neg\varphi\vee\varphi} \; (\neg I_m)$$

In the other direction, $(\neg I_m)$ is derivable from (LEM).

$$\frac{\dfrac{}{\varphi\vee\neg\varphi}\,(LEM) \quad \dfrac{\overset{[\varphi]_1}{\vdots}}{\dfrac{\chi}{\neg\varphi\vee\chi}}\,(\vee_2 I) \quad \dfrac{[\neg\varphi]_2}{\neg\varphi\vee\chi}\,(\vee_1 I)}{\neg\varphi\vee\chi}\,(\vee E^{1,2}) \qquad (4.5.168)$$

The upshot of the rules proposed by Milne is ([115], p. 514) the following observation

> ... introduce a connective into a position that may not be dominant, that may be subordinate to an occurrence of $\vee$.

Milne reduces the difference between Intuitionistic and Classical logic to this observation. This observation explains also the difference between the two logics in their sequent-calculus presentations, where Classical Logic requires multiple conclusion (succedent of size larger than one), while Intuitionistic Logic allows at most one formula in the succedent of a sequent (a phenomenon referred to as a "seemingly magical fact" by Hacking [69]).

Milne observes that the proposed rules induce the following property of derivability as a consequence.

$$\vdash\Gamma, \varphi : \psi \quad \text{iff} \quad \vdash\Gamma : \neg\varphi\vee\psi \qquad (4.5.169)$$

This consequence is related to the well-known ability of the expressions of one connective by others (e.g. $\varphi\rightarrow\psi$ as $\neg\varphi\vee\psi$), an ability possessed by Classical Logic, but not by Intuitionistic Logic (for the standard connectives).

Regarding the (intrinsic) harmony of the suggested system, Milne proposes the following reduction for negation and implication (in my notation).

Negation:

$$\frac{\overset{[\varphi]_1}{\mathcal{D}_1}\;\;\psi}{\neg\varphi\vee\psi}\,(\neg I_m^1) \quad \frac{\varphi\;\;[\neg\varphi]_2}{\xi_1}\,(efq) \;\ldots\; \frac{\varphi\;\;[\neg\varphi]_2}{\xi_n}\,(efq) \quad \frac{\overset{\mathcal{D}_2}{\chi}\quad\overset{[\psi]_3}{\underset{\chi}{\mathcal{D}_3}}}{\chi}\,(\vee E^{2,3}) \;\leadsto_r\; \frac{\overset{\varphi}{\mathcal{D}_1}\;\;\mathcal{D}_3[\psi := \psi]}{\chi}$$

Milne doesn't relate to the closure of his system under derivation composition. In his notation, the reduced derivation is

$$\begin{array}{c}\varphi\\\mathcal{D}_1\\\psi\\\mathcal{D}_3\\\chi\end{array}$$

Disjunction:

$$
\dfrac{\begin{array}{c}[\varphi]_1\\ \mathcal{D}_1\\ \varphi\vee\xi\end{array}}{(\varphi\to\psi)\vee\chi}\,(\to I_m^1)
\qquad
\dfrac{\dfrac{\varphi\quad[\varphi\to\psi]_2}{\psi}\,(\to E)\quad\begin{array}{c}[\xi]_3\\ \mathcal{D}_2\\ \xi\end{array}\quad\begin{array}{c}\mathcal{D}_3\\ \chi\end{array}}{\xi}\,(\vee E^{2,3})
\quad\leadsto_r\quad
\dfrac{\begin{array}{c}\varphi\\ \mathcal{D}_1\\ \varphi\vee\chi\end{array}\quad\begin{array}{c}[\varphi]_2\\ \mathcal{D}_2\\ \xi\end{array}\quad\begin{array}{c}[\chi]_3\\ \mathcal{D}_3\\ \xi\end{array}}{\xi}\,(\vee E^{2,3})
$$

Note that in those reductions, "everything lives in the scope of disjunction".

Finally, Milne notes that his proposed ND-system has the following features.

- The system is *not* conservative over the (implicational fragment of) Intuitionistic Logic, as it can derive classical strength witnesses like Peirce's rule or (DNE).

- The separation property holds only under a very liberal conception of it. Suppose φ contained only the operators in $C = \{c_1, \cdots, c_m\}$. When deriving φ, one can use not only the rules for the connective in C, i.e., *actually* occurring in φ, but also the rules pertaining to connectives in C.

  This "liberal" view of separability is accompanied by another "liberal" view of a traditional concept, that of compositionality. According to his view, in addition to the "standard" way of deriving an operator compositionally, there may be exist auxiliary definitions of that operator, *not considered canonical*, as long as these auxiliary definitions do not conflict with the compositional definition. Milne views his system as such. This view is a counterargument to Tennant's objection (in [215], pp. 314-15) to the use of impure rules.

  Milne relates his view to Dummett's distinction between assertoric content and ingredient sense (cf. p. 13), seeing the canonical derivation yielding content, while the non-canonical (auxiliary) rules yielding ingredient sense. For example (Milne [115], p. 529), for the conclusion $(\varphi\to\psi)\vee\chi$ of $(\to I_m)$, having $\varphi\to\psi$ as a sub-sentential component, the contribution of $\varphi\to\psi$ is its ingredient sense, either a derivation of χ, or a derivation of $\varphi\to\psi$ *according to its contents*, a canonical derivation of $\varphi\to\psi$, i.e., a derivation of ψ under a discharged assumption φ.

4.5.4 Other arguments against the rejection of Classical Logic

In Kürbis [97], while retaining the main principles of Dummett's theory of meaning, Kürbis raises two main argument against rejecting Classical Logic. His final conclusion is that PTS cannot decide between Classical Logic and Intuitionistic Logic.

Complexity, (LEM) and (RAA): Kürbis rejects the view that those classical characteristic rules violate the complexity argument as stated in Section 3.7.2. According to his argument, the *semantic* complexity of φ and $\neg\varphi$ differs from their syntactic complexity and *is the same*. In the particular case of (LEM), the resources needed to comprehend $\neg(\varphi\vee\neg\varphi)$ are the same as those needed to comprehend $\varphi\vee\neg\varphi$, namely, comprehending disjunction and negation. In particular, the derivation of (LEM) in (2.2.17) should not be viewed as violating compositionality. For the same token, $\neg\neg\varphi$ and φ also have the same complexity. Kürbis sees Geach's argument (in [60], stated in terms of negating explicit predication, not sentential negation) as an argument to the same effect. Note that this argument does not depend in any way on the logic being Classical Logic, and applies equally well to Intuitionistic Logic.

Circularity, $\bot$ and $\neg$: Kürbis, after analyzing the various views of $\bot$, reaches the conclusion that there is an *unavoidable* circularity in the rules defining them. His conclusion is that negation, *whether in Intuitionistic Logic or in Classical Logic* is undefinable merely by rules. Thus, while accepting Dummett's theory of meaning in general, Kürbis excludes negation from its scope.

4.6   Summary

In this chapter, I have presented several attempts to justify Classical Logic, rejecting the claim that it is not amenable to proof-theoretic means of conferring meaning. The arguments range from finding different presentations of CL (such as $MCND$ or bilateralism) that do meet the required criteria for qualifying as meaning conferring, through revising the conception of those criteria, up to restricting the scope of applicability of those criteria.

My own position in this dispute is the following. As long as the goal is the justification of Classical Logic (which anyway contains disjunction), I find the arguments acceptable. Furthermore, I find the general use of impure ND-rules acceptable for the justification of *any* logic containing all the operators mentioned in the impure rules. However, when the goal is the (schematic) definition of negation, when added to an *arbitrary* object language (or a fragment thereof, not necessarily containing negation), I find this use of impure rules for defining negation unacceptable. After all, harmony (here achieved via the use of impure rules) is not a goal by itself – it only serves to rule out "bad" definition that do satisfy the more immediate demands of purity and separability. Certainly, a schematic extension of (fragments of) object languages fits better my aim of applying PTS to natural language, as described in part II.

Chapter 5

Proof-Theoretic Semantic Values

5.1   Introduction

In this chapter, I present in detail an explicit definition of meaning for the logical language as determined by a meaning-conferring natural-deduction proof-system, coined before as reified meaning.   Recall that the sentential meaning is associated with canonical derivations, as described in Section 5.2. For sub-sentential phrases (here, the logical constants themselves), this process involves a proof-theoretic reinterpretation of types (Section 5.3.2), by means of which *contribution* to sentential meaning is defined (Section 5.3.5) based on Frege's *context principle* (CP) (see Frege [56], section 62) and on a *type-logical grammar* (TLG) (see Moortgat [123]) for the logical language.

As already mentioned, a parallel reification of semantic values can be done based on the E-view, where reified sentential meaning is determined by the E-rules. Similarly, another parallel presentation is in terms of bilateral PTS. Such parallel presentations require only the change of the proof-theoretic interpretation of type t (explained below). The main presentation is in terms of a I-based, unilateral PTS. The adaptations needed for the incorporation of an E-based approach and a bilateral approach to PTS are indicated where appropriate.

5.2   Proof-Theoretic Reified Sentential Meanings

I start by assigning reified proof-theoretic meaning to formulas of a logical object language, say L, assumed to be recursively defined over a collection of basic formulas (usually, atomic propositions), according to an ND-system taken to be meaning-conferring for L.

The definition of meaning corroborates, and makes precise, a common observation about PTS (cf. for example, Pagin [133] and Peregrin [141] for recent discussions): sentential meanings, while being compositional, are not *directly* compositional. The reason *here* is, that on the I-based approach, an I-canonical derivation has as an *immediate* sub-derivations of the essential application of an I-rule an *arbitrary* derivations of the premises, not just a canonical derivation thereof; similarly, the E-canonical derivation have arbitrary sub-derivations arising from the immediate conclusion. Due to the recursive nature of L, where sentences may be constituents of other sentences, there is an "interfacing" function applied to the meaning of a sentence, a collection of I-canonical derivations (or E-canonical, on the E-based approach), yielding the collection of *all* the derivations (from the same context) of that sentence. However, this does not mean that sentential meanings do not depend on their component phrases: only that the dependence is somewhat indirect. This is reflected here by having the ND-rules determining *two* semantic values (both for sentences and for sub-sentential components, such as the connectives themselves), matching Dummett's distinction between assertoric content and ingredient sense (cf. p. 13). For an expression ξ (either a whole sentence or a sub-sentential component), the one semantic value, corresponding to Dummett's assertoric content, is its *contributed* semantic value $[[\xi]]$, serving also as its *meaning*; the other, auxiliary, semantic value, corresponding to Dummett's ingredient sense, is its *contributing* semantic value $[[\xi]]^*$, used when ξ is part of some larger expression ξ'. Thus, the function-argument relationships, induced by the given grammar of the logical language, plays a central role in obtaining semantic values. Those two semantic values might be seen as also counterparts of what is referred to in Peregrin [141] as the *inferential potential (ip)* and the *inferential role (ir)* of ξ, respectively. As shown below, $[[\xi]]^*$ can be recovered from $[[\xi]]$, but the presentation seems clearer if the two are thought of as if independent.

To emphasize the point of the need for two semantic values, consider the following example from $L_{propmin}$:

$$\varphi = \varphi_1 \wedge \varphi_2 \tag{5.2.1}$$

The (unilateral) I-based meanings of φ_1 and φ_2, $[[\varphi_1]]$ and $[[\varphi_2]]$, respectively, are (approximately, made precise below) the collections of I-canonical derivations of the two sentences φ_1 and φ_2 (say in NJ). Yet the unilateral, I-based meaning of the conjunction, $[[\varphi_1 \wedge \varphi_2]]$, depends on the collections of *all* derivations of the two sentences, as each I-canonical derivation of $\varphi_1 \wedge \varphi_2$ in NJ has as its premises arbitrary NJ-derivations of φ_1 and φ_2, to which the I-rule $(\wedge I)$ of NJ is applied.

Analogously, according to the E-view, the meaning of (5.2.1), is the collection of its E-canonically derived conclusions (from a given Γ). Every such conclusion, say ψ, is the conclusion of an arbitrary derivation (not necessarily canonical) either of φ_1 or of φ_2.

Thus, I take unilateral, I-based, sentential meanings (i.e., contributed sentential semantic values) of compound sentences in L to originate from I-canonical $\mathcal{N}$-derivations (or E-canonical, for the E-based approach). On the other hand, the contributing semantic sentential value (for both atomic and compound sentences) is taken as the collection of *all* (not just canonical) $\mathcal{N}$-derivations. I emphasize once again, that it is explicit *sentential meanings* that are *directly* defined by the ND-system, , and are the primary meaning carriers, whereas meanings of connectives are *extracted* from compound sentential meanings as shown in Section 5.3, and are secondary meaning carriers. In the sequel, I refer to functions from contexts to collections of ND-derivations as *contextualized* functions. To avoid cluttering the notation, I overload in the sequel the meaning of $[\![\cdots]\!]$, leaving it to context to determine the variant meant.

Definition 5.2.37 (reified sentential meaning). *Let* $\varphi \in L.$

1. *The contributed semantic value, i.e. its* meaning $[\![\varphi]\!]$, *is given, according to the different views of PTS, as follows.*

   Unilateral, I-based:

   $$[\![\varphi]\!]^{U,I} = \lambda\Gamma.[\![\varphi]\!]^{Ic}_\Gamma \tag{5.2.2}$$

   Unilateral, E-based:

   $$[\![\varphi]\!]^{U,E} = \lambda\Gamma.[\![\varphi]\!]^{Ec}_{\Gamma.} \tag{5.2.3}$$

   Bilateral, I-based:

   $$[\![\varphi]\!]^{B,I} = \lambda\Gamma.\langle[\![+\varphi]\!]^{Ic}_\Gamma, \; [\![-\varphi]\!]^{Ic}_\Gamma\rangle \tag{5.2.4}$$

   Bilateral, E-based:

   $$[\![\varphi]\!]^{B,E} = \lambda\Gamma.\langle[\![+\varphi]\!]^{Ec}_\Gamma, \; [\![-\varphi]\!]^{Ec}_\Gamma\rangle \tag{5.2.5}$$

   Thus, bilateral reified sentential meanings reflect also canonical denials, in addition to canonical assertions. Recall that atomic sentences are considered here as endowed too with canonical derivations, the identity derivation.

2. *The* contributing semantic value *is given as follows.*

   Unilateral:

   $$[\![\varphi]\!]^* = \lambda\Gamma.[\![\varphi]\!]^*_\Gamma \tag{5.2.6}$$

   Bilateral:

   $$[\![\varphi]\!]^* = \langle[\![+\varphi]\!]^*_\Gamma, \; [\![-\varphi]\!]^*_\Gamma\rangle \tag{5.2.7}$$

The following immediately holds, relating the auxiliary semantic value to meanings, allowing to hold only meanings in the full lexicon of a TLG for L (such a TLG grammar is shown in detail below).

$$[\![\varphi]\!]^* = \lambda\Gamma.[\![\rho([\![\varphi]\!])]\!]^*_\Gamma \tag{5.2.8}$$

(recall the definition of ρ on p. 31). I use the notation $\mathbf{ex}([[\varphi]])$ for the r.h.s., referring to it as the *exportation* (of the collection of I/E-canonical derivations to that of all derivations). Note that the "denotational" meaning of φ is *a proof-theoretic object*, a contextualized function from contexts to the collection of (I/E-canonical) derivations from that context, not to be confused with model-theoretic denotations (of truth-values, in this case).

To realize the role of canonicity in the definition of reified proof-theoretic meanings (according to the I-view), consider the following example derivation.

$$\frac{\Gamma : \alpha \quad \Gamma : (\alpha \rightarrow (\phi \wedge \psi))}{\Gamma : \phi \wedge \psi} \ (\rightarrow E)$$

$$(5.2.9)$$

This is a derivation of a conjunction – but not a canonical one, as it does not end with an application of $(\wedge I)$. Thus, the conjunction here was *not* derived according to its meaning! As far as this derivation is concerned, it could mean anything, e.g., disjunction. On the other hand, the following example derivation *is* according to the conjunction's meaning, being canonical.

$$\frac{\dfrac{\Gamma : \alpha \quad \Gamma : \alpha \rightarrow \varphi}{\Gamma : \varphi} \ (\rightarrow E) \quad \dfrac{\Gamma : \beta \quad \Gamma : \beta \rightarrow \psi}{\Gamma : \psi} \ (\rightarrow E)}{\Gamma : \varphi \wedge \psi} \ (\wedge I)$$

$$(5.2.10)$$

Similar considerations justify the definition of reified proof-theoretic meanings according to the other views of PTS.

We now can see the importance of the second clause in the definition of an I-canonical derivation from open assumptions (Definition 1.5.9). Consider the meaning of disjunction in minimal propositional logic. Pre-Theoretically, one might wish to say that the *commutativity* and *associativity* of disjunction, as expressed by the derivability claims $\vdash (\varphi \vee \psi) : (\psi \vee \varphi)$ and $\vdash (\varphi \vee (\psi \vee \chi)) : ((\varphi \vee \psi) \vee \chi)$, are parts of the meaning of disjunction, and therefore there should be I-canonical derivations establishing them. The standard derivations of the sequents expressing commutativity and associativity are, respectively,

$$\frac{(\varphi \vee \psi) \quad \dfrac{[\varphi]_1}{(\psi \vee \varphi)} \ (\vee_2 I) \quad \dfrac{[\psi]_2}{(\psi \vee \varphi)} \ (\vee_1 I)}{(\psi \vee \varphi)} \ (\vee E^{1,2})$$

$$(5.2.11)$$

and

$$\frac{(\varphi \vee (\psi \vee \chi)) \quad \dfrac{[\varphi]_1}{(\varphi \vee \psi)} \ (\vee I_1)}{((\varphi \vee \psi) \vee \chi)} \ (\vee I_1) \quad \dfrac{[(\psi \vee \chi)]_2 \quad \dfrac{\dfrac{[\psi]_3}{(\varphi \vee \psi)} \ (\vee I_2)}{((\varphi \vee \psi) \vee \chi)} \ (\vee I_1) \quad \dfrac{[\chi]_4}{((\varphi \vee \psi) \vee \chi)} \ (\vee I_2)}{((\varphi \vee \psi) \vee \chi)} \ (\vee E^{3,4})}{((\varphi \vee \psi) \vee \chi)} \ (\vee E^{1,2})$$

$$(5.2.12)$$

Both these derivations exhibit the same problematic character. By the conception of I-canonicity of derivations from open assumptions as based only on 'the last rule applied', these derivations are *not* I-canonical, as the last rule applied in each one of them is $(\vee E)$. Furthermore, *there are no I-canonical derivations* (under this conception of I-canonicity) for the above sequents establishing commutativity and associativity of disjunction. This non-existence is related to the well-known failure of Dummett's *Fundamental assumption (FA)* (see Section 1.5.6) for derivations with open assumptions, even in intuitionistic logic, while it does hold for proofs (with no open assumptions) in this logic. See, for example, Kürbis [96]. Thus, there would be a conflict between the pre-theoretic conception of the meaning of disjunction and its formal definition had we restricted the definition of I-canonicity to its first clause only. A similar situation emerges when passing to first-order intuitionistic logic and considering the rules for the existential quantifier. Adding the second clause to the definition resolves this conflict as both derivations (5.2.11) and (5.2.12) *are I-canonical* by the full definition.

5.2.1 Grounds for assertion/denial

Next, I define the *grounds for assertion/denial* of a sentence, as well as its *outcomes of assertion/denial*. Grounds and outcomes are needed for connecting meaning to the speech acts of assertion and denial. They play a central role in the definition of proof-theoretic consequence presented below. They also have a secondary use for imposing equivalence on sentential meanings, discussed below. Note that for atomic sentences, both their grounds for assertion and outcomes of assertion depend on the underlying (*assumed*) proof-theoretic meanings.

The notion of grounds considered here is different, though in the same spirit, from the grounds considered by Prawitz in [153]. The grounds here are formal entities, collections of sentences canonically deriving a sentence. On the other hand, Prawitz considers grounds as *mental* counterparts, associated with *possession* of the formal grounds and justifying the *epistemic acts* of inference, assertion and denial.

Definition 5.2.38 (grounds and outcomes). *Let $\varphi \in L$.*

grounds for assertion:

    Unilateral:

$$GA[\![\varphi]\!] = \{\Gamma \mid \vdash^{Ic}\Gamma : \varphi\} \tag{5.2.13}$$

    Bilateral:

$$GA[\![\varphi]\!] = \{\Gamma \mid \mathcal{C}(\Gamma) \text{ and } \vdash^{Ic}\Gamma : +\varphi\} \tag{5.2.14}$$

grounds for denial:

$$GD[\![\varphi]\!] = \{\Gamma \mid \mathcal{C}(\Gamma) \text{ and } \vdash^{Ic}\Gamma : -\varphi\} \tag{5.2.15}$$

outcomes from assertion:

    Unilateral:

$$OA[\![\varphi]\!] = \{\langle \Gamma, \psi \rangle \mid \vdash^{Ec}\Gamma, \varphi : \psi\} \tag{5.2.16}$$

Bilateral:

$$OA[\![\varphi]\!] = \{\langle \Gamma, \sigma \rangle \mid C(\Gamma, +\varphi) \text{ and } \vdash^{Ec}\Gamma, +\varphi : \sigma\} \qquad (5.2.17)$$

outcomes from denial:

$$OD[\![\varphi]\!] = \{\langle \Gamma, \sigma \rangle \mid C(\Gamma, -\varphi) \text{ and } \vdash^{Ec}\Gamma, -\varphi : \sigma\} \qquad (5.2.18)$$

Example 5.2.37 (Limit cases). *Consider two limit cases of this definition when applied to intuitionistic logic.*

-
$$GA[\![\bot]\!] = \varnothing \qquad (5.2.19)$$

  There are no grounds for asserting $\bot$, as expected, as there is no I_c-canonical derivation of $\bot$ from any Γ.

-
$$OA[\![\bot]\!] = \{\langle \Gamma, \psi \rangle \mid \psi \in L\} \qquad (5.2.20)$$

  Every $\psi \in L$ is an outcome of asserting $\bot$, obtained by applying ($\bot E$) to $\bot$ for any Γ.

Equating grounds for assertion with (possibly empty) collections of sentences (syntactic elements, members of L) might be an over-simplification, but it suffices for our immediate need of connecting meaning to the speech act of assertion. A fuller picture is presented in Prawitz [153], that aims for attributing *epistemic significance* to the ND-rules, to valid inferences carried out using them and to valid arguments formulated by them.

According to this fuller picture, basic sentences have, again, given grounds. An I-rule of $\mathcal{N}$ induces a *grounds propagation operation*, converting grounds for the premises (including the discharged assumptions) to a ground for the conclusion. Similarly, E-rules propagate outcomes. Thus, grounds and outcomes are not syntactic objects at all. I assume here a primitive relationship of *possession* of some grounds (under this simplified view) by an agent, that "hides" the epistemic issues involved in arriving at such grounds. Thus, possession of some $\Gamma \in GA[\![\varphi]\!]$ is considered (in Dummett's terminology) a warrant for justified assertion of φ, and $\Gamma \in GD[\![\varphi]\!]$ is considered a warrant for the *denial* of φ.

The next task is to generalise the notion of grounds for assertion from a single sentence to a collection, say Δ, of sentences, namely $GA[\![\Delta]\!]$. The presentation follows Francez [47].

There are two[1] natural possibilities here, distinguished by the way assumptions are combined.

---

[1] In contrast to MTS, being traditionally the only way, the conjunctive way, of combining assumptions. However, see Read [165] for a discussion of fusing assumptions also in a model-theoretic setting.

common grounds:

$$GA_c[\![\Delta]\!] \stackrel{\text{df.}}{=} \cap_{\psi \in \Delta} GA[\![\psi]\!] \tag{5.2.21}$$

This is a *conjunctive* combination of the individual assumptions in Γ, the same as the one in the MTS case.

joint grounds:

$$GA_j[\![\Delta]\!] \stackrel{\text{df.}}{=} \circ_{\psi \in \Delta} GA[\![\psi]\!] \tag{5.2.22}$$

where '$\circ$' is *fusion*, known also as intensional conjunction. Recall that the difference between conjunction and fusion originates in the I-rules for conjunction being *additive* (or shared context), while the I-rules for fusion are *multiplicative* (context free), as evident from their I-rules.

$$\frac{\Gamma : \varphi \quad \Gamma : \psi}{\Gamma : \varphi \wedge \psi} \; (\wedge I) \qquad \frac{\Gamma_1 : \varphi \quad \Gamma_2 : \psi}{\Gamma_1 \Gamma_2 : \varphi \circ \psi} \; (\circ I) \tag{5.2.23}$$

Recall also that conjunction has two general-elimination rules (GE-rules), given by

$$\frac{\Gamma : \varphi \wedge \psi \quad \Gamma, \varphi : \chi}{\Gamma : \chi} \; (\wedge GE_1) \qquad \frac{\Gamma : \varphi \wedge \psi \quad \Gamma, \psi : \chi}{\Gamma : \chi} \; (\wedge GE_2) \tag{5.2.24}$$

that can be simplified (assuming the structural rule of *weakening*) to the more familiar

$$\frac{\Gamma : \varphi \wedge \psi}{\Gamma : \varphi} \; (\wedge E_1) \qquad \frac{\Gamma : \varphi \wedge \psi}{\Gamma : \psi} \; (\wedge E_2) \tag{5.2.25}$$

Thus, conjunction projects each conjunct *separately*.

On the other hand, fusion has a single GE-rule, given by

$$\frac{\Gamma : \varphi \circ \psi \quad \Gamma, \varphi, \psi : \chi}{\Gamma : \chi} \; (\circ GE) \tag{5.2.26}$$

which does not simplify, hence projecting *both* conjuncts *simultaneously*. When I want to remain neutral regarding this difference in combining grounds, I will speak generically of "collective grounds", with a generic notation $GA[\![\Delta]\!]$ (without a qualifying superscript). A similar generalization applies (in principle[2]) to $GD[\![\Delta]\!]$, the collective grounds for denial.

This distinction plays a major role in defining proof-theoretic consequence in the next section.

---

[2] Substructural bilateral ND-systems have not been investigated in the literature, to the best of my knowledge.

5.2.2   Proof-theoretic consequence

As mentioned in the Introduction chapter, the main "visible" semantic property of (sentential) meaning is that of *entailment*. Recall that in MTS, entailment is defined in terms of truth conditions in models, namely by propagation of truth in models. A sentence φ entails a sentence ψ iff for every (suitable) model $\mathcal{M}$, if $\mathcal{M} \vDash \varphi$ then $\mathcal{M} \vDash \psi$. This is naturally extended (conjunctively) to $\Gamma \vDash \psi$. My aim here is to present a proof-theoretic counterpart of entailment, or consequence.

Based on the definitions of collective grounds, I define a proof-theoretic notion of *consequence*. Most notably, proof-theoretic consequence need not, in general, coincide with derivability! Once again, the presentation follows Francez [47], where a rational for this definition is argued for.

It is here that the assumption of atomic canonicity is helpful, allowing for a uniform treatment of consequence for both atomic and compound sentences, both in premises and in conclusions.

Definition 5.2.39 (proof-theoretic consequence). *ψ is a* proof-theoretic consequence *(PT-consequence) of Γ ($\varphi \Vdash \psi$) iff*

Unilateral consequence: $GA[[\Gamma]] \subseteq GA[[\psi]]$.

Bilateral Consequence: $GA[[\Gamma]] \subseteq GA[[\psi]]$ *and* $GD[[\psi]] \subseteq GD[[\Gamma]]$.

Thus, (unilateral) PT-consequence is based on *grounds propagation*: every Γ that derives canonically ψ, already derives canonically φ. By this definition, φ is a PT-consequence of Γ *according to its meaning* as it involves canonical derivability.

By specifying the way grounds are combined, two distinct PT-consequence can be distinguished (I show the unilateral case only).

conjunctive PT-consequence: ψ is a *conjunctive proof-theoretic consequence* of Γ ($\Gamma \Vdash_c \psi$) iff $GA_c[[\Gamma]] \subseteq GA[[\psi]]$.

fused PT-consequence: ψ is a *fused proof-theoretic consequence* of Γ ($\Gamma \Vdash_j \psi$) iff $GA_j[[\Gamma]] \subseteq GA[[\psi]]$.

Proposition 5.2.10 (properties of conjunctive PT-consequence). *Conjunctive PT-consequence enjoys the following properties.*

1. *$\varphi \Vdash_c \varphi$*

2. *If $\Gamma \Vdash_c \varphi$ and $\varphi, \Gamma' \Vdash_c \psi$ then $\Gamma, \Gamma' \Vdash \psi$.*

3. *For every Γ', if $\Gamma \Vdash_c \varphi$ then also $\Gamma, \Gamma' \Vdash_c \varphi$.*

In other words, conjunctive PT-consequence is reflexive and transitive since inclusion is. In addition, conjunctive PT-consequence is monotonic.

As for the fused PT-consequence, while also being reflexive and transitive, it is *not monotonic*. For example, $\varphi \Vdash_j \varphi$, but $\varphi, \varphi \nVdash_j \varphi$, since 'o' is known not to be idempotent.

As is expected from a notion of consequence, PT-consequence has a modal force of necessity, as manifested by the universal quantification over grounds, the counterpart of the universal quantification over models in the MTS definition of consequence.

Example 5.2.38 (PT-consequence:). *I show that in intuitionistic propositional logic*

$$\varphi \to (\psi \to \chi) \Vdash \varphi \wedge \psi \to \chi \tag{5.2.27}$$

As we are dealing with a single assumption, the mode of combining assumptions does not matter here.

Suppose that $\Gamma \in GA[\![\varphi \to (\psi \to \chi)]\!]$. *A canonical derivation of* $\varphi \to (\psi \to \chi)$ *from* Γ *ends with* ($\to I$). *Therefore,*

$$\vdash \Gamma, \varphi : \psi \to \chi \tag{5.2.28}$$

which, in turn, implies

$$\vdash \Gamma, \varphi, \psi : \chi \tag{5.2.29}$$

But then, $\vdash \Gamma, \varphi \wedge \psi : \chi$ *too, by which* $\vdash^{I_c} \Gamma : \varphi \wedge \psi \to \chi$, *implying* $\Gamma \in GA[\![\varphi \wedge \psi \to \chi]\!]$; *i.e. the grounds* Γ *have been propagated from assumption to conclusion, establishing PT-consequence.*

A simple example of fused PT-consequence is the following.

Example 5.2.39 (fused PT-consequence:).

$$\varphi, \psi \Vdash_j \varphi \circ \psi \tag{5.2.30}$$

If $\Gamma_1 \in GA_j[\![\varphi]\!]$ *and* $\Gamma_2 \in GA_j[\![\psi]\!]$, *then* $\Gamma_1 \circ \Gamma_2 \in GA_j[\![\varphi, \psi]\!]$. *But* $\Gamma_1 \circ \Gamma_2 \in GA_j[\![\varphi \circ \psi]\!]$ *by one application of* ($\circ I$) *(in (5.2.23)).*

It is important to note that by definition, PT-consequence *is not* derivability, though they sometimes coincide. In Francez [47], I present the rational for this distinction and I show the coincidence in the case of intuitionistic logic.

5.2.3   Connectives as internalization of structural properties

The proof-theoretic view of meaning of logical connectives can be seen as fully in concert with Došen's characterization of connectives as *internalizing structural properties* (of '$\vdash$') , as presented in Došen, [28]. The latter is, in Došen's terms, less congenial to the model-theoretic view of the meaning of the connectives. In Došen, [28] (p. 366), the following assumption **[C]** is made.

Any constant of the object language on whose presence the description of a nonstructural formal deduction depends can be ultimately analyzed in structural terms.

While Došen views this assumption as a characterization for logicality of the constant, here my concern is more to the connection of this ultimate analysis and the constant's (proof-theoretic) meaning. Došen explains 'ultimate analysis' by means of *double-line* rules (an abbreviation device) using multiple-conclusion sequents, A similar view can be read-off the ND-rules taking part in the determination of the constant's proof-theoretic meaning. The 'ultimate analysis' means expressing the deductive role of a constant in structural terms, *without mentioning the constant*. Such schematic sequents (not mentioning any constant of the object language) are referred to by Došen as *structural sequents*. The following examples (due to Došen) will illustrate the approach.

Example 5.2.40. *The most clearly visible internalization is in the proof-theoretic meaning of implication, where the meaning of* $\vdash \Gamma \; : \; \varphi \to \psi$ *captures the structurally-expressible claim that* $\vdash \Gamma, \varphi \; : \; \psi$*, which is just* $(\to I)$*. The various variants of implication (classical, intuitionistic, relevant, ...) are obtained by varying the structural assumptions underlying the internalized derivation.*

Example 5.2.41. *The meaning of the conjunction in* $\vdash \Gamma \; : \; \varphi \wedge \psi$ *(i.e., to the right of* ':'*), captures the structurally-expressed claim of the existence of two derivations* $\vdash \Gamma \; : \; \varphi$ *and* $\vdash \Gamma \; : \; \psi$*, which is just* $(\wedge I)$*. Again, variants of conjunction, e.g., fusion, are obtained by varying the structural assumptions underlying the two internalized derivations.*

On the other hand, the meaning of a conjunction to the left of ':' *can be viewed as an internalization of the structural comma (separating assumptions in the antecedent of a sequent). This internalization allows for the reduction of the notion of a derivation from a finite number of assumptions to a derivation from a single assumption (formed by conjoining all those assumptions).*

Example 5.2.42. *The meaning of the universal quantifier in* $\vdash \Gamma \; : \; \forall x.\varphi$ *captures a derivation* $\vdash \Gamma \; : \; \varphi$ *in which* x *does not occur free; again, this is embodied in* $(\forall I)$*. It reflects the closure of derivations under (capture free) variable substitution. Došen suggests viewing this property as a structural rule (not mentioned by Gentzen).*

$$\frac{\Gamma : \varphi}{\Gamma[x := y] : \varphi} \qquad\qquad (5.2.31)$$

5.2.4   The granularity of the reified meaning

The presentation in this section is based on Francez [50]. Clearly, the reified meanings obtained as above are very *fine-grained*. For example (under a unilateral definition),

$$[\![\varphi \wedge \varphi]\!]^{I_c} \neq [\![\varphi]\!]^{I_c}, \quad [\![\varphi \wedge \psi]\!]^{I_c} \neq [\![\psi \wedge \varphi]\!]^{I_c}, \quad [\![((\varphi \wedge \psi) \wedge \xi)]\!]^{I_c} \neq [\![(\varphi \wedge (\psi \wedge \xi))]\!]^{I_c}$$
$$(5.2.32)$$

The first inequality in (5.2.32), for example, obtains because every derivation which is the image, for some Γ, of a member of $[[\varphi \wedge \varphi]]^{I_c}$ ends with an application of $(\wedge I)$, while derivations in $[[\varphi]]^{I_c}$, for any Γ, end with an application of the I-rule of the main operator of φ.

In particular, note that it is not the case that all tautologies (or all contradictions) share the same meaning like in MTS. For example,

$$[[\varphi \to \varphi]]^{I_c} \neq [[(\varphi \to (\psi \to \varphi))]]^{I_c} \tag{5.2.33}$$

Next, I turn to the imposition of an equivalence relation on semantic values, avoiding at least some of the difficulties raised by their fine-grainedness. This fine-grainedness can be somewhat relaxed by imposing equivalence relations on meanings, and identifying sameness of meaning with such an equivalence.

As an example, I base an equivalence of sentences on *sameness of grounds* to obtain coarser reified meanings.

Definition 5.2.40 (*G*-**equivalence**). *For two sentences* φ, ψ, $\varphi \equiv_G \psi$ *iff*

Unilateral *G*-equivalence: $GA[[\varphi]] = GA[[\psi]]$.

Bilateral *G*-equivalence: $GA[[+ \varphi]] = GA[[+ \psi]]$, *and* $GD[[- \varphi]] = GD[[- \psi]]$.

Thus, the unilateral meanings of two sentences having the same grounds (i.e., assertable under the same condition) are equated. In the bilateral case, the two sentences have to be also deniable under the same conditions.

Let us return to the example mentioned above in (5.2.32). Note that

$$GA[[\varphi \wedge \psi]] = GA[[\psi \wedge \varphi]] = GA[[\varphi]] \cup GA[[\psi]] \tag{5.2.34}$$

So, it is easy to see that, in the unilateral case, '$\equiv_G$' "forgets" the difference in meaning between φ and $\varphi \wedge \varphi$, namely

$$GA[[\varphi \wedge \varphi]] = GA[[\varphi]] \tag{5.2.35}$$

hence

$$\varphi \wedge \varphi \equiv_G \varphi \tag{5.2.36}$$

similarly

$$\varphi \wedge \psi \equiv_G \psi \wedge \varphi \tag{5.2.37}$$

Finally, since

$$GA([[((\varphi \wedge \psi) \wedge \xi)]]) = GA([[(\varphi \wedge (\psi \wedge \xi))]]) = GA[[\varphi]] \cup GA[[\psi]] \cup GA[[\xi]] \tag{5.2.38}$$

we get

$$((\varphi \wedge \psi) \wedge \xi) \equiv_G (\varphi \wedge (\psi \wedge \xi)) \qquad (5.2.39)$$

The equivalence (5.2.37) and (5.2.39) justify the common view of conjunction as having an arbitrary arity, based on its commutativity and associativity. The grounds for asserting such a general conjunctive sentence are the union of the grounds for assertion of the separate conjunct. This situation is analogous to the situation in MTS, where the truth-condition of conjunction, conceived as having *arbitrary* arity (due to its commutativity and associativity), is the truth (in the same model) of all conjuncts.

An analogous refinement of $[\![\varphi]\!]^{E_c}$ as an equivalence based on sameness of outcomes of assertion can be defined too.

An additional issue regarding granularity is also resolved by G-equivalence. Under the unilateral meaning definition, all the *contradictions* have identical reified meanings, as they have *no* canonical derivations from any Γ. However, by appealing to bilateral G-equivalence, the negations of inequivalent tautologies have different meanings, as they may differ on their ground for denial.

5.3   Extracting Lexical Meanings for the Connectives and quantifiers

In this section, I show how to extract, based on their contribution to sentential semantic values, phrasal, including lexical, semantic values – also proof-theoretic objects. In particular, we obtain semantic values for the connectives and the quantifiers, reifying their meaning as induced by the meaning-conferring ND-system, taken here as NJ. I emphasize the dependency of the extracted meaning of the connectives and the quantifiers on the following.

- The object language and the specific TLG-grammar chosen for it. Taking a different TLG-grammar may yield different extracted semantic values, by determining different contributions of the operators to the sentential meanings.

- The particular ND-system chosen as meaning-conferring, generating the sentential meanings from which the connectives/quantifiers' meanings are extracted.

5.3.1   A type-logical grammar a fragment of the propositional calculus

In this section, I define a simple **L**-based type-logical grammar (see Appendix A) $G_{minprop}$ for a small fragment[3] $L_{minprop}$ of the propositional Minimal Logic (MiL),

---

[3]The extension to a propositional language having more connectives is straightforward.

$G_{minprop}$:

lexeme	category
$p \in P$	s
(	l
)	r
$\wedge$	$(((s \rightarrow (l \rightarrow s)) \leftarrow r) \leftarrow s)$
$\Rightarrow$	$(((s \rightarrow (l \rightarrow s)) \leftarrow r) \leftarrow s)$

Figure 5.1: The propositional-logic syntactic lexicon

with conjunction ('$\wedge$') and implication ('$\Rightarrow$') only, to be used for determining the contribution of the operators (connectives) to sentential meanings, in accordance to Frege's context principle (see 5.3.4). In Figure 5.1 I present the (syntactic) lexicon exhibiting the alphabet and assigning syntactic categories to its elements. I use the following collection $\mathcal{B}_{minprop}$ of basic categories: $\mathcal{B}_{minprop} = \{s, l, r\}$. Here s is the designated category of a sentence, and l, r are the categories for left and right parentheses, respectively. Note that the same symbols are used for parentheses in $L_{minprop}$ (the defined object language) and within compound functional categories. This should cause no confusion. Below is a sample derivation of $(p \wedge (p \Rightarrow q))$.

$$(5.3.40)$$

It is easy to verify that $L[[G_{minprop}]]$ defines the considered fragment of $L_{minprop}$.

5.3.2 Proof-Theoretic Type Interpretation

To express those intended contributed semantic values (meanings), I first introduce *proof-theoretic type interpretations*, the counterpart of the prevailing type interpretation (in models) for MTS, used by TLG ever since Montague's seminal work. What makes those type interpretations proof-theoretic is that their domains of interpretation are not arbitrary set-theoretic (or domain-theoretic) models; rather, they are collections of ND-derivations and contextualized functions therein. Since we are dealing with a *propositional* language, there is no use of an analog of the Montagovian basic type e (entities, in the model-theoretic interpretation). I only keep the *reinterpreted* basic type t, (for sentential meanings) – contributed semantic values. The meta-variable τ (possibly indexed) ranges over types in $\mathcal{T}_{minprop}$. With each type $\tau \in \mathcal{T}_{minprop}$, there is associated a domain D_τ of proof-theoretic objects, of possible denotations of the type. Here t is interpreted as contextualized functions from contexts to the collections of I-canonical derivations (from the context) of formulae. In other words, the proof-theoretical interpretation of type t consists in the collection of *all sentential meanings* (for all sentences in the fragment).

Definition 5.3.41 (propositional proof-theoretic types interpretation). *The set of types $\mathcal{T}_{minprop}$ is the least set satisfying the following clauses.*

1. *$t \in \mathcal{T}_{prop}$ is a (basic) type, with $D_t = \{[[\varphi]] \mid \varphi \in L_{minprop}\}$. Thus, by varying the sentential meanings (E-based, or bilateral), the proof-theoretic denotation of t varies accordingly. Type t "hides" the underlying notion of sentential meanings – making, as seen below, the extraction of sub-sentential meanings independent of the actual sentential meanings used.*

2. *If $\tau_1, \tau_2 \in \mathcal{T}_{minprop}$ are types, so is $(\tau_1, \tau_2) \in \mathcal{T}_{minprop}$, called a* functional type. *We put $D_{(\tau_1, \tau_2)} = D_{\tau_2}^{D_{\tau_1}}$, the set of* contextualized *functions.*

Remarks:

1. Note that all the type denotations are *contextualized*, in being dependent on some context Γ.

2. For $z \in D_t$, $\rho(z)$ is a singleton; if $z = \lambda\Gamma.[[\varphi]]_\Gamma^{Ic}$, we get $\rho(z) = \{\varphi\}$, and by abuse of notation, we let $\rho(z) = \varphi$.

Finally, we define the category-to-type mapping.

$$CtoT(\mathbf{c}) = \left\{ \begin{array}{ll} t & \mathbf{c} = s \\ (t, t) & \mathbf{c} \in \{l, r\} \end{array} \right. \tag{5.3.41}$$

5.3.3   Introduction and Elimination Functions

It is convenient to "hide" applications of I/E-rules to sub-derivations inside functions associated with the rules for operators. These functions hide also the uncovering of meanings from collections of arbitrary derivations. Thus, with each operator, say '$*$', are associated two functions, I_* and E_*, capturing the applications of the respective rules $(*I)$ and $(*E)$ in the ND-system employed, $\mathcal{N}$. As an example, in Figure 5.2 are the functions for Minimal Logic conjunction ('$\wedge$') and implication ('$\rightarrow$'), The explanation of the auxiliary retrieval functions $\rho, \gamma, \gamma_1, \gamma_2$ is as follows.

- $\rho(\mathcal{D})$: retrieves the root formula of the derivation $\mathcal{D}$ (its conclusion). So, for example, in $I_\wedge$, $\rho(\mathcal{D}_1)$ is the first premise, say φ; $\rho(\mathcal{D}_2)$ is the second premise, say ψ, so $\rho(\mathcal{D}_1) \wedge \rho(\mathcal{D}_2)$ is the conclusion $\varphi \wedge \psi$ of the application of $(\wedge I)$.

$$I_\wedge = \lambda z_2^t \lambda z_1^t \lambda \Gamma. \left\{ \frac{\mathcal{D}_1 \quad \mathcal{D}_2}{\Gamma : \rho(\mathcal{D}_1) \wedge \rho(\mathcal{D}_2)} \, {}^{(\wedge I)} \,\Big|\, \mathcal{D}_1 \epsilon \mathbf{ex}(z_1)(\Gamma) \& \mathcal{D}_2 \epsilon \mathbf{ex}(z_2)(\Gamma) \right\}$$

$$I_\to = \lambda z_2^t \lambda z_1^t \lambda \Gamma. \left\{ \frac{\mathcal{D}}{\Gamma : (\rho(z_1) \to \rho(\mathcal{D}))} \, {}^{(\to I)} \,\Big|\, \mathcal{D} \epsilon \mathbf{ex}(z_2)(\Gamma, \rho(z_1)) \right\}$$

$$E_{\wedge,1} = \lambda z^t \lambda \Gamma. \left\{ \frac{\mathcal{D}}{\Gamma : \gamma_1(\rho(z))} \, {}^{(\wedge_1 E)} \,\Big|\, \mathcal{D} \epsilon \mathbf{ex}(z)(\Gamma) \right\}$$

$$E_{\wedge,2} = \lambda z^t \lambda \Gamma. \left\{ \frac{\mathcal{D}}{\Gamma : \gamma_2(\rho(z))} \, {}^{(\wedge_2 E)} \,\Big|\, \mathcal{D} \epsilon \mathbf{ex}(z)(\Gamma) \right\}$$

$$E_\to = \lambda z_2^t \lambda z_1^t \lambda \Gamma. \left\{ \frac{\mathcal{D}_1 \quad \mathcal{D}_2}{\Gamma : \gamma(\rho(\mathcal{D}_1))} \, {}^{(\to E)} \,\Big|\, \mathcal{D}_1 \epsilon \mathbf{ex}(z_1)(\Gamma) \& \mathcal{D}_2 \epsilon \mathbf{ex}(z_2)(\Gamma) \right\}$$

Figure 5.2: I/E-functions for Minimal Logic conjunction and implication

- For an implication, $\gamma(\varphi \to \psi)$ is the consequent ψ.

- Similarly, for a conjunction, $\gamma_1(\varphi \wedge \psi)$ is the first conjunct φ, and $\gamma_2(\varphi \wedge \psi)$ is the second conjunct ψ.

The type of the I/E-functions in those examples is $(t, (t, t))$. Note that the arguments of the I/E functions, yielding I/E-canonical derivations, are first expanded via **ex**, and the actual application of an I/E-rule is to the expanded arguments, yielding arbitrary derivations.

In general, for an operator '$*$', we thus have the functions I_* and E_* (or several, if there is more than one I-rule or E-rule for '$*$'), of arities depending on the number of premises of the corresponding I/E-rules. The result of the extraction process (described in details below) is the following reified meaning for for a general '$*$', assuming both $(*I)$ and $(*E)$ have two premises.

$$[\![*]\!]^I = \lambda z_2^t \lambda z_1^t . I_*(z_2)(z_1) \tag{5.3.42}$$

and

$$[\![*]\!]^E = \lambda z_2^t \lambda z_1^t . E_*(z_2)(z_1) \tag{5.3.43}$$

Those meanings satisfy the following expected equalities.

$$[\![*]\!]^{I_c}([\![\varphi_2]\!]^{I_c})([\![\varphi_1]\!]^{I_c}) = [\![\varphi_1 * \varphi_2]\!]^{I_c} \tag{5.3.44}$$

$$[\![*]\!]^{E_c}([\![\varphi_2]\!]^{E_c})([\![\varphi_1]\!]^{E_c}) = [\![\varphi_1 * \varphi_2]\!]^{E_c} \tag{5.3.45}$$

Those equalities generalize to arbitrary arities.

5.3.4   Frege's Context Principle and its use for sub-sentential meaning extraction

The original formulation of Frege's *Context Principle (CP)* ([56], p. 116) is:

> We must never try to define the meaning of a word in isolation, but only as it is used in the context of a proposition.

The *CP* can be generalized from words[4] to arbitrary sub-sentential phrases, yielding[5] ([78]):

Principle F: The meaning of a phrase w is the *contribution* that w makes to the meanings of phrases u containing w.

Contributions: Frege did not give any explicit definition of 'contribution'. I propose one here. The general method of identifying contributions is to decompose (in accordance with the given TLG-categorization) the semantic values, starting with sentential meanings as determined by the meaning-conferring ND-system $\mathcal{N}$ (NJ in the example below), into *function-argument structures*. Suppose that by the TLG-categorization, a phrase w_1 of some functional category is adjoined to a phrase w_2 of the suitable argument category (w.l.o.g, we consider the case where the string categorized as a function precedes the string categorized as an argument). We have the following decomposition principle:

$$\delta : \ [[w_1 w_2]] = [[w_1]]([[w_2]]) \tag{5.3.46}$$

Now, the sub-phrase w_1 contributing the function can obtain its contributed semantic value by abstraction over the argument's[6] contributing semantic value (see (5.3.48) below). Thus, transitively, sub-sentential meanings are characterized by their contribution to (the independently defined) sentential meanings, where the extraction of contributions is driven by the TLG, which assigns syntactic categories to phrases.

We can observe here an important difference in the role TLG plays in our PTS, as opposed to its traditional role in MTS. In the latter, (5.3.46) is viewed from right to left, determining $[[w_1 w_2]]$ based on $[[w_1]]$ and $[[w_2]]$. Here, (5.3.46) is viewed from left to right, using $[[w_1 w_2]]$ (given by the ND proof-system conferring meaning on full sentences) to yield $[[w_1]]$ and $[[w_2]]$. Recall that the recursive decomposition terminates when $[[P]]$ is reached, as meanings of atomic propositions is given externally. The ability to read (5.3.46) from left to right depends on compositionality, but also on the uniqueness of meaning with which sentences in this language are endowed. In case

---

[4]The original formulation takes 'word' in its natural-language use; here, we refer by 'word' to any item in the lexicon of a TLG.

[5]I consider 'sub-sentential phrase' where Hodges considers 'term' in his more Algebraic setting.

[6]Note that there is a strong dependency of the sub-sentential meanings obtained on the categorization used by the grammar; changing it may lead to a change in the sub-sentential meanings obtained from the *same* sentential meanings.

of semantic ambiguity, the whole process needs to be adapted accordingly. Since this does not arise here, I skip the details. An example is presented in Chapter 9, where ambiguous NL sentences are considered. Note that this general conception is geared towards its use for natural language (in Part II of the book). For L_{prop}, the only interesting application is to the connectives themselves (viewed as "words"). For some recent philosophical debate about CP see Barth [6] (and references therein).

The whole approach is only *indirectly* in concert with Frege's *compositionality principle*: The extraction of word-meanings and word contributing values occurs *once only*; but when the extracted word contributing semantic values (though *not* contributed semantic values) are combined – per (successfully) parsed sentence – according to the grammar, they reconstruct compositionally its "original" sentential meaning (see Proposition (5.3.11) below). An example is presented below. This view of reconciling the two of Frege's principles, Context and compositionality, shows that difficulties like those pointed out in Pagin [133] are partially overcome-able in a TLG framework.

5.3.5   Contributions and meaning extraction

The discussion here follows the I-view of PTS. In an analog treatment based on the E-view, E-functions would be used instead of the use of I-functions below.

To identify the contributions of a (binary) connective $* \in \{\wedge, \Rightarrow\}$, we inspect the derivation, in $G_{minprop}$, of sentences of the form $\varphi = (\varphi_1 * \varphi_2)$, which have '$*$' as their main connective. We obtain extensions of the two evaluation functions: $[[\cdots]]$ and $[[\cdots]]^*$, to sub-sentential phrases. Such a derivation in $G_{minprop}$ has the following form.

$$\dfrac{(\ \dfrac{\varphi_1}{l:x_1}\quad \dfrac{\dfrac{s:z_1}{\qquad}\ \dfrac{\dfrac{\dfrac{*}{(((s \to (l \to s)) \leftarrow r) \leftarrow s):y}\ \dfrac{\varphi_2}{s:z_2}}{((s \to (l \to s)) \leftarrow r):(yz_2)}(\leftarrow E)\ \dfrac{)}{r:x_2}}{\dfrac{(s \to (l \to s)):(x_2(yz_2))}{(l \to s):((x_2(yz_2))z_1)}(\to E)}(\leftarrow E)}{s:(x_1((x_2(yz_2))z_1))}(\to E)$$

$$(5.3.47)$$

In contrast to natural language sentences, logical formulae do not have "interesting" non-recursive sub-sentential phrases, except for the connectives themselves, which are the focus of my interest here. We may assume that parentheses, having an atomic category, do not contribute to meaning, so their lexical semantic value can be chosen as the identity function of the appropriate type (the contribution of "pure arguments", of atomic category, is discussed in more detail in Chapter 9, in a natural language context). To simplify the presentation, I shall just ignore them[7], and take the final meaning term to be just $M = ((yz_2)z_1)$. From the structure of M, we see that the

---

[7]One might choose a simpler syntax, using infix, parentheses-free (Polish) notation, avoiding the treatment of the meaning of parentheses altogether.

contribution of '$*$' is a function, applied (Curried) to the two semantic values of the two sub-formulas φ_1, φ_2, say M_1, M_2, respectively. Thus, we get[8]

$$[[\varphi]] = [[((yM_2)M_1)]] = [[*]]([[\varphi_2]])([[\varphi_1]]) \; [= \lambda\Gamma.[[\varphi]]_\Gamma^{Ic}] = I_*([[\varphi_2]])([[\varphi_1]]) \tag{5.3.48}$$

To solve for $[[*]]$ in (5.3.48), we need to abstract over the two arguments, first replacing them by variables. We thus obtain the following lexical meaning for a (binary) connective '$*$', extracted, by principle $\mathbf{F}$, from sentential meanings in which it occurs and to the meaning of which it contributes.

$$[[*]] = \lambda z_2^t \lambda z_1^t . I_*(z_2)(z_1) \; [= I_*] \tag{5.3.49}$$

Thus, we see that the I-function corresponding to a connective constitutes the reification of the semantic value of that connective.

The non-direct-compositionality of meaning is "hidden", since

$$[[\varphi_1 * \varphi_2]] = [[*]]([[\varphi_2]])([[\varphi_1]]) \tag{5.3.50}$$

Recall that I_* extends the meanings to the contributing semantic value via an application of the function $\mathbf{ex}$.

As a final note, we observe that the grammar will assign meaning (semantic value) also to "uninteresting" sub-phrases, like '$\wedge\psi)$'. This is a result of the way the machinery of TLG works. This does no harm, of course. If one is interested only in the meaning of well-formed sentences (and not in intermediate meanings generated during their computation), one can confine attention only to sub-sentential phrases of category s. I skip further details as the topic is inessential for the general program, being an artefact of the formalism used.

To emphasize the point of the need for two semantic values, consider the following example from $L_{propmin}$:

$$\varphi = (\varphi_1 \wedge (\varphi_2 \wedge \varphi_3)) \tag{5.3.51}$$

The contribution of the outer '$\wedge$' is its meaning $[[\wedge]]$ (see below), depending on canonical derivations of φ (from φ_1 and $(\varphi_2\wedge\varphi_3)$); on the other hand, the internal '$\wedge$' contributes to $[[\varphi]]$ an *arbitrary* derivation of $(\varphi_1\wedge\varphi_2)$, obtained by using $[[\wedge]]^*$.

Proposition 5.3.11 (sentential meaning). *For $\varphi \in L_{propmin}$:*

- *The syntactic type of φ is s.*

- *the semantic type of φ is t, with $[[\varphi]] = \lambda\Gamma.[[\varphi]]_\Gamma^{Ic}$*

The proof is by induction on the derivation of φ in the TLG.

---

[8]Note that within λ-terms, we employ the traditional notation (fA) (applying f to A), while at the meta-level we employ the usual mathematical notation $f(A)$.

$G_{minprop}$:

lexeme	category	PT – meaning	type
$p \in P$	s	$[[p]]^{I_c}$	t
$($	l		
$)$	r		
$\wedge$	$(((s{\rightarrow}(l{\rightarrow}s)) \leftarrow r) \leftarrow s)$	$\lambda z_2^t \lambda z_1^t.I_\wedge(z_2)(z_1)$	$(t,(t,t))$
$\Rightarrow$	$(((s{\rightarrow}(l{\rightarrow}s)) \leftarrow r) \leftarrow s)$	$\lambda z_2^t \lambda z_1^t.I_\Rightarrow(z_2)(z_1)$	$(t,(t,t))$

Figure 5.3: The propositional-logic full lexicon

For some contexts Γ, $[[\cdots]]_\Gamma^*$ can be recovered from $[[\cdots]]_\Gamma^{I_c}$, but not always. In contrast to *proofs* (derivations from no open assumptions), the existence of a derivation does not imply the existent of a (normal) canonical derivation (from the same assumptions). Thus, for $\Gamma = \{(\varphi{\rightarrow}\psi), \varphi\}$, we have $[[\psi]]_\Gamma^* \neq \varnothing$, but $[[\psi]]_\Gamma^{I_c} = \varnothing$. In such cases the recovery fails.

Below I present the *full* lexicon for $L_{propint}$, including semantic values and their types.

5.3.6   Extracting Meanings for Quantifiers

In this section, I extend the ideas presented above to (a fragment of) minimal first-order logic. Let L_{FOL} be the fragment of L_{foint} (Section 2.1.4) containing, for simplicity, only unary and binary predicate symbols, conjunction and implication as the sole connectives, and the universal and existential quantifiers $\forall$, $\exists$, respectively. Recall that no individual constants are included (nor function symbols in general). For convenience, I rename the collection X of variables to V, and we take the meta-variables ranging over V, as X, X_i (capitals). This enables a simple way of distinguishing those variables from the xs in the meaning expression language. Let $\mathcal{P}, \mathcal{R}$ be the finite collections of unary and binary predicate symbols, ranged over, by P, P_i, R, R_i, respectively. The main interest here is to capture proof-theoretically the semantic values of the quantifiers, again based on their contributions to sentential meanings. Towards that end, I first introduce a TLG G_{FOL}. I use the following collection $\mathcal{B}_{FOL}$ of basic categories: $\mathcal{B}_{FOL} = \{s, v, l, r, c\}$. Here s is again the designated category of a sentence (including open sentences), and l, r are the categories for left and right parentheses, respectively. The following additional base categories are used: c (for comma) and v (for variable). The (syntactic) lexicon is presented in Figure 5.4. Below I present a sample derivation establishing sentencehood of $\forall X(P_1(X){\wedge}P_2(X))$. It is easy to verify that $L[[G_{FOL}]] = L_{FOL}$ in general. For clarity, the two sub-derivations for $P_1(X)$ and $P_2(X)$ to s have been factored out and displayed separately, omitting intermediate semantic terms. Once again, parentheses

G_{FOL} :

lexeme	category
$P \in \mathcal{P}$	$(((s \leftarrow r) \leftarrow v) \leftarrow l)$
$R \in \mathcal{R}$	$((((s \leftarrow r) \leftarrow v) \leftarrow c) \leftarrow v) \leftarrow l)$
(	l
)	r
,	c
$X \in V$	v
$\wedge, \Rightarrow$	$(((s \rightarrow (l \rightarrow s)) \leftarrow r) \leftarrow s)$
$\forall, \exists$	$((s \leftarrow s) \leftarrow v)$

Figure 5.4: The 1st-order logic syntactic lexicon

are considered semantically transparent, as well as the coma (separating the arguments of a binary relation symbol).

$$\frac{\dfrac{\wedge}{\dfrac{\dfrac{(((s \rightarrow (l \rightarrow s)) \leftarrow r) \leftarrow s) : y \quad s : z_2}{((s \rightarrow (l \rightarrow s)) \leftarrow r) : (yz_2)} \ (\leftarrow E) \quad \dfrac{)}{r : u_2}}{\dfrac{(\forall}{((s \leftarrow s) \leftarrow v) : z \quad \dfrac{X}{v : x}}{\dfrac{(s \leftarrow s) : (zx)}{}} (\leftarrow E) \quad \dfrac{l}{l : u_1} \quad \dfrac{s : z_1}{} \quad \dfrac{((s \rightarrow (l \rightarrow s)) : ((yz_2)u_2)}{\dfrac{(l \rightarrow s) : (((yz_2)u_2)z_1)}{s : ((((yz_2)u_2)z_1)u_1)}} (\leftarrow E)}{s : ((zx)((((yz_2)u_2)z_1)u_1))}} (\leftarrow E)$$

$$\frac{\dfrac{\dfrac{P_1 \quad \dfrac{(}{l}}{((s \leftarrow r) \leftarrow v) \leftarrow l)} \ \dfrac{((s \leftarrow r) \leftarrow v)}{(s \leftarrow r)} (\leftarrow E) \quad \dfrac{X}{v}}{(s \leftarrow r)} (\leftarrow E) \quad \dfrac{)}{r}}{s : z_1} (\leftarrow E)$$

$$\frac{\dfrac{\dfrac{P_2 \quad \dfrac{(}{l}}{((s \leftarrow r) \leftarrow v) \leftarrow l)} \ \dfrac{((s \leftarrow r) \leftarrow v)}{(s \leftarrow r)} (\leftarrow E) \quad \dfrac{X}{v}}{(s \leftarrow r)} (\leftarrow E) \quad \dfrac{)}{r}}{s : z_2} (\leftarrow E)$$

In order to obtain sentential meanings, I again use NJ as the meaning-conferring ND-system, repeated here in (Figure 5.5). Meta-variables φ, ψ range again over sentences (including open ones). Here, what is assumed to be given is the meaning for *atomic* sentences and of variables (these are the two recursion endings). For a recent view about the form such given meanings may have, see Więckowski [236]. For variables, we take their proof-theoretic meaning as themselves: $[\![X]\!] = X, X \in V$. As before, I set the proof-theoretical meaning of a compound $\varphi \in L_{FOL}$ as a function mapping contexts Γ to the I-canonical derivations of φ from Γ, where the derivations are again in NJ.

In order to extract meanings (in accordance with principle **F**), I introduce again a suitable collection of FOL proof-theoretically interpreted types.

$$\frac{}{\Gamma, \varphi : \varphi} \; (Ax)$$

$$\frac{\Gamma : \varphi}{\Gamma : \forall X(\varphi)} \; (\forall I) \qquad \frac{\Gamma : \forall X(\varphi)}{\Gamma : \varphi[X := Y]} \; (\forall E)$$

$$\frac{\Gamma : \varphi[X := Y]}{\Gamma : \exists X(\varphi)} \; (\exists I) \qquad \frac{\Gamma : \exists X(\varphi) \quad \Gamma, [\varphi[X := Y]]_i : \psi}{\psi} \; (\exists E^i)$$

where X is fresh for Γ in $(\forall I)$ and Y fresh for Γ, ψ in $(\exists E)$.

Figure 5.5: NJ-rules for intuitionistic quantifiers

Definition 5.3.42 (FOL proof-theoretic type interpretation). *The set of FOL types* $\mathcal{FT}_{FOL}$ *is the least set satisfying the following clauses.*

1. *$t \in \mathcal{FT}_{FOL}$ is a (basic) type, with[9] $D_t = \{\lambda \Gamma. [\![\varphi]\!]_\Gamma^c \mid \varphi \in L_{FOL}\} = \{[\![\varphi]\!] \mid \varphi \in L_{FOL}\}$. All sentential meanings, again!*

2. *$V \in \mathcal{FT}_{FOL}$ is a basic type (of variables), with[10] $D_V = V$.*

3. *If $\tau_1, \tau_2 \in \mathcal{FT}_{FOL}$ are types, so is the functional type $(\tau_1, \tau_2) \in \mathcal{FT}_{FOL}$, with $D_{(\tau_1, \tau_2)} = D_{\tau_2}^{D_{\tau_1}}$, the set of contextualized functions.*

The category-to-type mapping is as follows.

$$CtoT(\mathbf{c}) = \left\{ \begin{array}{ll} t & \mathbf{c} = s \\ V & \mathbf{c} = v \\ (t, t) & \mathbf{c} \in \{l, r, c\} \end{array} \right. \tag{5.3.52}$$

The additional I-functions for the quantifiers, based on ND_1 I-rules, are as follows.

$$I_\forall = \lambda z_2^V \lambda z_1^t \lambda \Gamma. \left\{ \frac{\mathcal{D}}{\Gamma : \forall z_2(\rho(\mathcal{D}))} \; (\forall I) \; : \; \mathcal{D} \in \mathbf{ex}(z_1)(\Gamma), z_2 \text{ fresh for } \Gamma \right\}$$

$$I_\exists = \lambda z_2^V \lambda z_1^t \lambda \Gamma. \left\{ \frac{\mathcal{D}}{\Gamma : \exists z_2(\rho(\mathcal{D})[Y := z_2])} \; (\exists I) \; : \; \mathcal{D} \in \mathbf{ex}(z_1)(\Gamma), z_2 \notin Fr(\rho(\mathcal{D})) \right\}$$

To identify the contribution of a quantifier $q \in \{\forall, \exists\}$, we inspect the derivation, in G_{FOL} of sentences $\varphi = qX\psi$, which have 'q' as their main quantifier. Such a derivation has the following form.

$$\frac{\dfrac{\overset{q}{((s \leftarrow s) \leftarrow v) : u_1} \quad \overset{X}{v : u_2}}{(s \rightarrow s) : (u_1 u_2)} \; (\leftarrow E) \quad \overset{\psi}{s : z}}{s : ((u_1 u_2) z)} \; (\leftarrow E) \tag{5.3.53}$$

---

[9]For the sake of this definition, the (given) meanings of atomic sentences are regarded as canonical derivations, even though not strictly introduced via an I-rule.

[10]I abuse notation by using V both as type and as its domain of interpretation – the collection of variables.

We thus get, by taking $z = M$,

$$[[\varphi]] = [[((u_1 u_2)z)]] = [[qX]]([[\psi]]) = I_q([[X]])([[\psi]]) \qquad (5.3.54)$$

Solving (5.3.54) for the quantifier-phrase meaning $[[qX]]$, we get by abstracting over the argument:

$$[[qX]] = \lambda z_2^t . I_q([[X]])(z_2) \qquad (5.3.55)$$

Also, from the above derivation in the grammar, we have that $[[qX]] = (u_1 u_2) = [[q]]([[X]])$, from which, since $[[X]] = X$,

$$[[q]] = \lambda z_1^V \lambda z_2^t . I_q(z_1)(z_2) \; [= I_q] \qquad (5.3.56)$$

In a similar way, starting from the derivation of atomic sentences in the grammar, we obtain the meanings of predicate symbols (again, simplified by ignoring parentheses and the comma).

$$[[P]] = \lambda x^v . [[P(x)]], \quad [[R]] = \lambda x^v \lambda y^v . [[R(x,y)]] \qquad (5.3.57)$$

where the r.h.s. expressions are atomic sentences with given meanings.

Note again, that the grammar will assign meanings also to "uninteresting" sub-phrases such as $\wedge P(X))$; again, I skip the details.

5.4   Explaining non-eliminability

I now return to the issue of *non-eliminability* mentioned at the end of Section 1.6. In [19], Contu rejects both the I- and the E-views, with attributing a certain advantage to the latter, lacking a circularity he finds in the former. He favors the (combined) C-view, argued for by the non-eliminability from $\mathcal{N}$ of either the I-rules or the E-rules. Here I propose another explanation of this non-eliminability, based on the *reification of (proof-theoretical) meaning*. The argument was first presented in Francez [46]. This argument leaves the choice of the preferred view of PTS to be determined independently of this non-eliminability.

I will not go in detail into Contu's analysis of the meaning of rules, only in as much as relevant to the claim I intend to refute. The following is a quotation from Contu ([19], Section 6, pp.583-4).

> Insofar as we are concerned with the formal meaning of logical signs, i.e., the content that can be read off from the rules of the formal system they belong to, it is clear that if *if any rule is not eliminable, then it is essential to determining the formal meaning* [my emphasis, N.F.].

Thus, according to Contu, non-eliminability of either the E-rules (in the I-view) or the I-rules (in the E-view) is a reason to reject those views, the latter regarding meaning conferred by one group, which also justifies the other, non-eliminable group. In passing, I would like to mention that regarding the I-view, Contu arrives at this conclusion based on the use of E-rules to analyze assumptions in a derivation, the "place holder" view of assumptions discussed in Section 1.5.2.1.

I am now in a position to *explain* the non-eliminability of the ND-rules of the justified class (E-rules in the I-view and I-rules in the E-view), thereby refuting Contu's claim that this non-eliminability undermines the PTS approach where meaning is based on one group of ND-rules, the defining group, the other group being justified by the defining group. The explanation is as follows. Reified meanings are defined by *canonical* derivations, where canonicity is relative to the defining group. However, as Contu is obviously also aware of, canonical derivations may, in general, have non-canonical sub-derivations! Those are possible only because the non-defining group of rules is present.

Let me exemplify the situation, in the I-view, with an example from Minimal Logic, using the rules in Figure 2.1 (in their simple presentation). The following simply presented I-canonical derivation $\mathcal{D}$ is part of the reified meaning of $\varphi \wedge \psi$ (and indirectly of the I-meaning of '$\wedge$').

$$\mathcal{D}: \quad \dfrac{\dfrac{\chi \to \varphi \quad \chi}{\varphi} \; (\to E) \quad \dfrac{\xi \to \psi \quad \xi}{\psi} \; (\to E)}{\varphi \wedge \psi} \; (\wedge I) \tag{5.4.58}$$

Here $\varphi \wedge \psi$ is I-canonically derived, according to its meaning. However, both of the two premises from which it was derived, via $(\wedge I)$, are *not* I-canonically derived, that is, not according to their meanings, via $(\to E)$. Note that

$$\Gamma = \chi \to \varphi, \chi, \xi \to \psi, \xi \in G[\![\varphi \wedge \psi]\!] \tag{5.4.59}$$

If, in the I-view, all E-rules were eliminated, $\mathcal{D}$ (and many more derivations) would not be part of the reified meaning of $\varphi \wedge \psi$, not allowing the I-rules to exhaust the *full* meaning of the (minimal) conjunction. An analogous situation pertains regarding the E-view, where sub-derivations of an E-canonical derivation need not themselves be E-canonical. Consider the following derivation $\mathcal{D}'$.

$$\mathcal{D}': \quad \dfrac{\dfrac{\dfrac{\varphi \wedge \psi}{\varphi} \; (\wedge_1 E) \quad \varphi \to \alpha}{\alpha} \; (\to E) \quad \dfrac{\dfrac{\varphi \wedge \psi}{\varphi} \; (\wedge_1 E) \quad \varphi \to \beta}{\beta} \; (\to E)}{\alpha \wedge \beta} \; (\wedge I) \tag{5.4.60}$$

For $\Gamma' = \varphi \to \alpha, \varphi \to \beta$, we have

$$\langle \Gamma', \alpha \wedge \beta \rangle \in O(\varphi \wedge \psi) \tag{5.4.61}$$

if, on the E-view, all I-rules were eliminated, $\mathcal{D}'$ (and many other derivations) would not be part of the reified meaning of $\varphi \wedge \psi$, thereby prohibiting the E-rules to exploit the full meaning of the minimal '$\wedge$'.

Of course, this argument leaves open the question as to which of the two views, the
I-view and the E-view, or one of the other views, is preferable. Contu might have a
case here, preferring the the E-view over the I-view. However, his preference of the
C-view, based on non-eliminability, I see as unwarranted.

5.5 Summary

The main concern of this chapter was the attribution of explicit semantic values, reified
meanings, both to sentences and parts their-of. The chapter elaborated on the process
of abstraction leading, based of Frege's context principle, from sentential meanings to
the subsentential ones. This is done within the framework of a type-logical grammar
for the object language, by means of which contributions of constituents are explicated
and made precise. A major concept, that of proof-theoretic semantics (different from
derivability!), was defined in terms of grounds for assertion — a byproduct of meanings. The issue of the fine granularity of PT-meanings, was considered and seen as an
advantage, not assigning the same reified meanings to logically equivalent formulas.

Chapter 6

Other Developments

In this chapter I present some further developments falling within the PTS programme.

6.1 Higher-Level Rules an derivations

In this section, I present another approach to the form rules should take to be harmonious in form, originating in Schroeder-Heister [183], called *higher-level rules* (*HL*-rules).

6.1.1 Higher-level rules

While standard GE-rules have formulas (or sequents, depending on the presentation mode – I'll focus on simple presentation here), HL-rules may have *rules* as dischargeable assumptions. Such rules are another way of expressing a premise that is a sub-derivation, possibly discharging assumptions. To see the difference between standard GE-rules and HL-rules, consider the case of implication. First, the I-rule and standard GE-rule are repeated below.

$$\frac{\genfrac{}{}{0pt}{}{[\varphi]_i}{\genfrac{}{}{0pt}{}{\vdots}{\psi}}}{\varphi\to\psi}\ (\to I^i) \qquad \frac{\varphi\to\psi \quad \varphi \quad \genfrac{}{}{0pt}{}{[\psi]_i}{\genfrac{}{}{0pt}{}{\vdots}{\chi}}}{\chi}\ (\to GE^i) \qquad\qquad (6.1.1)$$

Next, suppose the availability of a derivation of ψ from φ, guaranteed by the premise of the $(\to I)$-rule, is recorded as a "temporary" *rule*, to be used by the E-rule to derive

205

an arbitrary conclusion. Denote this rule as[1] $\varphi \Rightarrow \psi$. The $(\to I)$ can be rewritten as $\dfrac{\varphi \Rightarrow \psi}{\varphi \to \psi}$ $(\to I)$. The HL-rule for the implication can be now presented as

$$\frac{\begin{array}{cc} & [\varphi \Rightarrow \psi]_i \\ & \vdots \\ \varphi \to \psi & \chi \end{array}}{\chi} \ (\to HL^i) \qquad\qquad (6.1.2)$$

To understand the way such a rule functions, consider the following derivations, showing that $(\to GE)$ and $(\to HL)$ are mutually inter-derivable. The proof is from Schroeder-Heister [194].

From $(\to GE)$ to $(\to HL)$: Suppose we have derivations for the premises of $(\to GE)$, namely

$$\begin{array}{ccc} & & \psi \\ \mathcal{D}_1 & \mathcal{D}_2 & \mathcal{D}_3 \\ \varphi \to \psi, & \varphi, & \chi \end{array}$$

With theses, we can form

$$\begin{array}{c} \mathcal{D}_2 \\ \varphi \\ \hline \psi \\ \mathcal{D}_3 \\ \chi \end{array} \ (\varphi \Rightarrow \psi)$$

which is the required minor premise of $(\to HL)$. Combining with the major premise, we get

$$\frac{\begin{array}{cc} & \begin{array}{c} \mathcal{D}_2 \\ \varphi \\ \hline \psi \\ \mathcal{D}_3 \end{array} \ (\varphi \Rightarrow \psi)_1 \\ \begin{array}{c} \mathcal{D}_1 \\ \varphi \to \psi \end{array} & \chi \end{array}}{\chi} \ (\to HL^1)$$

Note that the assumed rule $(\varphi \Rightarrow \psi)$ is discharged by the application of $(\to HL)$.

From $(\to HL)$ to $(\to GE)$: Suppose we have derivations for the premises of $(\to HL)$, namely

$$\begin{array}{cc} & \varphi \Rightarrow \psi \\ \mathcal{D}_1 & \mathcal{D}_2 \\ \varphi \to \psi, & \chi \end{array}$$

Next, consider an application of $\varphi \Rightarrow \psi$ within $\mathcal{D}_2$, having the form $\begin{array}{c} \mathcal{D}_3 \\ \varphi \\ \hline \psi \end{array} (\varphi \Rightarrow \psi)$. Replace this application of $\varphi \Rightarrow \psi$ with

$$\frac{\begin{array}{ccc} \mathcal{D}_1 & \mathcal{D}_3 & \\ \varphi \to \psi & \varphi & [\psi]_1 \end{array}}{\psi} \ (\to GE^1)$$

---

[1] Schroeder-Heister refers to '$\Rightarrow$' as a "structural implication", analog to the comma separating formulas in a sequent being a "structural conjunction".

In this way, the conclusion χ of $(\rightarrow HL)$ of $\mathcal{D}_2$ is preserved.

As emphasized by Schroeder-Heister [194], This proof relies on ψ (the arbitrary conclusion) being *schematic*. It was instantiated as ψ in the proof. Thus, the rules $(\rightarrow HL)$ and $(\rightarrow GE)$ are not equivalent per instance. In this sense, $(\rightarrow HL)$ is stronger than $(\rightarrow GE)$.

One might be tempted to consider replacing a derivation

$$
\begin{array}{c}
\mathcal{D}_3 \\
\varphi \\
\hline
\psi
\end{array} (\varphi \Rightarrow \psi)
$$

$$
(*) \quad \begin{array}{c} \mathcal{D}_4 \\ \chi \end{array}
$$

with

$$
(**) \quad \cfrac{\begin{array}{ccc} \mathcal{D}_1 & \mathcal{D}_3 & \overset{[\psi]_1}{\mathcal{D}_4} \\ \varphi\rightarrow\psi & \varphi & \chi \end{array}}{\chi} (\rightarrow GE^1)
$$

leaving χ uninstantiated. This, however, would be wrong, since in in $(*)$ assumptions may be discharged by $\mathcal{D}_4$; those are not available for discharge by the uncoupled $\mathcal{D}_4$ $(**)$.

Clearly, there is nothing special about implication, and the idea is applicable to arbitrary n-ary connectives (and to quantifiers, not discussed here.)

Suppose some object language is given.

Definition 6.1.43 (rule levels). *Rule levels are defined inductively.*

1. *A formula φ (in the given object language) is a rule of level 1.*

2. *A rule $\rho_1, \cdots, \rho_n \Rightarrow \psi$ is of level $l + 1$, provided the maximal level of all the $\rho_i s$ is l.*

6.1.2 Higher-level derivations

The possibility of having HL-rules as assumptions requires a certain modification of the definition of a derivation, to allow applications of open assumption-rules within a derivation. I will present only the simply-presented case (cf. Definition 1.5.2). The underlying system $\mathcal{N}$ is left implicit. To be able to relate to open rule-assumptions, Let $\mathbf{d}_{\varphi,\mathcal{D}}^r \subseteq \mathbf{d}_{\varphi,\mathcal{D}}$ be the collection of open rule-assumptions (of level greater than 1) on which φ depends (in $\mathcal{D}$).

Definition 6.1.44 (HL-Derivations). Derivations *are inductively defined as follows.*

- *Every assumption φ is a derivation $\overset{\mathcal{D}}{\varphi}$, with $\mathbf{d}_{\varphi,\mathcal{D}} = \{\varphi\}$, and if φ is a rule-assumption, also $\mathbf{d}_{\varphi,\mathcal{D}}^r = \{\varphi\}$.*

$$[\Sigma_j]$$
$$\mathcal{D}_j$$

- If ψ_j, for $1 \le j \le p$, are derivations with dependency sets $\mathbf{d}_{\psi_j, \mathcal{D}_j}$ with $\Sigma_j \subseteq \mathbf{d}_{\psi_j, \mathcal{D}_j}$, and if

$$
\cfrac{[\Sigma_1]_1 \qquad [\Sigma_p]_p}{\begin{matrix}\vdots & & \vdots \\ \psi_1 & \cdots & \psi_p\end{matrix}} \quad (R^{\bar{i}}) \atop \psi
$$

is a rule in $\mathcal{R}_{\mathcal{N}} \cup_{1 \le j \le p} \mathbf{d}^r_{\psi_j, \mathcal{D}_j}$, then

$$
\cfrac{\mathcal{D}}{\psi} =_{\mathrm{df.}} \quad \cfrac{\begin{matrix}[\Sigma_1]_1 & & [\Sigma_p]_p \\ \mathcal{D}_1 & & \mathcal{D}_p \\ \psi_1 & \cdots & \psi_p\end{matrix}}{\psi} \quad (R^{\bar{i}})
$$

is a derivation with $\mathbf{d}_{\psi, \mathcal{D}} = \cup_{1 \le j \le p} \mathbf{d}_{\psi_j, \mathcal{D}_j} - \cup_{1 \le j \le p} \Sigma_j$ and $\mathbf{d}^r_{\psi, \mathcal{D}} = \cup_{1 \le j \le p} \mathbf{d}^r_{\psi_j, \mathcal{D}_j} - \cup_{1 \le j \le p} \Sigma^r_j$. Here $\Sigma^r \subseteq \Sigma$ is the sub-collection of the discharged assumptions that are assumption-rules. The derivations $\mathcal{D}_1, \cdots, \mathcal{D}_p$ are the direct sub-derivations of $\mathcal{D}$.

Thus, at each stage in a derivation, either a basic rule of $\mathcal{N}$ is applied, or an open (at that stage) rule-assumption is applied.

Remarks:

1. In a tree-like depiction of a derivation, open rule-assumptions are leafs, like ordinary formulas.

2. In the logistic notation $\vdash_{\mathcal{N}} \Gamma : \psi$, Γ may contain HL-rules too, in addition to formulas.

Note that in derivations as defined above only formulas may appear as conclusions. However, in order to define derivation composition (and closure under it), the definition of derivations need to be extended. Recall that in the definition of composition of derivations without assumption-rules (cf. Definition 1.5.6), composition involved replacing a leaf in one derivation by another derivation having this leaf as its root. To apply the idea in the current set-up, there is a need to replace a leaf that is an assumption-rule by a derivation having that assumption-rule as a leaf. This means that there is a need for a definition of deriving assumption-rules from other rules. Such a definition is given[2] by Schroeder-Heister [183].

Definition 6.1.45 (*HL-rules derivability*). *A rule $\langle \Gamma_1 \Rightarrow \varphi_1, \cdots, \Gamma_n \Rightarrow \varphi_n \rangle \Rightarrow \psi$ is derivable from Γ iff*

$$\vdash \Gamma, \Gamma_1 \Rightarrow \varphi_1, \cdots, \Gamma_n \Rightarrow \varphi_n : \psi \qquad (6.1.3)$$

---

[2]The actual definition in Schroeder-Heister [183] is a little different, and equivalence with the definition here is proven as a lemma. I find this characterization clearer and have chosen *it* as the definition.

Thus, derivation of a rule (from other rules) is reduced to a derivation of the rule's conclusion from the other rules and the rule's premises as additional assumptions.

Now composition can be defined as before, by replacing a rule-assumption leaf node ρ in one derivation, say $\mathcal{D}_1 : \Delta, \rho : \psi$, by another derivation $\mathcal{D}_2 : \Delta : \rho$ having that leaf as a root. However, the story here is more complicated. While in the ordinary derivation composition an occurrence of a formula as an assumption in $\mathcal{D}_1$ can serve only as a premise, so the assumption can be replaced by its re-derivation via $\mathcal{D}_2$, a rule-assumption is *not* a premise; rather it is *applied* within the derivation $\mathcal{D}_1$. So, when an assumption-rule ρ is replaced by some derivation $\mathcal{D}_2$ of it, all its applications within $\mathcal{D}_1$ have to be replaced accordingly. This replacement is defined by induction on the level of the replaced rule-assumption ρ.

- The basis of the induction (level 1) is obvious.

- Consider a level-2 ρ of the form $\langle \varphi_1, \cdots, \varphi_n \rangle \Rightarrow \psi$. In this case, every application of ρ is replaced by a derivation of ψ from $\Delta, \varphi_1, \cdots, \varphi_n$, existing by the definition of rule derivability.

- Suppose the replaced rule has the form $\langle \Gamma_1 \Rightarrow \varphi_1, \cdots, \Gamma_n \Rightarrow \varphi_n \rangle \Rightarrow \psi$ with level greater than 2. Consider the topmost application (nearest to the root) of ρ in $\mathcal{D}_1$. By the definition of rule derivability and the induction hypothesis, $\Delta, \Delta_i, \Gamma_i : \varphi_i$, where Δ_i are the rule-assumptions open above φ_i. Then the application of ρ can be replaced by the derivation of ψ from $\Delta, \Delta_1, \cdots, \Delta_n$.

Note that the similarity of HL-rules to implications (in the object language) is rather superficial. Syntactically, in a rule the conclusion is always a formula, never a rule of a higher-level.

Closure under such higher-level composition needs to be established for every particular system used, to allow reductions.

6.1.3 Harmony with higher-level rules

Consider first intrinsic harmony. As an example of a reduction involving a higher-level rule, consider the following example due to Read [171]. In order to express it, the notation of substitution into a derivation is extended with $\mathcal{D}_1[\rho : \mathcal{D}]$, denoting the replacement in $\mathcal{D}_1$ of every application of the assumption-rule ρ by the derivation $\mathcal{D}$ of the conclusion of ρ from its premises, according to the definition of rule derivability. Consider again the ($\rightarrow HL$)-rule in (6.1.2), giving rise to the following reduction.

$$\dfrac{\begin{array}{c}[\varphi]_1 \\ \mathcal{D} \\ \psi \end{array}}{\dfrac{\varphi \rightarrow \psi}{\chi} \; (\rightarrow I^1)} \qquad \dfrac{\begin{array}{c}[\varphi \Rightarrow \psi]_2 \\ \mathcal{D}_1 \\ \chi \end{array}}{} (\rightarrow HL^2) \qquad \rightsquigarrow_r \qquad \mathcal{D}_1[(\varphi \Rightarrow \psi) := \begin{array}{c}[\varphi]_1 \\ \mathcal{D} \\ \psi \end{array}] \qquad (6.1.4)$$

Next, consider harmony in form, this time with HL-rules. Suppose $*(\varphi_1, \cdots, \varphi_n)$ is an n-ary connective with m I-rules of the form

$$\frac{\Sigma_j(\varphi_1, \cdots, \varphi_n)}{*(\varphi_1, \cdots, \varphi_n)} \; (*I_j), \; 1 \leq j \leq m \tag{6.1.5}$$

Then, the *Harmoniously-Induced HL-rule* (implementing the inversion principle) is given by

$$\frac{*(\varphi_1, \cdots, \varphi_n) \qquad \begin{array}{c} [\Sigma_1(\varphi_1, \cdots, \varphi_n)]_{i_1} \\ \vdots \\ \chi \end{array} \qquad \cdots \qquad \begin{array}{c} [\Sigma_1(\varphi_1, \cdots, \varphi_n)]_{i_m} \\ \vdots \\ \chi \end{array}}{\chi} \; (*HL^{i_1, \cdots, i_m}) \tag{6.1.6}$$

Clearly, the level of $(*HL)$ is greater by 1 from the level of $(*I)$. The difference between this scheme and the one for the harmoniously-induced GE-rule (cf. Section 3.6.3) is that here the Σ_j can themselves be rule assumptions (of some level), like in (6.1.2).

As evident from the example of $(\rightarrow GE)$ and $(\rightarrow HL)$, HL-rules need not be equivalent to GE-rules (sometimes referred to as "flattened"), although equivalence might hold for certain instantiation of the schematic rules. The following example from Schroeder-Heister [194] shows that sometimes "flattening" a HL-rule may require more than one GE-rule.

Example 6.1.43. *Consider a ternary connective* $c(\varphi_1, \varphi_2, \varphi_3)$ *with the following I-rule (in simple presentation).*

$$\frac{\varphi_1 \qquad \begin{array}{c} [\varphi_2]_1 \\ \vdots \\ \varphi_3 \end{array}}{c(\varphi_1, \varphi_2, \varphi_3)} \; (cI^1) \tag{6.1.7}$$

The harmoniously-generated HL-rule (in simple presentation) is

$$\frac{c(\varphi_1, \varphi_2, \varphi_3) \qquad \begin{array}{c} [\varphi_1]_1, [\varphi_2 \Rightarrow \varphi_3]_2 \\ \vdots \\ \chi \end{array}}{\chi} \; (cHL^{1,2}) \tag{6.1.8}$$

There is no equivalent single GE-rule; however, the following pair of GE-rules is equivalent to the above HL-rule.

$$\frac{c(\varphi_1, \varphi_2, \varphi_3) \qquad \begin{array}{c} [\varphi_1]_1 \\ \vdots \\ \chi \end{array}}{\chi} \; (cHL_1^1) \qquad\qquad \frac{c(\varphi_1, \varphi_2, \varphi_3) \qquad \varphi_2 \qquad \begin{array}{c} [\varphi_3]_2 \\ \vdots \\ \chi \end{array}}{\chi} \; (cHL_2^2) \tag{6.1.9}$$

However, as the following example from Schroeder-Heister [194] shows, even this splitting strategy does not always suffice for flattening a HL-rule.

Example 6.1.44. *Consider a connective* $c(\varphi_1, \varphi_2, \varphi_3, \varphi_4)$, *with the following I-rules.*

$$
\cfrac{\begin{array}{c}[\varphi_1]_1\\\vdots\\\varphi_2\end{array}}{c(\varphi_1, \varphi_2, \varphi_3, \varphi_4)}\ (cI_1^1) \qquad \cfrac{\begin{array}{c}[\varphi_3]_2\\\vdots\\\varphi_4\end{array}}{c(\varphi_1, \varphi_2, \varphi_3, \varphi_4)}\ (cI_2^2) \qquad\qquad (6.1.10)
$$

(which is $(\varphi_1 \to \varphi_2) \vee (\varphi_3 \to \varphi_4)$*). The harmoniously-generated HL-rule (in simple presentation) is*

$$
\cfrac{c(\varphi_1, \varphi_2, \varphi_3, \varphi_4) \qquad \begin{array}{c}[\varphi_1 \Rightarrow \varphi_2]_1\\\vdots\\\chi\end{array} \qquad \begin{array}{c}[\varphi_3 \Rightarrow \varphi_4]_2\\\vdots\\\chi\end{array}}{\chi}\ (cHL^{1,2}) \qquad\qquad (6.1.11)
$$

However, an attempt to flatten along the line of the previous examples would produce a GE-rule like the following.

$$
\cfrac{c(\varphi_1, \varphi_2, \varphi_3, \varphi_4) \quad \varphi_1 \quad \begin{array}{c}[\varphi_2]_1\\\vdots\\\chi\end{array} \quad \varphi_3 \quad \begin{array}{c}[\varphi_4]_2\\\vdots\\\chi\end{array}}{\chi}\ (cGE^{1,2}) \qquad\qquad (6.1.12)
$$

This rule is incorrect, as an assumption above $\varphi_1 \Rightarrow \varphi_2$ *or* $\varphi_3 \Rightarrow \varphi_4$ *is discharged below the respective assumed HL-rules, and would not be accessible for discharge in a decoupling. This happens to the assumption* ψ *in the derivation of* $(\psi \to \varphi_2) \vee (\psi \to \varphi_4)$ *from* $c(\varphi_1, \varphi_2, \varphi_3, \varphi_4)$, $\psi \to \varphi_2$ *and* $\psi \to \varphi_4$, *presented below. For brevity, the regular* $(\to E)$ *is used, not the* $(\to GE)$ *rule.*

$$
\cfrac{c(\varphi_1, \varphi_2, \varphi_3, \varphi_4) \qquad \cfrac{\cfrac{\cfrac{\psi \to \varphi_1 \quad [\psi]_3}{\varphi_1}\ (\to E)}{\cfrac{\varphi_2}{\psi \to \varphi_2}\ (\to I^3)}\ (\varphi_1 \Rightarrow \varphi_2)_1}{(\psi \to \varphi_2) \vee (\psi \to \varphi_4)}\ (\vee I) \qquad \cfrac{\cfrac{\cfrac{\psi \to \varphi_3 \quad [\psi]_4}{\varphi_3}\ (\to E)}{\cfrac{\varphi_4}{\psi \to \varphi_4}\ (\to I^4)}\ (\varphi_1 \Rightarrow \varphi_2)_2}{(\psi \to \varphi_2) \vee (\psi \to \varphi_4)}\ (\vee I)}{(\psi \to \varphi_2) \vee (\psi \to \varphi_4)}\ (cHL^{1,2})
$$

$$\qquad\qquad (6.1.13)$$

A natural question arises: does *every* connective have an *I*-rule expressible by the above scheme, to guarantee the existence of a Harmoniously-Induced HL-rule? Schroeder-Heister ([194]) argues that a ternary connective $c(\varphi_1, \varphi_2, \varphi_3)$, meaning the same as $\varphi_1 \to (\varphi_2 \vee \varphi_3)$ is not expressible only in term of its sub-formulas, that is, without using neither '$\to$' nor '$\vee$'.

6.2 Validity of arguments

Since PTS proposes an alternative theory of meaning, it also calls for a different notion of *validity* of an argument (captured as a derivation in the meaning-conferring ND-system). Proof-theoretic validity replaces the role of truth by canonical derivability.

Assume some informal, pre-theoretic notion of "*correctness*". The traditional view of validity of an argument is, in Schroeder-Heisters' terminology (e.g., [193], to be further elaborated below), *the transmission of correctness*. This view is inherently connected with the notion of (logical) *consequence*. An argument is valid if it transmits correctness from the assumptions to the conclusion: whenever all the assumptions are correct, so is the conclusion. There is an implicit universal quantification underlying this view, expressed by 'whenever'. The quantification varies with the semantic theory of meaning used.

In Section 6.3, the model-theoretic notion of correctness is reviewed.

6.3   Model-theoretic validity

First, the underlying notions of consequence are reviewed.

6.3.1   Model-Theoretic consequence

Recall that the basic model-theoretic notion is *truth in a model*, denoted $\mathcal{M} \vDash \varphi$, usually defined by recursion on φ. Such a definition captures, according to the model-theoretic theory of meaning, the meaning of φ. Truth of a formula in a model is naturally extended (i.e., conjunctively) to truth of a set Γ of formulas in a model (where here only finite sets are considered), denoted $\mathcal{M} \vDash \Gamma$, by

$$\mathcal{M} \vDash \Gamma \text{ iff } \mathcal{M} \vDash \psi \text{ for every } \psi \in \Gamma$$

Based on this notion, φ being a consequence of Γ, denoted $\Gamma \vDash \varphi$, is defined as follows.

Definition 6.3.46 (model-theoretic consequence).

$$\Gamma \vDash \varphi \text{ iff for every } \mathcal{M} : \text{ if } \mathcal{M} \vDash \Gamma \text{ then } \mathcal{M} \vDash \varphi \qquad (6.3.14)$$

Remarks:

1. The universal quantification expressing 'whenever' is over models.

2. In the order of explanation, the *categorical* concept of truth (in a model) is *prior* to the *hypothetical* concept of consequence. The latter is reducible (according to (6.3.14)) to the latter. This reduction expresses what Schroeder-Heister [192, 193] refers to as *dogmas of standard semantics*:

   • The priority of the categorical over the hypothetical.

   • The transmission view of consequence.

(yet another such dogma is discussed below). A criticism of those dogmas is expressed also by Došen in [31], with insights from category theoretical semantics.

3. As remarked by Schroeder-Heister [192], the notion of consequence in intuitionistic logic adheres to the same dogmas, only rendering the notion of truth with an extra *constructive* flavor.

4. The definition of truth (and, therefore, of consequence) can be parameterized by a parameter expressing explicitly a dependence on an *atomic base*, expressing the correctness (here – truth) of atomic formulas. In the definition above, this dependence is implicit, as a model comes with an interpretation function through which truth of atomic formulas is defined (the base of the recursion defining truth in a model).

An argument from assumptions Γ to a conclusion φ is, then, (model-theoretically) *valid* iff φ is a model-theoretic consequence of Γ.

6.4   Proof-Theoretic Validity

6.4.1   Proof-theoretic consequence

In this section, to be faithful to the literature, I revert back to the view of proof-theoretic consequence as derivability in some formal proof-system, and the desire is to define it without resorting to any model-theory. In the current set-up, we might assume a (single conclusion) natural-deduction proof-system. Suppose such a system $\mathcal{N}$ is given. The definition of φ being a proof-theoretic consequence of Γ, is cast as the $PT-validity$ of $\vdash_{\mathcal{N}} \Gamma : \varphi$. This constitutes yet another dogma of standard semantics.

• The identification of consequence with validity (correctness of inference).

Recall that the traditional justification of proof-theoretic consequence, where the model-theoretic consequence is the yardstick for correctness, is given by the *soundness* of $\mathcal{N}$.

Definition 6.4.47 (soundness). $\mathcal{N}$ *is sound iff*

$$\text{If } \vdash_{\mathcal{N}} \Gamma : \varphi, \text{then } \Gamma \vDash \varphi \tag{6.4.15}$$

A proof of the soundness of $\mathcal{N}$ is usually reduced, inductively, to the soundness of the individual rules of $\mathcal{N}$, that are shown to transmit model-theoretic consequence: If each premise of a rule is a (model-theoretic) consequence of the open assumptions it depends on, so is the conclusion of the rule.

A desirable additional property of $\mathcal{N}$, enhancing its attractiveness, is its completeness.

Definition 6.4.48 (completeness). $\mathcal{N}$ *is* complete *iff*

$$\text{If } \Gamma \vDash \varphi, \text{then } \vdash_{\mathcal{N}} \Gamma : \varphi \qquad\qquad (6.4.16)$$

The question now is, assuming a transmission-of-correctness view of validity is to be preserved, *can proof-theoretic validity be formulated by having a genuine proof-theoretic rendering of correctness?* I next attend to such a view.

6.4.2   The Dummett-Prawitz proof-theoretic validity

This view of proof-theoretic validity originates in Dummett [35] and Prawitz [148, 149, 152], and is elaborated upon in Schroeder-Heister [198]. As the central notion on which PTS is founded is canonical provability, proof-theoretic validity (PT-validity) is taken as the transmission of canonical derivability. This notion of PT-validity accords with the I-view of PTS (cf. Section 1.6.1).

6.4.2.1   A preparatory simplified definition

Before attending to a detailed full presentation, it is useful to start with a preparatory exposition of a simplified case, due to Humberstone ([80], §4.13), to get a feeling of what is involved. Consider an impoverished sublanguage of intuitionistic logic having only conjunction and disjunction as logical operators, with their I/E-rules from NJ (cf. Figure 2.1). Let '$\vdash$' indicate the derivability relation in this impoverished fragment. In view of the special role of I-rules in conferring meaning, consider a derivability relation $\vdash^{I}$, where $\vdash^{I}\Gamma : \varphi$ iff there is a derivation of φ from Γ *without using any E-rules*. In this case, φ is *I-entailed* by Γ. Note that $\vdash^{I}\varphi : \varphi$. Next, suppose some (finite) set of formulas Θ is distinguished so that I-entailment by Θ is taken as the correctness to be transmitted by valid arguments. Call Γ *correct* iff for every $\psi \in \Gamma$, $\vdash^{I}\Theta : \psi$ holds. Based on this, say that $\Gamma : \varphi$ is $PT - valid_0$ iff φ is correct when Γ is.

A natural question to ask at this point is how does $PT - validity_0$ relate to derivations using E-rules. Consider an application of $(\wedge E_1)$, $\dfrac{\Gamma : \varphi \wedge \psi}{\Gamma : \varphi}\ (\wedge E_1)$. Suppose Γ is correct. We would like to reason as follows. The only way to obtain $\vdash^{I}\Theta : \varphi \wedge \psi$ is to apply $(\wedge I)$ to the premises $\vdash^{I}\Theta : \varphi$ and $\vdash^{I}\Theta : \psi$, and hence φ is correct, and therefore $(\wedge E_1)$ transmits correctness. For $(\wedge E_2)$ the argument is similar.

However, there is a caveat here, since it might be the case that $\varphi \wedge \psi \in \Theta$, so that it vacuously I-entails itself, without applying any I-rule. A first reaction might be to exclude application of the identity axiom from counting as I-entailment, enforcing applications of I-rules. But this would exclude any correct argument with atomic conclusion, say $\vdash^{I}\Gamma : p$, from being correct, with the undesired consequence that any $\Gamma : \varphi$ is correct if Γ contains an atomic proposition. Therefore, we restrict Θ to have *atomic propositions only*; anticipating Prawitz's definition below, call Θ an *atomic base*. Members of Θ might be thought of as having proofs external to the ND-system considered.

This results in the following definition of $PT - validity_0$:
$\Gamma : \varphi$ is $PT - valid_0$ iff for every atomic base Θ under which Γ is correct, φ is correct too.
Note that the argument above involving $(\wedge E)$ is now restored. Now φ being a proof-theoretic consequence of Γ is identified with the $PT - validity_0$ of $\Gamma : \varphi$.

We are now in the position of establishing the correctness of *every* argument $\vdash_N \Gamma : \varphi$ (over the impoverished language), thereby abandoning soundness and completeness as the yardstick according to which derivability is justified.

Claim: If $\vdash_N \Gamma : \varphi$, then $\Gamma : \varphi$ is $PT - valid_0$.
Proof: Proceed by induction on the derivation of φ from Γ. Let an arbitrary atomic base Θ be given, and suppose Γ is correct.

identity axiom: Obviously, if $\vdash^I \Theta : \varphi$ then $\vdash^I \Theta : \varphi$.

$(\wedge I)$: Let

$$\mathcal{D} : \quad \frac{\begin{matrix} \mathcal{D}_1 & \mathcal{D}_2 \\ \Gamma : \varphi & \Gamma : \psi \end{matrix}}{\Gamma : \varphi \wedge \psi} \ (\wedge I)$$

By the induction hypothesis, both $\mathcal{D}_1$ and $\mathcal{D}_2$ are $PT - valid_0$. Thus, $\vdash^I \Theta : \varphi$ and $\vdash^I \Theta : \psi$. Therefore,

$$\frac{\Theta : \varphi \quad \Theta : \psi}{\Theta : \varphi \wedge \psi} \ (\wedge I)$$

establishes $\vdash^I \Theta : \varphi \wedge \psi$, the correctness of $\varphi \wedge \psi$, and, hence, the $PT - validity_0$ of $\mathcal{D}$.

$(\vee I)$: Similar.

$(\wedge E)$: Let

$$\mathcal{D} : \quad \frac{\begin{matrix} \mathcal{D}_1 \\ \Gamma : \varphi \wedge \psi \end{matrix}}{\Gamma : \varphi} \ (\wedge E_1)$$

By the induction hypothesis, $\mathcal{D}_1$ is $PT - valid_0$, and therefore $\vdash^I \Theta : \varphi \wedge \psi$. Therefore, $\vdash^I \Theta : \varphi$, establishing the correctness of φ and thereby the $PT - validity_0$ of $\mathcal{D}$. The proof for $(\wedge E_2)$ is similar.

$(\vee E)$: Let

$$\mathcal{D} : \quad \frac{\begin{matrix} \mathcal{D}_1 & \mathcal{D}_2 & \mathcal{D}_2 \\ \Gamma : \varphi \vee \psi & \Gamma, \varphi : \chi & \Gamma, \psi : \chi \end{matrix}}{\Gamma : \chi} \ (\vee E)$$

By the induction hypothesis, $\mathcal{D}_1, \mathcal{D}_2$ and $\mathcal{D}_3$ are $PT - valid_0$. By the $PT - validity_0$ of $\mathcal{D}^I$, $\vdash^I \Theta : \varphi \vee \psi$. W.l.o.g., suppose $\vdash^I \Theta : \varphi \vee \psi$ was inferred by

($\vee I_1$) from $\vdash^I \Theta : \varphi$. Hence φ is correct, and so is Γ, φ. By the $PT - validity_0$ of $\mathcal{D}_2$, χ is correct too, establishing the $PT - validity_0$ of $\mathcal{D}$.

The converse claim, completeness, can be proved[3] too.

$$\mathcal{D}$$
Claim: If an argument $\Gamma : \chi$ is $PT - valid_0$, then $\Gamma : \chi$ it is provable, namely $\vdash \Gamma : \chi$.
Proof: Let Γ be correct. The proof is by induction on the complexity of χ.

$\chi = p$, **atomic:** By the $PT - validity_0$ of $\mathcal{D}$, $\vdash^I \Theta : p$, implying, by atomicity, $p \in \Theta$. Therefore the argument is $p : p$, whereby $\vdash p : p$ as an identity axiom instance.

$\chi = \varphi \wedge \psi$: By $PT - validity_0$, $\vdash^I \Theta : \varphi \wedge \psi$. That is,

$$\frac{\vdash^I \Theta : \varphi \quad \vdash^I \Theta : \psi}{\vdash^I \Theta : \varphi \wedge \psi} \ (\wedge I)$$

By the induction hypothesis, both $\vdash \Gamma : \varphi$ and $\vdash \Gamma : \psi$. Hence,

$$\frac{\vdash \Gamma : \varphi \quad \vdash \Gamma : \psi}{\vdash \Gamma : \varphi \wedge \psi} \ (\wedge I)$$

$\chi = \varphi \vee \psi$: By $PT - validity_0$, $\vdash^I \Theta : \varphi \vee \psi$. W.l.o.g,

$$\frac{\vdash^I \Theta : \varphi}{\vdash^I \Theta : \varphi \vee \psi} \ (\vee I)$$

By the induction hypothesis, $\vdash \Gamma : \varphi$. Therefore,

$$\frac{\vdash \Gamma : \varphi}{\vdash \Gamma : \varphi \vee \psi} \ (\vee I)$$

Let us next consider what happens if this definition of $PT - validity_0$ is attempted to be applied to implication, using the same reasoning. Suppose that Γ is correct, and

$$\frac{\Gamma, \varphi : \psi}{\Gamma : \varphi \rightarrow \psi} \ (\rightarrow I)$$

($\rightarrow I$) is applied . Suppose further that $\Gamma, \varphi : \psi$ is $PT - valid_0$. We need to show $PT - validity_0$ of $\Gamma : \varphi \rightarrow \psi$. To establish the latter, we have to show $\vdash^I \Theta : \varphi \rightarrow \psi$. For that, apparently, $\vdash^I \Theta, \varphi : \psi$ should suffice. The trouble is, that in general φ *need not be atomic*, and Θ, φ need not be an atomic base!

The conclusion is, that a more general definition is needed in order to handle properly I-rules that discharge assumptions.

---

[3]Humberstone [80] proves completeness by a model-theoretic argument.

6.4.3 PT-validity

The main idea of Prawitz is to consider, for the sake of $PT - validity$ (expressing proof-theoretic consequence), open assumptions as *placeholders* for their closed derivations (proofs). By appealing to the reductions (viewed as justifications) and assuming a finite atomic base Θ, the following picture emerges, whereby the $PT - validity$ of an open derivation is perceived as the *transmission of PT - validity of closed derivations*. In the original definition of Prawitz, the atomic base is depicted as a production-system, having rules for deriving an atom from other atoms. I find Humberstone's collection of atomic formulas clearer and initially present the definition accordingly. The extension to general atomic bases, with its justification, is attended to below. Denote the fact that an atomic base Θ' *extends* Θ by $\Theta' \geq \Theta$.

Definition 6.4.49 (PT-validity). *For every atomic base Θ:*

- *A closed derivation of an atomic proposition p is $PT - valid$ iff $p \in \Theta$.*

- *A canonical closed derivation is $PT - valid$ iff its sub-derivations are $PT - valid$.*

- *A closed non-canonical derivation is $PT - valid$ if it reduces to a $PT - valid$ canonical derivation.*

- *An open derivation is $PT - valid$ if every closed derivation resulting from replacing every open assumption by a closed $PT - valid$ derivation thereof w.r.t. any $\Theta' \geq \Theta$ is $PT - valid$ w.r.t. Θ'.*

Returning to the implication and the $(\rightarrow\! I)$-rule, a derivation

$$\dfrac{\begin{array}{c}[\varphi]_i\\ \mathcal{D}\\ \psi\end{array}}{\varphi\rightarrow\psi}\ (\rightarrow\! I^i)$$

is $PT - valid$ iff for every $PT - valid$ closed derivation $\overset{\mathcal{D}_1}{\varphi}$ w.r.t. any $\Theta' \geq \Theta$, the derivation

$$\begin{array}{c}\mathcal{D}_1\\ \varphi\\ \mathcal{D}\\ \psi\end{array}$$

(in my notation, $\ \mathcal{D}[\varphi := \overset{\mathcal{D}_1}{\varphi}]\ \atop \psi$)

is $PT - valid$ w.r.t. Θ'. This amounts to the $PT - validity$ of the open subderivation

$$\begin{array}{c}\varphi\\ \mathcal{D}\\ \psi\end{array}$$

Note that φ itself may have occurrence of '$\rightarrow$', therefore $\mathcal{D}$ may need to apply ($\rightarrow E$); the reduction, that embodies the harmony of the $\rightarrow$-rules, takes care of such applications to derive only $\vdash^I$-justified conclusions.

Why does the definition of $PT-validity$ of open derivations have to appeal to extensions Θ' of the atomic base Θ?
The answer is, in order to avoid vacuous $PT-validity$ in case an open assumption has no $PT-valid$ closed derivation of it under the given atomic base. In that case, there is nothing to substitute for the open assumption, and the derivation is vacuously $PT-valid$, without having really been "tested". However, by extending the atomic base, $PT-valid$ closed derivations of the open assumptions may emerge, and once substituted for that assumption the whole derivation is put to real test.

Example 6.4.45. *Consider again the one-step derivation $\dfrac{\varphi\wedge\psi}{\varphi}$ ($\wedge E$) where $\varphi\wedge\psi$ is contingent (non-tautological). To test this derivation for $PT-validity$, all the closed derivations of $\varphi\wedge\psi$ need to be substituted for the open assumption, and the resulting closed derivation needs to be tested for $PT-validity$. Now, it may well be the case that under the given atomic base, there are no closed derivations for $\varphi\wedge\psi$. Assume temporarily that both φ and ψ are themselves atomic. By extending the atomic base with both, there are now closed derivations of $\varphi\wedge\psi$ to substitute for the open assumption.*

But what happens if φ and ψ are not atomic?
Then, the atomic base cannot be directly extended with φ and ψ. But, if we had production-rules allowing the derivation of both φ and ψ from some atomic sentences, then again by an extension a closed derivation of $\varphi\wedge\psi$ *inside the production-system* becomes available for substitution for the open assumption.

Example 6.4.46. *Suppose, in the above example, that φ is $p\rightarrow q$ and ψ is $r\rightarrow s$, where p,q,r,s are atomic. Then, to get the desired closed derivation for $\varphi\wedge\psi$, we can add production rules $\dfrac{p}{q}$ and $\dfrac{r}{s}$. The closed derivation is*

$$\dfrac{\dfrac{\dfrac{[p]_1}{q}}{p\rightarrow q}\ (\rightarrow I^1)\quad \dfrac{\dfrac{[r]_2}{s}}{r\rightarrow s}\ (\rightarrow I^1)}{(p\rightarrow q)\wedge(r\rightarrow s)}\ (\wedge I)$$

Denote by S an atomic base that may include also production-rules for atomic sentences. To achieve the above goal, we need to extend the first clause in the definition of $PT-validity$ as follows.

- Any[4] derivation in S is $PT-valid$.

---

[4] In the original definition by Prawitz, only *closed* S-derivations were considered $PT-valid$. This is clearly insufficient.

I have only considered the propositional case here. In the first-order case, open assumptions may contain free variables. Those are also seen as placeholders, this time for closed terms. The definition of $PT - validity$ in the first-order case involves substitutions of closed terms closing open formulas, similar to the substitution of closed derivations closing the open assumptions. An example of a rule in a first-order system is

$$\frac{F(x)}{G(x,x)}$$

where F is a monadic predicate symbol and G is dyadic predicate symbol.

I skip the technical details, but the alert reader should not miss the connection of the various freshness side-conditions on rules with free variables.

In general, $PT - validity$ is monotonic, in the sense that extending the atomic base Θ to a larger Θ' preserves $PT - validity$ w.r.t. Θ, but can render more derivations $PT - valid$.

To end this discussion, let me mention that Prawitz presents a certain generalization of $PT - validity$, pertaining to a more abstract notion of a derivation, called a *proof-structure*, that consists of trees of formulas not necessarily generated by a system of I/E-rules. Instead of reductions, it considers an abstract collection of $\mathcal{J}$ of justifications, giving rise to abstract "reductions". Definition 6.4.49 is then mimicked using those abstractions. As this abstraction is not directly concerned with meanings of logical operators, I will leave out further discussion of it, that may be found in Prawitz's papers cited about, and also in Schroeder-Heister, e.g., in [198] and [189].

6.4.4   Definitional reasoning

In a series of papers [71, 184, 186, 187, 188, 189, 193], Hallnäs and Schroeder-Heister introduce a different framework for defining proof-theoretic validity, escaping the three dogmas mentioned above. This is framework for *definitional reasoning*, of which logic can be viewed as a special case. While in the framework considered so far atomic sentences had an external source of proof-theoretic meaning, here atoms are internalized by having *definitions* within the framework itself. The presentation here follows the above references. I first present the general framework, and then the definition of the usual logical operators using this framework.

6.4.4.1   Presenting definitional reasoning

Let A, B (possibly indexed) range over a set of *atoms*, not further specified at this stage. Let Δ (possibly indexed) range over finite collections of atoms. A *clause* has the form[5] $A \Leftarrow \Delta$, where A is the *head* of the clause and Δ its *body*. A clause with an

---

[5]The direction of the arrow originates from a tradition in Logic Programming.

empty body is presented as $A \Leftarrow$.

Definition 6.4.50 (definitions).   • *A definition* **D** *has the form*

$$\mathbf{D} : \begin{cases} A_1 \Leftarrow \Delta_1 \\ \vdots \\ A_n \Leftarrow \Delta_n \end{cases} \tag{6.4.17}$$

The A_is are not necessarily distinct. Any Δ_i may be empty.

• *For an atom A, the collection* **D**(A) *of its* defining clauses *(in* **D***) is*

$$\mathbf{D}(A) =^{df\cdot} \{\Delta_i \mid A_i \leftarrow \Delta_i \in \mathbf{D} , A_i = A\} \tag{6.4.18}$$

Possibly **D**$(A) = \varnothing$.

Here atoms are defined in terms of other atoms. Note that an atom A may have more than one defining clause in **D**(A), understood disjunctively. A further generalization, allowing for some more complex objects taking part in the definition of an atom is considered later. The discussion here is on a "propositional level"; atoms with variables are considered later.

Equational reasoning is embodied in the following two principles of *definitional closure* and *definitional reflection*. A relation of *definitional consequence*, denoted by a sequent-like object $\Delta \Vdash_{\mathbf{D}} A$ is associated with a definition **D**, expressing that **D** can generate that sequent (in a way specified below). The use of '$\Vdash$' is to emphasize the difference between definitional consequence and the model-theoretic consequence '$\models$' and provability/derivability in a formal system '$\vdash$'. As usual, when **D** is clear from context, it is omitted. Note that as a special case, it is possible that $\Vdash_{\mathbf{D}} A$ (see Example 6.4.48 below). This is naturally extended to $\Vdash_{\mathbf{D}} \Gamma$ holding iff $\Vdash_{\mathbf{D}} A$ for every A in Γ.

For the formulation of definitional closure, the following abbreviation is useful. Suppose that $\Delta_i = A_{i_1}, \cdots, A_{i_m} \in \mathbf{D}(A)$. Abbreviate $\Delta \Vdash A_{i_1}, \cdots, \Delta \Vdash A_{i_m}$ as $\Delta \Vdash \Delta_i$, and $A_{i_1} \Vdash B, \cdots, AI_{m_i} \Vdash B$ as $\Delta_i \Vdash B$.

First, $A \Vdash A$ for every atom A.

Definition 6.4.51 (definitional closure).

$$\frac{\Delta \Vdash \Delta_i}{\Delta \Vdash A} \; (\Vdash \mathbf{D}), \; 1 \le i \le m \tag{6.4.19}$$

This principle means that A is a definitional consequence of Δ whenever each member of a defining clause of A is a definitional consequence of Δ.

Definition 6.4.52 (definitional reflection).

$$\frac{\{\Delta, \Delta_i \Vdash B : A \Leftarrow \Delta_i \in \mathbf{D}(A)\}}{\Delta, A \Vdash B} \; (\mathbf{D} \Vdash) \tag{6.4.20}$$

This principle means that whatever is a definitional consequence of the body of every defining clause of A (in a context Δ) is a definitional consequence of A itself (in the same context Δ). Note that in contradistinction from the ND-related "forward" reasoning (from assumptions to conclusions), here assumptions are introduced to the left of 'ⵏ⊢' during derivation. As its name suggests, the principle embodies a reflection over the whole definition (of a given atom A). It is in the spirit of the way inductive definitions are formulated, expressing exhausting *all* the defining clauses by "and nothing else ...". such an exhausting condition is a typical characteristic of definitions in general. Note that in case A has no defining clauses, namely $\mathbf{D}(A) = \emptyset$, what results is $A \Vdash B$ for any B, a kind of 'ex falso quodlibet'.

It is also possible to present the above rules for definitional reasoning in an ND-like presentation, introducing/eliminating atoms (e.g., as in Schroeder-Heister [184]), as shown below (in (6.4.27), (6.4.28), after the inclusion of variables).

Any resemblance to the inversion principle is not accidental ...

Note the symmetry inherent in those two principles, resembling the symmetric structure of structure in sequent calculi (cf. Section 6.5.2). As emphasized by Schroeder-Heister in [189]), definitional reasoning with those two principles is not "forward oriented", from assumptions to consequence; rather, assumptions (occurring to the left of 'ⵏ⊢') are added *during* reasoning. Furthermore, they are added *according to their meaning* as determined by their definitional clauses.

Thus, in definitional reasoning the dogma of viewing assumptions as placeholders is avoided!

Example 6.4.47. *Consider, for example, the following definition* $\mathbf{D}_0$. *Let* $\{a, b, c, d, e, f, g, h\}$ *be a collection of atoms.*

$$\mathbf{D}_0 : \begin{cases} a \Leftarrow b, c \\ b \Leftarrow \\ c \Leftarrow \\ d \Leftarrow a, b \\ e \Leftarrow a \\ e \Leftarrow b \\ f \Leftarrow a \\ g \Leftarrow a \\ g \Leftarrow b \end{cases} \qquad (6.4.21)$$

1.

Example 6.4.48. $\Vdash_{\mathbf{D}_0} d :$ *a closed derivation.*

$$\frac{\dfrac{\overline{b} \quad \overline{c}}{a} \ (\Vdash \mathbf{D}) \quad \overline{b}}{d} \ (\Vdash \mathbf{D})$$

2.

Example 6.4.49. $f, g \Vdash_{\mathbf{D}_0} e$: *an open derivation.*

$$\dfrac{\dfrac{\overline{b} \ \ \overline{c}}{a} \ (\Vdash \mathbf{D}) \quad f \quad \dfrac{g}{h} \ (\Vdash \mathbf{D})}{e} \ (\Vdash \mathbf{D})$$

3.

Example 6.4.50. *Note that* $\mathbf{D}_0(g) = \{g \Leftarrow a, \ g \Leftarrow b\}$.
$g \Vdash_{\mathbf{D}_0} e$:

$$\dfrac{\dfrac{a \Vdash a}{a \Vdash e} \ (\Vdash \mathbf{D}) \quad \dfrac{b \Vdash b}{b \Vdash e} \ (\Vdash \mathbf{D})}{g \Vdash e} \ (\mathbf{D} \Vdash)$$

Note that e is a definitional consequence of the body of every defining clause of g. The definitional reflection yields the result that e is a definitional consequence of g itself.

6.4.4.2   Definitional reasoning involving variables

So far, only propositional-level atoms and defining clauses have been considered. The ideas naturally extend to a 1st-order variant, involving variables in atoms and defining clauses, following [188].   Let x, y, z (possibly indexed) range over a countable collection of variables, and a, b, c (possibly indexed) over a countable collection of constants. It is also possible to include a countable collection of function symbols, ranged over by f, g, h (possibly indexed). If functions symbols are included, their arity is assumed given too. *terms* are defined as usual. Atoms now have the form $A(t_1, \cdots, t_m)$ (with each t_i a term), so the As are treated as predicate symbols (also with given arities). Atoms and terms are *ground* whenever they contain no variables; otherwise $var(A)$ denotes the collection of variables in A.

Definition 6.4.53 (substitutions).   • *A substitution σ is a partial mapping with a finite domain dom_σ, mapping variables to terms. I will display substitutions as finite collections in the form $\sigma = [x_1 := t_1, \cdots, x_s := t_s]$, where $dom_\sigma = \{x_1, \cdots, x_s\}$.*

- *If $dom_{\sigma_1} \cap dom_{\sigma_2} = \varnothing$, the* union *$\sigma_1 \cup \sigma_2$ is defined with $dom_{\sigma_1 \cup \sigma_2} = dom_{\sigma_1} \cup dom_{\sigma_2}$.*

- *If each t_i in the range of σ is itself a variable, σ is a* renaming>

- *A* variant *of a syntactic object (term, atom, clause or a whole definition) is the result of applying to it a renaming substitution. An* instance *of a syntactic object is the result of applying to it an arbitrary substitution.*

The definition of $\mathbf{D}(A)$ is revised as follows.

$$\mathbf{D}(A) =^{df\cdot} \{\Delta\sigma \mid B \Leftarrow \Delta \in \mathbf{D}, \, B\sigma = A\} \tag{6.4.22}$$

Next, the definition of definitional closure and definitional reflection are adapted to handle variables. The notations $\Delta \Vdash \Delta_i$, $\Delta_i \Vdash \Delta$ are naturally extended to $\Delta \Vdash \Delta_i\sigma$, $\Delta_i\sigma \Vdash \Delta$.

Definition 6.4.54 (definitional closure with variables).

$$\frac{\Delta \Vdash \Delta_i\sigma}{\Delta \Vdash A\sigma} \; (\Vdash \mathbf{D}) , \, 1 \leq i \leq m \tag{6.4.23}$$

Definition 6.4.55 (definitional reflection with variables).

$$\frac{\{\Delta, \Delta_i\sigma \Vdash B : \Delta_i \in \mathbf{D}(A), A = B\sigma\}}{\Delta, A\sigma \Vdash B} \; (\mathbf{D} \Vdash) \tag{6.4.24}$$

under the side-condition $\mathbf{D}(A\sigma) \subseteq (\mathbf{D}(A))\sigma$. The side-condition ensures that $(\mathbf{D} \Vdash)$ is applicable only in case the body of a clause does not contain variables in the head of that clause.

The side-condition ensures the closure under substitutions of $\mathbf{D} \Vdash)$. By specifying A further via applying additional substitutions, its definition does not extend. To see the effect of violating the side-condition, consider the following example.

Example 6.4.51. *Consider the definition*

$$\mathbf{D} : \begin{cases} P(b) \Leftarrow Q(x) \\ Q(a) \Leftarrow \end{cases} \tag{6.4.25}$$

The body of the first clause contains the variable x not present in the ground head of that clause. Ignoring the side-condition leads to the derivation

$$\frac{Q(b) \Vdash Q(b)}{P(b) \Vdash Q(b)} \; (\mathbf{D} \Vdash)$$

resulting in $P(b)$ being derivable in spite of $Q(b)$ not being derivable.

Schroeder-Heister [188] considers also a more general formulation of $(\mathbf{D} \Vdash)$, referred to as $(\mathbf{D} \Vdash)_\omega$, that allows universal generalization from ground instances.

Definition 6.4.56 (most general unifier). • *A unifier of two syntactic objects o_1 and o_2 is a substitution σ s.t. $o_1\sigma = o_2\sigma$.*

• *A unifier σ of o_1 and o_2 is* most general, *denoted by $mgu(o_1, o_2)$, if any other unifier σ' of o_1 and o_2 is a substitution instance of σ.*

Definition 6.4.57 (generalized definitional reflection with variables).

$$\frac{\{\Delta, \Delta_i\sigma \Vdash C\sigma : B \Leftarrow \Delta_i \in \mathbf{D}', \sigma = mgu(A, B)\}}{\Delta, A \Vdash C} \; (\mathbf{D} \Vdash)_\omega \tag{6.4.26}$$

where $\mathbf{D}'$ is a renamed variant of $\mathbf{D}$ and the same side-condition on variables as in the special case is kept. Note that $mgu(A, B)$ need not exist, in which case the rule is inapplicable.

Example 6.4.52. *From Schroeder-Heister [188]: Suppose the only defining clauses of a predicate P are*

$$\begin{cases} P(a) \Leftarrow Q(a) \\ P(b) \Leftarrow Q(b) \end{cases}$$

Then, $P(x) \Vdash Q(x)$ *can be inferred using* $(\mathbf{D} \Vdash)_\omega$ *but not using* $(\mathbf{D} \Vdash)$*. The two mgus applicable to the two premises are* $\sigma_1 = [x := a]$ *and* $\sigma_2 = [x := b]$*.*

For a more thorough study of this generalized rule see Schroeder-Heister [188].

As mentioned above, it is also possible to present the definitional reasoning rules in an ND-like system.

$$\frac{\Delta_i\sigma}{B\sigma}\,(A\,I), \quad A \Leftarrow \Delta_i \in \mathbf{D}(A),\ A = B\sigma \qquad (6.4.27)$$

$$\frac{A\sigma \quad \overset{[\Delta_1\sigma]_1}{\underset{B}{\vdots}} \quad \cdots \quad \overset{[\Delta_m\sigma]_m}{\underset{B}{\vdots}}}{B}\,(A\,E^{1,\cdots,m}), \quad A \Leftarrow \Delta_i \in \mathbf{D}(A), \qquad (6.4.28)$$

As stressed in Schroeder-Heister [184], the rules introduce and eliminate atoms, not predicates. Thus, if A is $p(t)$ (for some term t), it is $p(t)$ that is introduced and eliminated, not p.

6.4.4.3   Definitional reasoning with logical constants

Following Schroeder-Heister [193], it is possible to put the use the machinery of definitional reasoning to define logical constants and reasoning with them. The definition below defines the propositional Intuitionistic Logic.

$$\mathbf{D}_{log} : \begin{cases} p \Leftarrow p \\ \varphi \wedge \psi \Leftarrow \varphi, \psi \\ \varphi \vee \psi \Leftarrow \varphi \\ \varphi \vee \psi \Leftarrow \psi \\ \varphi \rightarrow \psi \Leftarrow \varphi \Rightarrow \psi \\ \neg\varphi \Leftarrow \varphi \mapsto \bot \end{cases} \qquad (6.4.29)$$

The way to think of the logical formulas as atoms in the current sense is to consider the main operator as mapping atoms to atoms. Thus, $\varphi \wedge \psi$ is better viewed as $\wedge(\varphi, \psi)$.

Note the absence of a defining clause for '⊥'. For atomic propositions p, the defining clause leaves their definition as unspecific. If this clause were absent, atomic propositions would be indistinguishable from '⊥'. For the sake of $PT - validity$, additional defining clauses are needed, in order to capture the atomic base.

There is a clear correspondence between instances of definitional closure based on $\mathbf{D}_{log}$ and the I-rules of NJ (cf. Section 2.1.5), as well as between definitional reflection instances and GE-rules for the intuitionistic constants (cf. Section 3.6.1.5). The former can be easily obtained via $\mathbf{D}\vdash$, while the latter, in GE-form, via $\vdash\mathbf{D}$.

6.4.5   Detaching consequence from validity

In [193], Schroeder-Heister proposes a different approach of validity and consequence (within the proof-theoretic approach), detaching the former from the latter and abandoning altogether the transmission view of consequence, the second and third dogmas mentioned above. The hypothetical notion of consequence is defined *directly* without reference to the categorical notion, thereby abandoning also the first dogma, the priority of the categorical over the hypothetical. Categorical validity will turn out to be no more than a limit case of hypothetical consequence, when there are no open assumptions. Yet another difference is the dependence on atomic bases, that is justified in logical systems, but not, as we saw, in general definitional reasoning systems. Thus, the view of consequence emerging here is more general than logical consequence, the latter being merely a special case (though a highly important one).

As the definition of consequence, definitional consequence $\Vdash_\mathbf{D}$ is chosen. Consequence and correctness of inference would be identified in case

$$\Gamma \Vdash_\mathbf{D} A \Leftrightarrow (\Vdash_\mathbf{D} \Gamma \text{ implies } \Vdash_\mathbf{D} A) \qquad (6.4.30)$$

holds.

- The definition $\mathbf{D}$ is *total* in case the direction

$$\Gamma \Vdash_\mathbf{D} A \Rightarrow (\Vdash_\mathbf{D} \Gamma \text{ implies } \Vdash_\mathbf{D} A) \qquad (6.4.31)$$

  holds. It means that consequence ensures validity, namely correctness of inference.

- The definition $\mathbf{D}$ is *complete* in case the direction

$$\Gamma \Vdash_\mathbf{D} A \Leftarrow (\Vdash_\mathbf{D} \Gamma \text{ implies } \Vdash_\mathbf{D} A) \qquad (6.4.32)$$

  holds. It means that validity (correctness of inference) establishes consequence.

By detaching validity from consequence, it is possible to formulate conditions under which totality or completeness hold, but neither need always hold.

A counterexamples to totality arises from non-well-founded (circular) definitions, leading to well-known *paradoxes*. Such an example from Schroeder-Heister [193] considers a definition

$$\mathbf{D}: \left\{ a \Leftarrow (a{\rightarrow}\bot) \right.$$

defining a by $\neg a$. It can be shown that both $\Vdash_{\mathbf{D}} a$ and $\Vdash_{\mathbf{D}} \bot$ hold, but this is impossible since '$\bot$' has no defining clauses. See Schroeder-Heister [193] for a discussion of two sufficient conditions for the totality of $\mathbf{D}$.

A simple counterexample to completeness, also from Schroeder-Heister [193], is the following definition.

$$\mathbf{D}: \begin{cases} A \Leftarrow A \\ B \Leftarrow B \end{cases}$$

Here $\Vdash_{\mathbf{D}} A$ implies $\Vdash_{\mathbf{D}} B$ holds vacuously, since $\Vdash_{\mathbf{D}} A$ is not derivable. However, $A \Vdash_{\mathbf{D}} B$ is not derivable. See Schroeder-Heister [193] for a discussion of the definition of Intuitionistic Logic in Section 6.4.4.3.

Consequently, validity and inference are mutually independent!

6.5   Basing Proof-Theoretic Semantics on sequent-calculi

6.5.1   Introduction

The development of PTS in its mainstream is based on ND as its underlying deductive formalism, reflecting "use". However, there were some other voices, and basing PTS on *Sequent Calculi* (SC), also invented by Gentzen [61], was also advocated, though not extensively, and not with fully worked out details. Clearly, if the underlying deductive system is changed, so are also the adequacy criteria for qualifying as meaning-conferring. In this section, I survey several such approaches. First, SCs are briefly reviewed.

6.5.2   Sequent Calculi

The object languages and sequent structure of SCs are the same as those for ND-systems, sequents having typically the form $\Gamma : \Delta$. However, the way rules are conceived is different. In contrast to the "forward" reasoning of ND-systems, from (possibly discharged) assumptions towards a (single) conclusion, where the rules are I/E-rules, in SCs the derivation is "inside-out", and the operational rules are only I-rules, but of two kinds:

- R-rules, that introduce a formula to the right of ':', as a conclusion. The R-rules naturally correspond to I-rules in ND-systems.

- L-rules, that introduce a formula to the left of ':', as an assumption. The L-rules correspond to "reversed" E-rules.

In addition, there are explicit structural rules (including (cut), desirably admissible), cf. Section 1.5.2, and the identity axiom, that in this setting guarantees that an operator dominating a formula has the same meaning whether positioned left of ':' (in a sequent), or right of ':'. As emphasized by Schroeder-Heister [190], L-rules introduce assumptions *according to their meaning*, in contrast to the uniform way assumptions are introduced in ND-systems.

Example 6.5.53. *Below are the SC-rules for implication for IL. In this SC, all the Δs (right contexts) contain at most one formula).*

$$\frac{\Gamma : \varphi \quad \Gamma, \psi : \chi}{\Gamma, (\varphi \to \psi) : \chi} \ (\to L) \quad \frac{\Gamma, \varphi : \psi}{\Gamma : (\varphi \to \psi)} \ (\to R) \tag{6.5.33}$$

Derivations are again trees of sequents and have no simple presentation, only logistic one. A notable difference between trees embodying ND-derivations and trees embodying SC-derivations is the form of the leaves.

ND: Leaves are (open) assumptions.

SC: Leaves are instances of identity axioms.

6.5.3 Motivation and adequacy criteria

In this section, I survey the proposals for basing PTS on SC rather than on ND, indicating the motivation for the proposal as well as its suggestion for adequacy for qualification as meaning-conferring. The need for such criteria is clear, as the "disturbing" *tonk* can be defined in a SC as follows, with a similar effect of trivializing the derivability relation.

$$\frac{\psi, \Gamma : \chi}{\varphi \ tonk \ \psi, \Gamma : \chi} \ (tonk \ L) \quad \frac{\Gamma : \varphi}{\Gamma : \varphi \ tonk \ \psi} \ (tonk \ R) \tag{6.5.34}$$

In multiple-conclusions SC, the definition is

$$\frac{\psi, \Gamma : \Delta}{\varphi \ tonk \ \psi, \Gamma : \Delta} \ (tonk \ L) \quad \frac{\Gamma : \varphi, \Delta}{\Gamma : \varphi \ tonk \ \psi, \Delta} \ (tonk \ R) \tag{6.5.35}$$

The trivializing derivation is

$$\frac{\dfrac{\overline{\varphi : \varphi} \ (id)}{\varphi : \varphi \ tonk \ \psi} \ (tonk \ R) \quad \dfrac{\overline{\psi : \psi} \ (id)}{\varphi \ tonk \ \psi : \psi} \ (tonk \ L)}{\varphi : \psi} \ (cut) \tag{6.5.36}$$

6.5.3.1   Cut-admissibility as the adequacy criterion

An early proposal to base PTS on SC rather than on ND is by Kramer [91]. Kramer rejects Dummett's and Prawitz's theory of meaning because of two main reasons.

- A theory of meaning should not be based either on grounds (for assertion) or on outcomes of assertion (using my previous nomenclature, being either an I-view or an E-view). Rather, such a theory should put those two aspects of use on equal footing (C-view, in my terminology).

- A theory of meaning should not attempt to settle the rivalry between different logics, seen by him as a pragmatic question. The bias of their theory towards IL is unwarranted.

The main adequacy criterion proposed by Kremer as embodying harmony – here between L-rules and R-rules – is is (cut)-admissibility, guaranteeing conservative extension. He relativizes this absolute notion of harmony by inducing a partial order on rules according to the class of languages they extend conservatively, opting for being maximally harmonious, rejecting the $tonk$-rules as minimally harmonious, conservatively extending the trivial derivability relation. Clearly, (cut) is not admissible in the presence of the $tonk$-rules. E.g., the application of (cut) in (6.5.35) cannot be eliminated.

An additional adequacy criterion on rules to qualify as meaning-conferring is the reducibility of the structural rules of Weakening and the identity axiom, when applied to formulas with the defined connective, to formulas without that connective. This idea originates from Hacking [69]. Such reducibility guarantees that the defined connective can only be introduced via its defining rules, so that those rules exhaust its meaning.

Another proposal for basing PTS on SC rather than on ND is by Wansing [231]. He also regards (cut)-admissibility and reducibility of the id-axiom as criteria for rules to qualify as meaning-conferring (though he doubts the role of (W)). Wansing also stresses about some properties of the form rules should have to count as adequate. In particular, they should be (stated in previous terms) *single ended* (cf. Section 1.5.2), separable (cf. Definition 1.5.5) and pure (cf. Section 1.5).

6.5.3.2   The substructurality effect

Paoli in [135, 136] advocates the use of SC as the underlying formal system for PTS for two main reasons, both related to structural rules and the "substructural era" in logic.

- The ability to recognize a sharpening of the meanings of the classical connectives, disambiguating them according to whether the rules are additive or multiplicative. Other names for this distinction are lattice vs. group operations, and extensional vs. intensional rules. This distinction is transparent in the form of the operational rules, a form rendering them inequivalent in the absence of certain structural rules. Since the latter are usually "absorbed" within the operational rules in ND-systems, this important distinction, while expressible, is much less transparent.

- Another issue made more transparent when working with SC is a distinction between *local meaning*, defined solely in terms of the operational rules, and *global meaning*, determined by the totality of derivable sequents, thereby depending also on the structural rules. This distinction underlies Paoli's [135] conception of rivalry between logics, whereby rival logics can share the local meanings of operators, differing in their global meanings.

6.5.3.3   Canonical SC-rules

In [3] and [2], Avron and Lev shifts the $tonk$-debate to SC, claiming it provides better means for determining meaning (to include also CL) better amenable to precise mathematical formulation of adequacy criteria for determining meaning. While the meaning determined is model-theoretic, there is still interest in those criteria as determining proof-theoretic meaning according to the PTS-programme.

The first step is to identify a special form of rules, referred to as a *canonical rules* by Avron and Lev [3], in terms of which the adequacy criterion is to be formulated. The original notation is slightly modified.

Definition 6.5.58 (clause).   *1. A sequent $\Gamma : \Delta$ is a* clause *iff all members of Γ, Δ are atomic propositions.*

2. A clause is a positive Horn clause *iff $\Delta = \{q\}$ (a singleton).*

3. A clause is a negative Horn clause *iff $\Delta = \varnothing$.*

Definition 6.5.59 (canonical rule).   *1. R for an operator '$*$' has the form*

$$\frac{\{\Gamma_i : \Delta_i\}_{1 \leq i \leq m}}{\psi} \ (R)_{, \ m \geq 0}$$

where:

- *The conclusion ψ has one of the following two forms:*
  - *(a)* $*(p_1, \cdots, p_n)$:
  - *(b)* : $*(p_1, \cdots, p_n)$
- *For each $1 \leq i \leq m$, $\Gamma_i : \Delta_i$ is a (non-empty) clause, with $\Gamma_i, \Delta_i \subseteq \{p_1, \cdots, p_n\}$.*

2. *An* application *of R when ψ is $*(p_1, \cdots, p_n)$: has the form*

$$\frac{\{\Gamma, \Gamma_i^* : \Delta_i^*, \Delta\}_{1 \leq i \leq m}}{\Gamma, *(\varphi_1, \cdots, \varphi_n) : \Delta} \ (R^*)$$

where Γ^*, Δ^* are obtained from Γ, Δ, respectively, by substituting φ_i for p_i. The application when ψ is $*(p_1, \cdots, p_n)$ is defined analogously.

From this point, single-conclusion sequents are assumed, taking Δ as $\{\psi\}$.

Definition 6.5.60 (canonical L/R-rules). *1. A canonical R-rule is a canonical rule with all premises positive Horn clauses and conclusion : $*(p_1, \cdots, p_n)$.*

2. *A canonical L-rule is a canonical rule with all premises Horn clauses (either positive or negative) and conclusion $*(p_1, \cdots, p_n)$:.*

3. *An SC is canonical iff all its operative rules are canonical.*

The notion of application of L/R-rules is the natural adaptation of that of a general canonical rule. Below are examples of canonical L/R-rules.

Example 6.5.54. Conjunction:

$$\frac{p_1, p_2 :}{p_1 \wedge p_2 :} \ (\wedge L_c) \qquad \frac{: p_1 \quad : p_2}{: p_1 \wedge p_2} \ (\wedge R_c)$$

with their (single-conclusion) applications (omitting the '$*$' on the rule's name)

$$\frac{\Gamma, \varphi_1, \varphi_2 : \psi}{\Gamma, \varphi_1 \wedge \varphi_2 : \psi} \ (\wedge L_c) \qquad \frac{\Gamma : \varphi_1 \quad \Gamma : \varphi_2}{\Gamma : \varphi_1 \wedge \varphi_2} \ (\wedge R_c)$$

Tonk:

$$\frac{p_2 :}{p_1 \ tonk \ p_2 :} \ (tonkL_c) \qquad \frac{: p_1}{: p_1 \ tonk \ p_2} \ (tonkR_c)$$

with the (single-conclusion) applications as in (6.5.34).

Now comes the definition of the adequacy criterion, called *coherence*.

Definition 6.5.61 (coherent SC). *A canonical SC is coherent iff for every two rules*

$$\frac{S_1}{*(p_1, \cdots, p_n) :} \ (*L_c) \qquad \frac{S_2}{: *(p_1, \cdots, p_n)} \ (*R_c)$$

with S_1, S_2 collections of clauses, $S_1 \cup S_2$ is (classically) inconsistent.

according to Avron, the coherence condition replaces the balance between I/E-rules in ND by a weaker *non-conflict* condition between L/R-rules in SC.

Example 6.5.55. *Returning to Example 6.5.54, it is easy to see that the rules for* $\wedge$ *are coherent, while those of tonk are not.*

$\wedge$**:** *The derivation of the empty clause:*

$$\cfrac{:p_2 \quad \cfrac{:p_1 \quad p_1,p_2:}{p_2:}\;(cut)}{:}\;(cut)$$

tonk: *The union of the premises is the (classically) consistent set of clauses* $\{p_2 :, : p_1\}$.

In Avron and Lev [3], coherence is proved as necessary and sufficient for (*cut*)-admissibility.

6.5.3.4 Bidirectionality

Schroeder-Heister [190] also advocates basing PTS on a formal system that is *SC*-like, though need not be an actual *SC*. He refers to such formalisms as *bidirectional*. In case the objects on which the formalism operates are sequents, the rules are *symmetrical*, as they introduce assumptions in both sides of ':'; more generally, assumptions are introducible in the various available positions.

The main advantage of a bidirectional proof-system is its more adequate treatment of hypothetical reasoning. Assumptions are not place-holders (for closed derivations – cf. Section 6.4.3); rather, in whichever position they are introduced, they are introduced *according to their meaning*.

The actual formalism Schroeder-Heister endorses is a *bidirectional ND*-system, in which all the *E*-rules are *GE*-rules with the following additional property:

- The major premise is always an assumption.

Adhering to this extra feature renders the *GE*-rules independent of the way such a formula is introduced. They are viewed as ways of introducing a complex assumption according to its meaning, forming a definitional step. This view imposes a symmetry on asserting and assuming. Note that this kind of consideration leads naturally to definitional reasoning (cf. Section 6.4.4).

As for the adequacy criteria for rules to qualify as meaning-conferring, the stress is on *locality* of such criteria, not depending on the whole proof-system, and not relying on reductions. Therefore, a criterion like conservative extension is rejected, while criteria like uniqueness (cf. Section 3.5), or its "cousin" local-completeness (local inversion) are kept.

Example 6.5.56. *Consider as an example the rule for conjunction as presented in Schroeder-Heister [190]. The standard SC L-rules for conjunction are the following.*

$$\frac{\Gamma, \varphi : \chi}{\Gamma, \varphi \wedge \psi : \chi} \ (\wedge_1 L) \qquad \frac{\Gamma, \psi : \chi}{\Gamma, \varphi \wedge \psi : \chi} \ (\wedge_2 L) \tag{6.5.37}$$

This rule can be interpreted as follows. If χ is derived by assuming either conjunct, then it is derived from assuming the conjunction, as if discharging the assumption of the conjunct. Recasting the same view as a GE-rule in a bidirectional ND-system leads to the following formulation.

$$\frac{\begin{matrix} & [\varphi]_i \\ & \vdots \\ \varphi \wedge \psi & \chi \end{matrix}}{\chi} \ (\wedge_1^i GE) \qquad \frac{\begin{matrix} & [\psi]_i \\ & \vdots \\ \varphi \wedge \psi & \chi \end{matrix}}{\chi} \ (\wedge_2^i GE) \tag{6.5.38}$$

6.6 Reductive harmony

6.6.1 Introduction

In a series of recent papers by Schroeder-Heister and Olkhovikov [132], [196] and [195], a new approach to the proof-theoretic semantics (PTS) for logical constant is advanced. This new approach to PTS is coined *reductive*, in contrast to the prevailing approach (as presented in Chapter 3), referred to as *foundational*. Two essential differences between the two approaches to PTS are described below.

Foundational: Characterized by:

    1. No logical constant is assumed to have a given, predetermined meaning; the meaning of every constant has its own, independent definition.

    2. Harmony and stability of the meaning-conferring natural-deduction (*ND*) system are defined in terms of *local-soundness* (the existence of *reduction*) and *local-completeness* (the existence of *expansions*), both explained below.

Reductive: Characterized by:

    1. The logical constants of second order propositional Intuitionistic Logic *PL2* (see below) are assumed to have a given, predetermined meaning, and the meaning of any newly defined logical constant is *in terms of PL2*; thus, any logic is reduced to *PL2*, hence the name of the approach.

    2. Harmony and stability of the meaning-conferring natural-deduction system are defined system are defined in terms of *inter-derivability* (in *PL2*) of two formulas, assigning propositional meaning to the *I/E*-rules of any defined constant.

In this section, the reductive approach is described.

The logic $PL2$ is IL (over '$\wedge$', '$\vee$', '$\rightarrow$' and '$\perp$') augmented with universal quantification over propositional variables '$\forall p$'.

6.6.2　I/E-Rules and I/E-meanings

The approach is only applicable to an object language freely generated from a collection of *propositional variables* (ranged over by p, p_i) by a set of logical constants. Let c range over the latter. The I/E-rules for an $n$ary constant c have the following (simply presented) rigid forms.

$$\frac{\begin{array}{ccc}[\Gamma_1]_{i_1} & & [\Gamma_n]_{i_n}\\ \vdots & & \vdots\\ s_1 & \cdots & s_n\end{array}}{c(p_1,\cdots,p_n)}\;(c\,I^{i_1,\cdots,i_n})\qquad \frac{c(p_1,\cdots,p_n)\quad \begin{array}{ccc}[\Gamma_1]_{i_1} & & [\Gamma_m]_{i_m}\\ \vdots & & \vdots\\ s_1 & \cdots & s_m\end{array}}{p}\;(c\,E^{i_1,\cdots,i_m})$$

(6.6.39)

Each Γ is a list of propositional variables, with $\wedge\Gamma$ the conjunction of all elements in Γ. In the I-rules, the propositional variables in each Γ_j and all s_j have to be among $p_{i_1},\cdots,p_{i_n}$. In the E-rules, the s variables are not restricted. The I-rules and E-rules are specified independently of each other.

A constant c with rules $(c\,I)$ and $(c\,E)$ is associated with two $PL2$-formulas c^I and c^E, said the express its I-*meaning* and E-*meaning*, respectively, defined as follows. Suppose c has k I-rules $(c\,I_j)$, $1 \le j \le k$. First, each $(c\,I_j)$-rule is associated with

$$c^{I_j} \text{ is } (\wedge\Gamma_{i_1}\rightarrow s_1)\wedge\cdots\wedge(\wedge\Gamma_{i_n}\rightarrow s_n)\qquad(6.6.40)$$

The I-meaning of c is given by

$$c^I \text{ is } c^{I_1}\vee\cdots\vee c^{I_k}\qquad(6.6.41)$$

Similarly, the E-meaning is associated with the E-rules, with a *unselective* universal quantification over all proposal variable in the E-rules which are not amongst $p_1,\cdots,p_n$. Suppose c has l E-rules.

$$c^{E_j} \text{ is } \forall(((\wedge\Gamma_{i_1}\rightarrow s_1)\wedge\cdots\wedge(\wedge\Gamma_{i_n}\rightarrow s_n))\rightarrow p)\qquad(6.6.42)$$

and

$$c^E \text{ is } c^{E_1}\wedge\cdots\wedge c^{E_l}\qquad(6.6.43)$$

The following two facts connect c with I^c, E^c. Let Γ and φ not contain c.

Fact 1:

$$\vdash_{PL+cI} : \Gamma : c(p_1,\cdots,p_n) \text{ iff } \vdash_{PL} \Gamma : c^I\qquad(6.6.44)$$

Fact 2:

$$\vdash_{PL2+cE} c(p_1,\cdots,p_n),\Gamma : \varphi \text{ iff } \vdash_{PL2} c^E,\Gamma : \varphi\qquad(6.6.45)$$

233

6.6.3   Reductive harmony

The new notion of harmony expresses another kind of balance between I-rules and E-rules for c.

Definition 6.6.62 (reductive harmony). *The I-rules and E-rules for c are* reductively harmonious *iff* $c^I \vdash_{PL2} c^E$ *and* $c^E \vdash_{PL2} c^I$.

This definition naturally raises the question of the existence of flat reductively-harmonious E-rules for arbitrary I-rules (adhering to the strict format restrictions required). A negative result is given in Olkhovikov and Schroeder-Heister [132] by showing that the ternary connective with I-rules given in (3.6.61) has no reductively harmonious E-rules. Instead, higher-level reductively-harmonious E-rules are show always to exist.

This certainly contrasts the situation discussed in Section 3.6.3. Thus, some explanatory notes are in order.

- The rule format assumed by the reductive approach is very restrictive, almost "tailored" for representability in $PL2$. Even intuitionistic 1st-order rules are not representable.

- In addition to the above, the strict format excludes any rules having *side-conditions*, a frequent characteristic of *ND I-rules*. For example, the rules for the *Relevant Logic* **R** (even its implicational fragment) are not representable in $PL2$ by the representation method used, as they impose a side-condition on $(\to I)$, namely, a use of the assumption φ as a condition for its discharge. See Francez [42] for a study of foundational harmony and stability in **R**.

- Still more, the strict format of the I-rules does not suit the application of PTS to natural language, as advocated in Francez and Dyckhoff [53] and Ben-Avi and Francez [52]. Here there are no propositional variables in the object language to start with.

- Yet another domain of application excluded by the reductive approach is that of *labeled ND*-systems. Such systems are used in the study of *Modal Logic*, e.g. by Read [161] or Negri [129].

6.7   Summary

In this chapter, I have collected several issues important to the understanding of PTS, but in one way or another did not fit into the main line of presentation. While some of the topics discussed date to an earlier period, others are recent and even very recent, manifesting the livelihood of PTS as a vivid area of research.

Part II

Proof-Theoretic Semantics for Natural Language

Chapter 7

Introduction II

In this part of the book, I present a more ambitious project, expanding the PTS approach to meaning to cover fragments of natural language (exemplified here by declarative sentences in English, in indicative mood). This approach is intended to overcome the many criticisms of the MTS approach to meaning, mentioned in Introduction I (cf. Section 1.2). The following citation from Schroeder-Heister [85] (p. 525) emphasizes the need for such a development.

> Although the *"meaning as use"* approach has been quite prominent for half a century now and provided one of the cornerstones of philosophy of language, in particular of ordinary language philosophy, it has never become prevailing in the *formal* semantics of artificial and natural languages. In formal semantics, the *denotational* approach which starts with interpretations of singular terms and predicates, then fixes the meaning of sentences in terms of truth conditions, and finally defines logical consequence as truth preservation under all interpretations, has always dominated.

Chapters 8 and 9 are based on Francez and Dyckhoff [53] and Francez, Dyckhoff and Ben-Avi [55], respectively. The proposed semantics is intended to constitute an alternative to the traditional *model-theoretic semantics (MTS)*, originating in Montague's seminal work [223], especially as used in TLG, my choice of a computational formalism for expressing natural language grammars (both syntax and semantics). See Lappin [101] for an indication of the dominance of MTS in formal semantics of NL, and for a variety of further references to MTS treatment of various semantic issues.

Because of the complication of the syntax and semantics of natural language, as compared with the formalized object languages for logic considered in part I of the book, some enhancement of the approach is needed.

- The syntax of logical calculi is usually *recursive*, in that each operator (connective, quantifier) is applied to (one, two or more) formulae of the calculus, to yield another formula. Thus, there is a natural notion of the *dominant* (or *main*) operator which is introduced into/eliminated from a formula, and, typically, this operator is a *sentence forming* operator. In the NL fragments considered here, on the other hand, there is no such notion (in general) of a dominant operator. So, there is a somewhat more general notion of *what* is being introduced/eliminated by the ND-rules, and *where*.

  where: There is a general notion of *locus of introduction (loi)*, the position in a sentence into which an expression is introduced. Naturally, the same position is the position from which expressions are eliminated.

  what: Many of the expressions that are introduced/eliminated are *term-forming* expressions (examples follow) like determiner-phrases[1] (abbreviated *dps*). The proposed PTS adheres to the principle that term-forming expressions (known also as *subnectors*) are always considered within sentences containing them, known also as contexts. This is very much akin Russell's *contextual definition* of the definite description operator, considered in Section 12.5. Other forms of introduction/elimination involve no *overt* operator at all, like (intersective) adjective modification. Both adjectival modification and relative-clause formation are noun-forming, not sentence-forming, operators, so again we see more general notions of introduction and elimination.

  Certain expressions (constituting operators) are introduced as if *according to their grammatical function*; for example, *dps* with determiners such as 'every' or 'some' may be introduced either into the subject or into the object of a transitive verb, or into both, or into various positions in relative clauses. In addition, there are introductions of non-overt operator, like in the case of (intersective) adjectives (see Section 8.1.3.2).

- Formal calculi are usually taken to be *semantically unambiguous*, while the fragments considered here (and NL in general) has semantically ambiguous sentences. In a PTS, the semantic ambiguity manifests itself via different derivations (from the same assumptions). This will be exemplified below by showing how traditional *quantifier-scope ambiguity* manifests itself (see Section 8.3.2).

- Unlike the various logics, which have well-known presentations in terms of ND-systems, there is no such well-established ND-system for natural language, and a "dedicated" one needs to be developed. Also unlike logic, the language over which the meaning-conferring ND-system is defined is not *directly* the NL the meaning of which is being defined; rather, it is a certain *extension* thereof, which, amongst other tasks, *disambiguates* the NL sentences. A particular feature of this extension is, following Gentzen, the inclusion of (individual and, later, notional) *parameters* in the proof-language, extending the resources used in formulating proofs (derivations in the dedicated ND-system). The proof-theoretic use of such parameters, while very natural, is not amenable

---

[1] Known also as noun-phrases (*nps*).

to an obvious model-theoretic interpretation. See Milne [116] for a discussion of this difficulty when applied to logic.

- Formal logical calculi usually have *(formal) theorems* (or *theses*), having a *proof*, i.e. a (closed) derivation from no open assumptions, namely, no assumptions that have not been discharged. In natural language (and in particular in the fragment we consider here) there are hardly any formal theorems. Typically, sentences are contingent and their derivations rely on open (undischarged) assumptions (but see Section 8.3.6.3). This difference has a direct influence on the conception of PTS-meanings (of sentences), and PTS-validity (of arguments, or derivations).

I am interested here not merely with NL meanings in abstracto; rather, I am interested in the incorporation of meanings in (preferably computational) grammars for natural language, exhibiting a "well-behaved" syntax-semantics interface. TLG is the grammatical framework of choice. For that, meanings (both sentential and subsentential) need to be *reified*, in having for each NL phrase an abstract object, denoted by $[[\cdots]]$, the meaning of that phrase.

In the next chapter, I survey (gradually) the various fragments for which a PTS is proposed. The fragments are mostly extensional (but see Chapter 11 for a discussion of an intensional construction). Then, in Section 8.2, the formal extension of the fragments into the corresponding proof-languages is presented, together with ND-rules proposed as being meaning conferring for the various fragments considered. Some properties of the proof-system are presented in Section 8.3.6.

Further applications of the PTS for NL introduced in this part of the book provide more in-depth studies, showing its advantage over the prevailing MTS approaches.

- In Chapter 10, I develop, based on Francez and Ben-Avi [52], a proof-theoretic theory of *determiners*, *proving* their conservativity based on their proof-theoretic type. This contrast with the approach *stipulating* this property for generalized quantifiers serving as model-theoretic denotations of NL determiners. In addition, *monotonicity* properties of determiners are expressed proof-theoretically.

- In Chapter 11 I develop a proof-theoretic semantics for *intensional transitive verbs*, a notoriously difficult problem in MTS. The central proof-theoretic tool is the addition of another kind of parameter, without any ontological commitments about what populates models.

- In Chapter 12, I study two of the major phenomena of NL related to their *contextually varying* meaning. I present an *explicit* definition of a context supporting the following proof-theoretically specified contextual variations.

    - Contextual quantifier domain restriction, where quantifiers range over a contextually-varying domain of quantification.
    - Contextual definiteness, where the existence and uniqueness associated with the meaning of the (in English) is restricted by any given context.

All the displayed natural language expressions are presented in a sans-serif font, and are always mentioned, not used.

Digression: a brief manifesto about formal semantics for NL

I would like to keep apart two facets of sentential meaning in natural language. While there may not be a sharp borderline between the two, I am working at a level of abstraction aiming at the first and avoiding the second. This choice is even more justified on case of the fragments I consider, which lack any pronouns, or other indexical expressions.

What a sentence S means: Here, I am following my own understanding of Montague's dictum [122] (though meant by him to apply to MTS):

> [...] There is in my opinion no important theoretical difference between natural languages and the artificial languages of logicians; indeed, I consider it possible to comprehend the syntax and semantics of both kinds of language within a single natural and mathematically precise theory

Assimilating natural and formal languages means, as I understand it, detaching the study of the former from actual circumstances of its use. This certainly is an idealisation, but one most suitable for mathematical study.

What a speaker means using S: Here, a user may express something different than the literal meaning of S, depending on such circumstances as context of utterance[2], speakers intentions, plans, awareness state and the like. In actual communication activities, the emerging difference mat turn huge, as for saying 'no' to mean 'yes' ...

I see this level of study as belonging to (possibly formal) pragmatics, the study of language in (some) use.

Thus, I exclude from consideration examples such as the following (from Stanley and Gendler Szabó [205])

$$\text{Fred is a good friend} \tag{7.0.1}$$

uttered by a speaker in some circumstances to express that Fred is, actually, a terrible friend. I do not take this interpretation of (7.0.1) as a *meaning* of (7.0.1) in any sense of 'meaning' that semantics is concerned with.

---

[2]For some way of incorporated explicit context dependencies into meaning see Chapter 12.

Chapter 8

Proof-Theoretic Sentential Meanings

8.1 The NL fragments

In this chapter, I gradually present a sequence of fragments of English, of growing level of complication, for which a PTS is developed. Those fragments (actually their extensions in the proof-languages) play the same role as the various object languages for the logics considered in part I of the book. Anticipating the presentation of a PTS for contextually-varying meanings in Chapter 12, I emphasise here that all the meanings considered in this chatter are taken as independent of any context.

8.1.1 The core fragment

I start by considering the core fragment, referred to as E_0^+, with sentences headed by intransitive and transitive verbs, and determiner phrases with a (singular, count) noun and a positive determiner (hence the '+' in E_0^+) (where the positive/negative classification of determiners is defined in Chapter 10, Definition 10.1.66). In addition, there is the copula. This is a typical fragment of many NLs, syntactically focusing on *subcategorization*, and semantically focusing on *predication* and *quantification*. Some typical sentences are listed below.

$$\text{every/some girl smiles} \tag{8.1.1}$$

$$\text{every/some girl is a student} \tag{8.1.2}$$

$$\text{every/some girl loves every/some boy} \tag{8.1.3}$$

241

Note the absence of *proper names*, to be added below in Section 8.1.3.1, and *negative determiners* like no, included in Chapter 10. I refer to expressions such as every girl, some boy as dps (determiner-phrases, known also as nps, noun-phrases). Every position that can be filled with a dp is a *locus of introduction* (of the quantifier corresponding to the determiner of the introduced dp). As already mentioned, this is a major source of *ambiguity* in E_0^+, known as quantifier-scope ambiguity, treated in Section 8.2.3. As will be seen in the next section, this ambiguity is removed from the proof-system, giving rise to a *unique* "next" locus of introduction, replacing the role played by the notion of a main operator in a logic formula, performing the same task of defining a locus of introduction. The same will hold true for eliminations too.

I note that this core fragment contains only two determiners, 'every' and 'some', each treated in a sui generis way. Again, this is just setting the stage and illustrating the treatment of determiners (and quantifiers based on them). In a Chapter 10, as already mentioned, I present a *general* treatment of determiners (and dps) in PTS, providing, for example, proof-theoretic characterization of their monotonicity properties, and capturing proof-theoretically their conservativity, traditionally expressed in model-theoretic terms. Also, a deeper study of negative determiners such as 'no', is added to the fragment in a later chapter.

8.1.2   Relative Clauses: Long-Distance Dependencies

I next add relative clauses (rcs) to the fragment. This fragment transcends the locality of subcategorization in E_0^+, in having *long-distance dependencies*. I refer to this (still positive) fragment as E_1^+. Typical sentences include the following.

every boy/some boy loves every/some girl who(m) smiles/

loves every flower/some girl loves   (8.1.4)

every girl/some girl is a girl who loves everyboy   (8.1.5)

some boy loves every girl who loves every boy who smiles (nested relative clause)
                                                                    (8.1.6)

So, girl who smiles and girl who loves every boy are *compound nouns*. I treat the case of the relative pronouns somewhat loosely, in the form of who(m), abbreviating either who or whom, as required. Note that E_1^+, by its nesting of rcs, expands the stock of available positions for dp-introduction/elimination. Thus, in (8.1.6), 'every boy who smiles' is the object of the relative clause modifying the object of the matrix clause. In addition, new scope relationships arise among the multitude of dps present in E_1^+ sentences. Island conditions, preventing some of the scopal relationships, are ignored here.

8.1.3   More Extensions

I present here two more simple extensions, proper names and intersective adjectives, that can be parts of either E_0^+ or E_1^+, as found convenient. They do not add any essential complication.

8.1.3.1   Adding Proper Names

Proper names such as Rachel, Jacob occur in dp positions. As will be seen later, their meaning *is not* the analog of a constant in logical object languages that have constants. Typical sentences are:

$$\text{Rachel is a girl} \tag{8.1.7}$$

$$\text{Rachel smiles} \tag{8.1.8}$$

$$\text{Rachel loves every/some boy} \tag{8.1.9}$$

$$\text{every boy loves Rachel} \tag{8.1.10}$$

I retain the same fragment names for the fragments augmented with proper names.

8.1.3.2   Adding Intersective Adjectives

I augment $E_{0/1}^+$ with sentences containing *adjectives*, schematized by A. The focus of the PTS presented is on *adjectival modification*, not on lexical semantics for the adjectives themselves. Both *attributive* use for modification and *predicative* use are considered. For the former, I consider here only what is known as *intersective adjectives*. A more comprehensive treatment of adjectival modification is presented in Francez [48]. The uses are distinguished syntactically by different uses of the copula. Attributive: isa, predicative: is. Typical sentences are:

$$\text{Rachel is a beautiful girl/clever beautiful girl/clever beautiful red} - \text{headed girl}$$
$$\tag{8.1.11}$$

$$\text{Rachel/every girl/some/girl is beautiful} \tag{8.1.12}$$

$$\text{Rachel/every beautiful girl/some beautiful girl smiles} \tag{8.1.13}$$

$$\text{Rachel/every beautiful girl/some beautiful girl loves Jacob/}$$

$$\text{every clever boy/some clever boy} \tag{8.1.14}$$

A noun preceded by an adjective is also a (compound) noun (the syntax is treated more precisely once the grammar is presented (see Section 9.2). Denote these extensions again by $E_{0/1}^+$.

8.2   The Proof Language

In this chapter, I introduce the (slightly) extended proof languages (extending the NL fragments), to be used for formulating the meaning-conferring ND-system. I refer to the extension of E_i^+ as L_i^+, respectively. This is a technical Section, introducing a lot of notation. There is a natural temptation to simplify the syntax of L_i^+, abstracting over some of the baroqueness of the NL syntax. For example, one might express both copular predication and intransitive verb predication in the usual FOL notation $P(x)$. I shall resist this temptation to emphasize the fact that I view the proof-system *directly* applied to slightly extended E_i^+-sentences.

8.2.1   Meta-variables

I start by the introduction of meta-variables, ranging over the various types of NL-expressions present in the fragments. Note that the proof-languages have *no individual constants*! As I do not consider models with entities, there is no need for any means of referring to elements of a domain.

***nouns*:** Nouns like girl, student and the like (all singular, count nouns) are ranged over by X. For the core language L_0^+, only *lexical* nouns are considered. For E_1^+, and for the extension with adjectives, *compound* nouns like girl who loves every boy, or beautiful girl, are considered too to fall under X.

***verbs*:** Intransitive verbs like smile are ranged over by P. Transitive verbs like love are ranged over by R. A verb-phrase headed by a transitive verb, where an object has been already incorporated, like loves some girl is treated as an intransitive verb.

***intersective adjectives*:** Intersective adjectives like beautiful are ranged over by A.

***proper names*:** Names like Rachel, Jacob are ranged over by N.

***sentences*:** Sentences are ranged over by S, giving rise to various additional notations specified below.

***individual parameters*:** The proof languages contain a (infinitely countable) set **P** of individual parameters (see Section 8.2.2 below), ranged over by bold-face variables **j**, **k**. These individual parameters can occupy any dp position in a sentence, and are viewed syntactically as dps.

8.2.2   Individual Parameters

As stated above, the proof languages contain a set **P** of individual parameters. Their main purpose is to express *predication*, like in **j** isa girl, **j** smiles or **j** loves every girl.

It is important to note that individual parameters are *artefacts* of the proof language, and cannot be used in assertions, the latter confined to proper E_i^+ sentences. Their functioning is intermediate between constants and free variables, as they can be bound (see below) on the one hand, but can be thought also as *arbitrary* representatives, as will be clear from their use in the I/E-rules for restricted quantification. In particular, they play an important role in discharged assumptions in the I/E-rules.

An expression of the form $S[\mathbf{j}]$ means a sentence with a parameter $\mathbf{j}$ occupying some distinguished dp-position in S. This can then be used by an I-rule to introduce a dp with a quantifier to that distinguished position.

Note that individual parameters, being syntactic objects, artefacts of the proof language, do not carry any ontological burden, no ontological commitment, as do elements in models in the use of MTS as the underlying theory of meaning.

Since at this stage no other parameters are considered, I relate to individual parameters just as 'parameters'.

8.2.2.1   Sentence Classification

Based on the presence or absence of parameters in the various dp positions, the following classification of sentences is induced.

pseudo-sentences: Schematic sentences containing at least one occurrence of a parameter are referred to as *pseudo-sentences*, emphasizing their distinction from assertible proper sentences (in the fragments themselves). They are ranged over by S too.

ground pseudo-sentences: Ground pseudo-sentences contain *only parameters* in every dp-position. Ground pseudo-sentences in E_0^+ (i.e., without any introduced relativization or adjectival modification) play the role of atomic sentences in logic. Their sentential meaning is assumed to be given externally (see below), and general sentential meanings will be defined *relative* to the given meanings of ground pseudo-sentences.

Example 8.2.57.
$$\mathbf{j} \text{ smiles/loves every girl} \qquad (8.2.15)$$
are pseudo-sentences, the first of which is ground.

8.2.3   Quantifier-Scope Disambiguation

As is well-known, the fragment E_0^+ exhibits a phenomenon known as *quantifier scope ambiguity*. Sentences like

$$\text{every girl loves some boy} \qquad (8.2.16)$$

are considered ambiguous between two readings, attributing two different relative scopes to the two quantifiers involved. In MTS, the two truth-conditions are usually expressed by the following two translations into 1st-order logic.

Subject wide-scope (sws):

$$\forall x.\mathbf{girl}(x) \rightarrow \exists y.\mathbf{boy}(y) \land \mathbf{love}(x, y) \qquad (8.2.17)$$

Subject narrow-scope (sns):

$$\exists y.\mathbf{boy}(y) \land \forall x.\mathbf{girl}(x) \rightarrow \mathbf{love}(x, y) \qquad (8.2.18)$$

In Section 8.3.2, I show how this ambiguity is treated in PTS. Here, I just take care of introducing a disambiguation notation[1] by using an *explicit scope indicator*.

For any *dp*-expression D having a quantifier, I use the notation $S[(D)_n]$ to refer to a sentence S having the designated position filled by D, where n is the *scope level* (*sl*) of the quantifier in D. In case D has no quantifier (i.e., it is a parameter), $sl = 0$. The higher the *sl*, the higher the scope. For example, $S[(\text{every } X)_1]$ refers to a sentence S with a designated occurrence of every X of the lowest scope. (every $X)_1$ loves (some $Y)_2$, representing in L_0^+ the object wide-scope reading of the E_0^+ sentence (8.2.18), while (every $X)_2$ loves (some $Y)_1$ represents the (8.2.17) reading. When sentential meanings are defined (in Section 8.5), disambiguation is assumed to have taken place. Each reading obtains its own meaning, and the original sentence, before disambiguation, has indeed two (or more) meanings.

When the *sl* can be unambiguously determined it is omitted. I use $r(S)$ to indicate the *rank* of S, the highest *sl* on a *dp* within S. In E_0^+, there are only two scope level. However, in E_1^+, where arbitrarily nested relative clauses can occur, there is an unbounded number of *dp* positions available, and hence an arbitrary number of scope levels.

8.3   The proof Systems

In this section, I gradually introduce the meaning-conferring, dedicated ND-systems, through which sentential meanings for sentences in the fragments are defined.

The presentation is again in Gentzen's "Logistic"-style *ND*, with shared contexts, single succedent, set antecedent sequents $\Gamma : S$, formed over contexts of schematic L_i^+ sentences. As is traditional in the simple presentation, I enclose *discharged assumptions* in square brackets and index them, using the index to mark the rule-application

---

[1]Note that I am not concerned here with the actual *pragmatic* disambiguation process, by which a "correct" reading is chosen. I only provide means for expressing the various readings.

$$\overline{\Gamma, S : S} \;\; (Ax)$$

$$\frac{\Gamma, \mathbf{j} \text{ isa } X : S[\mathbf{j}]}{\Gamma : S[(\text{every } X)_{r(S[\mathbf{j}])+1}]} \;\; (eI) \qquad \frac{\Gamma : \mathbf{j} \text{ isa } X \quad \Gamma : S[\mathbf{j}]}{\Gamma : S[(\text{some } X)_{r(S[\mathbf{j}])+1}]} \;\; (sI)$$

$$\frac{\Gamma : S[(\text{every } X)_{r(S[\mathbf{j}])+1}] \quad \Gamma : \mathbf{j} \text{ isa } X \quad \Gamma, S[\mathbf{j}] : S'}{\Gamma : S'} \;\; (eE) \qquad \frac{\Gamma : S[(\text{some } X)_{r(S[\mathbf{j}])+1}] \quad \Gamma, \mathbf{j} \text{ isa } X, S[\mathbf{j}] : S'}{\Gamma : S'} \;\; (sE)$$

where $\mathbf{j}$ is fresh for Γ, $S[\text{every } X]$ in (eI), and for Γ, $S[\text{some } X]$, S' in (sE).

Figure 8.1: The meta-rules for N_0^+

responsible for the discharge. There are I-rules and E-rules for each determiner forming a dp, the latter indexed for its scope level.

The definitions of a derivation and a canonical (here – only I-canonical) derivation are carried over from Section 1.5.

8.3.1   The core ND-system N_0^+

I start by presenting N_0^+, an ND-system formulated in L_0^+, intended to confer meaning on the sentences in the core fragment E_0^+.

In the rule names, I abbreviate 'every' and 'some' to 'e' and 's', respectively. The meta-rules for N_0^+ are presented in Figure 8.1. In addition, the structural rule of *contraction*, namely

$$\frac{\Gamma, S', S' : S}{\Gamma, S' : S} \;\; (C)$$

is assumed, allowing multiple uses of assumptions (example below).

Before attending to derivations in N_0^+, here is some commentary on the rules themselves.

Axioms: The axioms are the usual identity axioms in ND-systems, incorporating the structural rule (W) of weakening.

I-rules: The I-rules have in "intuitionistic flavour", explainable akin the famous BHK explanation of intuitionistic logic.

every (**Evidence transforming**): A proof of $S[(\text{every } X)]$ is a function mapping each proof of $\mathbf{j}$ isa X (for an arbitrary fresh parameter $\mathbf{j}$) into a proof of $S[\mathbf{j}]$. For example, a proof of every X P is a function mapping each proof of $\mathbf{j}$ isa X to a proof of $\mathbf{j}$ P. A proof of every X R $\mathbf{r}$ is a function mapping each proof of $\mathbf{j}$ isa X to a proof of $\mathbf{j}\,R\,\mathbf{r}$. Similarly, a proof of $\mathbf{j}\,R$ every Y is a function mapping each proof of $\mathbf{r}$ isa Y to a proof of $\mathbf{j}\,R\,\mathbf{r}$.

some (**Evidence combining**): A proof of $S[(\text{some } X)]$ is a pair[2] of proofs, one of $\mathbf{j}$ isa X and the other of $S[\mathbf{j}]$, for some parameter $\mathbf{j}$. For example, a proof of some X P is a pair of proofs, one of $\mathbf{j}$ isa X, and the other for $\mathbf{j}\,P$. A proof of some X R $\mathbf{r}$ is a pair of proofs, one of $\mathbf{j}$ isa X and the other for $\mathbf{j}\,R\,\mathbf{r}$. Similarly, a proof of $\mathbf{j}\,R$ some Y is a pair of proofs, one of $\mathbf{k}$ isa Y and the other for $\mathbf{j}\,R\,\mathbf{k}$.

Another word of explanation about the I-rules is due. The scope-level $r(S[j])$ is the highest scope of a quantifier already present in $S[\mathbf{j}]$. When a new dp is introduced into the position currently filled by $\mathbf{j}$, it obtains the scope level $r(S[j]) + 1$. Thereby its quantifier becomes the one with the highest scope in the resulting sentence. Thus, scope is represented by the *order of introduction* of dps. This order will play more roles below.

E-rules: The E-rules are all in the GE form, drawing arbitrary conclusions, thereby guaranteeing harmony (in form), discussed in Section (3.6.3). The E-rules always eliminate the quantifier with the highest scope.

The following is a convenient *derived* E-rule, that will be used to shorten derivations.

$$\frac{\Gamma : S[(\text{every } X)_{r(S[\mathbf{j}])+1}] \quad \Gamma : \mathbf{j} \text{ isa } X}{\Gamma : S[\mathbf{j}]} \ (e\hat{E}) \qquad\qquad (8.3.19)$$

Its derivability is shown by

$$\frac{\Gamma : S[(\text{every } X)_{r(S[\mathbf{j}])+1}] \quad \Gamma : \mathbf{j} \text{ isa } X \quad \Gamma, S[\mathbf{j}] : S[\mathbf{j}]}{\Gamma : S[\mathbf{j}]} \ (eE)$$

That is, I take the arbitrary consequence to be $S[\mathbf{j}]$ itself.

The usual notion of (tree-shaped) derivation is assumed.

Example 8.3.58. *Below is an example derivation (where I suppress the Γ, using only the succedent, to save space), establishing*

$$\vdash_{N_{0^+}} \text{some } U \text{ isa } X, \ (\text{every } X)_2 \ R \ (\text{some } Y)_1, \text{ every } Y \text{ isa } Z \ : \ (\text{some } U)_1 \ R \ (\text{some } Z)_2$$
$$(8.3.20)$$

---

[2]Strictly speaking, such a proof consists of an ordered *triple*, the first member of which is a parameter, say $\mathbf{j}$, and the other two members are the two above-mentioned proofs. As the parameter is trivially retrievable from the pair of proofs in N_0^+, I avoid this extra pedantry.

The derivation is

$$
\cfrac{
\cfrac{
\text{some } U \text{ isa } X \quad
\cfrac{
[\mathbf{k} \text{ isa } U]_1 \quad
\cfrac{(\text{every } X)_2 \; R \,(\text{some } Y)_1 \quad [\mathbf{k} \text{ isa } X]_2}{\mathbf{k} \; R \text{ some } Y} \,(e\hat{E})
}{(\text{some } U)_2 \; R \,(\text{some } Y)_1} \,(sI)
}{(\text{some } U)_2 \; R \,(\text{some } Y)_1} \,(sE^{1,2})
\qquad
\cfrac{
[\text{some } U \; R \,\mathbf{j}]_3 \quad
\cfrac{\text{every } Y \text{ isa } Z \quad [\mathbf{j} \text{ isa } Y]_4}{\mathbf{j} \text{ isa } Z} \,(e\hat{E})
}{(\text{some } U)_1 \; R \,(\text{some } Z)_1} \,(sI)
}{(\text{some } U)_1 \; R \,(\text{some } Z)_2} \,(sE^{3,4})
$$

$$(8.3.21)$$

Without the freshness side-condition, the following unwarranted derivations would be available (ignoring scope).

$$
\cfrac{
[\mathbf{j} \text{ isa } X]_1 \quad
\cfrac{\mathbf{j} \text{ isa } Y \quad \text{every } Y \text{ smiles}}{\mathbf{j} \text{ smiles}} \,(e\hat{E})
}{\text{every } X \text{ smiles}} \,(eI^1)
\qquad
\cfrac{
\cfrac{\mathbf{j} \; R \,\mathbf{j} \quad [\mathbf{j} \text{ isa } X]_1}{(\text{every } X)_1 \; R \,\mathbf{j}} \,(eI^1) \quad [\mathbf{j} \text{ isa } X]_2
}{(\text{every } X)_1 \; R \,(\text{every } X)_2} \,(eI^2)
$$

$$(8.3.22)$$

To see the need for contraction, consider the following: $\mathbf{j}$ isa X, every X isa Y, every X isa $Z \vdash$ some Y isa Z.

The assumption $\mathbf{j}$ isa X has to be used twice, to eliminate both occurrences of every.

$$
\cfrac{
\cfrac{\mathbf{j} \text{ isa } X \quad \text{every } X \text{ isa } Y}{\mathbf{j} \text{ isa } Y} \,(e\hat{E})
\qquad
\cfrac{\mathbf{j} \text{ isa } X \quad \text{every } X \text{ isa } Z}{\mathbf{j} \text{ isa } Z} \,(e\hat{E})
}{\text{some } Y \text{ isa } Z} \,(sI)
$$

$$(8.3.23)$$

8.3.2   Interlude: Quantifier-Scope Ambiguity

In order to better understand the PTS of E_0^+, consider one of its well-known features: *quantifier-scope ambiguity*. Consider again the ambiguous E_0^+ sentence in (8.2.16).

Under the proposed PTS, the difference in its (disambiguated) meanings reflects itself by the two readings having *different grounds for assertion*. This is manifested in derivations by different *order of introduction* of the subject and object *dps*.

Subject wide-scope (sws):

$$
\cfrac{
\cfrac{
[\mathbf{r} \text{ isa girl}]_i \\
\cfrac{\mathcal{D}_1}{\mathbf{r} \text{ loves } \mathbf{j}} \quad \cfrac{\mathcal{D}_2}{\mathbf{j} \text{ isa boy}}
}{\mathbf{r} \text{ loves (some boy)}_1} \,(sI)
}{(\text{every girl})_2 \text{ loves (some boy)}_1} \,(eI^i)
$$

$$(8.3.24)$$

Subject narrow-scope (sns):

$$
\cfrac{
\cfrac{
\cfrac{
[\mathbf{r} \text{ isa girl}]_i \\
\cfrac{\mathcal{D}_1}{\mathbf{r} \text{ loves } \mathbf{j}}
}{(\text{every girl})_1 \text{ loves } \mathbf{j}} \,(eI^i)
\quad
\cfrac{\mathcal{D}_2}{\mathbf{j} \text{ isa boy}}
}{(\text{every girl})_1 \text{ loves (some boy)}_2} \,(sI)
$$

$$(8.3.25)$$

Note that there is no way to introduce a dp with a narrow-scope where the dp with the wider-scope has already been introduced. In the N_0^+ calculus, only disambiguated sentences participate.

This way of capturing the source of quantifier-scope ambiguity, by means of order of application of I-rules, is a major element in the PTS tool-box. We shall encounter another use of this tool below, in the treatment of opaque transitive verbs. Its usefulness is a direct result of the way I-rules take part in conferring meaning. Obviously, this tool is unavailable to MTS.

8.3.3   Adding identity and Proper Names

Proper names, ranged over by N, M, are strictly distinct from (individual) parameters in the way they function in the proof-system, as explained below. I retain the name L_0^+ for this (minor) extension. I add to L_0^+ pseudo-sentences such as

$$\mathsf{j} \text{ is } N, \quad N \text{ is } \mathsf{j}, \quad \mathsf{j} \text{ is } \mathsf{k}, \quad N \text{ is } M \tag{8.3.26}$$

Note that pseudo-sentences having a proper name in any dp-position *are not ground!*

First, I add I-rules and E-rules for is (a disguised identity). I adopt a version of the rules in Read [166], as presented in [169].

$$\frac{\Gamma, S[\mathsf{j}] : S[\mathsf{k}]}{\Gamma : \mathsf{j} \text{ is } \mathsf{k}} \ (isI) \qquad \frac{\Gamma : \mathsf{j} \text{ is } \mathsf{k} \quad \Gamma : S[\mathsf{j}] \quad \Gamma, S[\mathsf{k}] : S'}{\Gamma : S'} \ (isE) \tag{8.3.27}$$

where S does not occur in Γ. By choosing the arbitrary conclusion S' in (isE) as $S[\mathsf{k}]$ itself, we again have the following simplified rule $(is\hat{E})$.

$$\frac{\Gamma : \mathsf{j} \text{ is } \mathsf{k} \quad \Gamma : S[\mathsf{j}]}{\Gamma : S[\mathsf{k}]} \ (is\hat{E}) \tag{8.3.28}$$

From these, I can derive rules for reflexivity $(is - refl)$, symmetry $(is - sym)$ and transitivity $(is - tr)$. For shortening the presentation of derivations, combinations of these rules are still referred to as applications of (isE).

$$\frac{}{\Gamma : \mathsf{j} \text{ is } \mathsf{j}} \ (is-refl) \quad \text{derived by} \quad \frac{\dfrac{}{\Gamma, S[\mathsf{j}] : S[\mathsf{j}]} \ (Ax)}{\Gamma : \mathsf{j} \text{ is } \mathsf{j}} \ (isI) \tag{8.3.29}$$

$$\frac{\Gamma : \mathsf{j} \text{ is } \mathsf{k}}{\Gamma : \mathsf{k} \text{ is } \mathsf{j}} \ (is-sym) \quad \text{derived by} \quad \frac{\dfrac{}{\Gamma : \mathsf{j} \text{ is } \mathsf{j}} \ (is-refl) \quad \Gamma : \mathsf{j} \text{ is } \mathsf{k}}{\Gamma : \mathsf{k} \text{ is } \mathsf{j}} \ (is\hat{E}) \tag{8.3.30}$$

$$\frac{\Gamma : \mathsf{j} \text{ is } \mathsf{k} \quad \Gamma : \mathsf{k} \text{ is } \mathsf{l}}{\Gamma : \mathsf{j} \text{ is } \mathsf{l}} \ (is-trans) \quad \text{derived by} \quad \frac{\Gamma : \mathsf{k} \text{ is } \mathsf{l} \quad \Gamma : \mathsf{j} \text{ is } \mathsf{k}}{\Gamma : \mathsf{j} \text{ is } \mathsf{l}} \ (is\hat{E}) \tag{8.3.31}$$

The harmony (local-soundness) of the identity rules is established by the following reduction.

$$
\begin{array}{c}
[S[\mathbf{j}]]_i \\
\mathcal{D}_1 \\
\dfrac{S[\mathbf{k}]}{\mathbf{j} \text{ is } \mathbf{k}} \, (isI^i) \quad \dfrac{\mathcal{D}_2}{S[\mathbf{j}]} \\
\dfrac{\phantom{S[\mathbf{k}]}}{S[\mathbf{k}]} \, (isE)
\end{array}
\quad \leadsto_r \quad
\begin{array}{c}
\mathcal{D}_2 \\
\mathcal{D}_1[S[\mathbf{j}] := S[\mathbf{j}]] \\
S[\mathbf{k}]
\end{array}
\tag{8.3.32}
$$

Local-completeness is established by the following expansion.

$$
\begin{array}{c}
\mathcal{D} \\
\mathbf{j} \text{ is } \mathbf{k}
\end{array}
\quad \leadsto_e \quad
\begin{array}{c}
\dfrac{\mathcal{D} \qquad [S[\mathbf{j}]]_i}{\mathbf{j} \text{ is } \mathbf{k}} \\[2pt]
\dfrac{S[\mathbf{k}]}{\mathbf{j} \text{ is } \mathbf{k}} \, (isI^i)
\end{array}
\qquad \quad S \text{ fresh}
\tag{8.3.33}
$$

Next, I incorporate I-rules and E-rules of proper names into dp-positions, letting names function similarly to determiner-headed dps, fitting their MTS view as generalized quantifiers (GQs) (here viewed proof-theoretically). In the MTS, where an intransitive verb has the predicate type (e,t), a proper name has the GQ-type $((e,t),t)$. So, it is not the meaning of the verb applied to the meaning of a proper name; rather, the meaning of a proper name is applied to the meaning of the verb.

$$
\dfrac{\Gamma : \mathbf{j} \text{ is } N \quad \Gamma : S[\mathbf{j}]}{\Gamma : S[N]} \, (nI)
\qquad
\dfrac{\Gamma : S[N] \quad \Gamma, \mathbf{j} \text{ is } N, S[\mathbf{j}] : S'}{\Gamma : S'} \, (nE)
, \quad \mathbf{j} \text{ fresh for } \Gamma
\tag{8.3.34}
$$

Below are two example derivations.

$\vdash_{N_0^+}$ Rachel isa girl, every girl smiles : Rachel smiles: Note that Rachel is not a parameter, and $(e\hat{E})$ is not *directly* applicable.

$$
\dfrac{\text{Rachel isa girl} \quad \dfrac{[\mathbf{r} \text{ is Rachel}]_1 \quad \dfrac{[\mathbf{r} \text{ isa girl}]_2 \quad \text{every girl smiles}}{\mathbf{r} \text{ smiles}} \, (e\hat{E})}{\text{Rachel smiles}} \, (nI)}{\text{Rachel smiles}} \, (nE^{1,2})
\tag{8.3.35}
$$

$\vdash_{N_0^+}$ Rachel isa girl, Rachel smiles : some girl smiles: Again, since Rachel is not a parameter, (sI) is not *directly* applicable.

$$
\dfrac{\text{Rachel isa girl} \quad \dfrac{\text{Rachel smiles} \quad \dfrac{\dfrac{[\mathbf{r}_1 \text{ is Rachel}]_1 \quad [\mathbf{r}_2 \text{ is Rachel}]_3}{\mathbf{r}_1 \text{ is } \mathbf{r}_2} \, (isE) \quad \dfrac{[\mathbf{r}_1 \text{ isa girl}]_2}{\mathbf{r}_2 \text{ isa girl}} \, (isE) \quad [\mathbf{r}_2 \text{ smiles}]_4}{\text{some girl smiles}} \, (sI)}{\text{some girl smiles}} \, (nE^{3,4})}{\text{some girl smiles}} \, (nE^{1,2})
\tag{8.3.36}
$$

8.3.4 Adding intersective adjectives

I next augment N_0^+ with ND-rules for (intersective) adjectives. A more comprehensive treatment of adjectival modification (for other classes of adjectives too) is presented in [48]. First, an additional ground pseudo-sentence is introduced, of the form $\mathbf{j}$ is A, A an intersective adjective. Note that a different copula, namely is, is used to express adjectival predication. The rules are:

$$\frac{\Gamma : \mathbf{j} \text{ isa } X \quad \Gamma : \mathbf{j} \text{ is } A}{\Gamma : \mathbf{j} \text{ isa } A\ X} \ (adjI) \qquad \frac{\Gamma : \mathbf{j} \text{ isa } A\ X \quad \Gamma, \mathbf{j} \text{ isa } X, \mathbf{j} \text{ is } A : S'}{\Gamma \vdash S'} \ (adjE)$$

$$(8.3.37)$$

Let the resulting system be $N_{0,adj}^{+}$.

Again, we can obtain the following *derived* elimination rules, used to shorten presentations of example derivations.

$$\frac{\Gamma : \mathbf{j} \text{ isa } A\ X}{\Gamma : \mathbf{j} \text{ isa } X} \ (adj\hat{E}_1) \qquad \frac{\Gamma : \mathbf{j} \text{ isa } A\ X}{\Gamma : \mathbf{j} \text{ is } A} \ (adj\hat{E}_2) \qquad\qquad (8.3.38)$$

Note that the intersectivity here is manifested by the rules themselves (embodying an "invisible" conjunctive operator), at the sentential level. These rules induce intersectivity as a lexical property of (some) adjectives by the way lexical meanings are extracted from sentential meanings, as shown in Chapter 9.

The following sequent, the corresponding entailment of which is often taken as the definition of intersective adjectives, is derivable in $N_{0,adj}^{+}$: $\vdash_{N_{0,adj}^{+}} \mathbf{j}$ isa $A\ X$, $\mathbf{j}$ isa Y : $\mathbf{j}$ isa $A\ Y$, as shown by

$$\frac{\mathbf{j} \text{ isa } Y \quad \dfrac{\mathbf{j} \text{ isa } A\ X}{\mathbf{j} \text{ is } A} \ (adj\hat{E}_2)}{\mathbf{j} \text{ isa } A\ Y} \ (adjI) \qquad\qquad (8.3.39)$$

As another example of derivations using the rules for adjectives, consider the following derivation for

$$\vdash_{N_{0,adj}^{+}} \mathbf{j} \text{ loves every girl} : \mathbf{j} \text{ loves every beautiful girl} \qquad (8.3.40)$$

In model-theoretic semantics terminology, the corresponding entailment is a witness to the *downward monotonicity* of the meaning of every in its second argument. I use an obvious schematization.

$$\frac{\mathbf{j}\ R \text{ every } Y \quad \dfrac{[\mathbf{r} \text{ isa } A\ Y]_1}{\mathbf{r} \text{ isa } Y} \ (adj\hat{E})}{\dfrac{\mathbf{j}\ R\ \mathbf{r}}{\mathbf{j}\ R \text{ every } A\ Y}} \ (e\hat{E}) \atop (eI^1)} \qquad (8.3.41)$$

A proof-theoretic reconstruction of monotonicity is presented in Chapter 10.

Under this definition of the meaning of intersective adjectives, such adjectives are also *extensional*, in the sense of satisfying the following entailment: $\vdash_{N_{0,adj}^{+}}$ every X isa Y :

every A X isa A Y, as shown by the following derivation:

$$\cfrac{\text{every } X \text{ isa } Y \quad \cfrac{\cfrac{[\text{j isa } A\ X]_1}{\text{j isa } X}\ (adj\hat{E}_1)}{\text{j isa } Y}\ (e\hat{E}) \quad \cfrac{[\text{j isa } A\ X]_1}{\text{j is } A}\ (adj\hat{E}_2)}{\cfrac{\text{j isa } A\ Y}{\text{every } A\ X \text{ isa } A\ Y}\ (eI^1)} \quad (adjI)$$

$$(8.3.42)$$

8.3.5   The N_1^+ system: Adding Relative Clauses

The ND-system N_1^+ extends N_0^+ by adding the following I-rules and E-rules. For their formulation, I extend the distinguished position notation with $S[-]$, indicating that the position is *unfilled*. For example, loves every girl and every girl loves have their subject and object dp positions, respectively, unfilled.

$$\cfrac{\Gamma : \text{j isa } X \quad \Gamma : S[\text{j}]}{\Gamma : \text{j isa } X \text{ who } S[-]}\ (relI) \qquad \cfrac{\Gamma : \text{j isa } X \text{ who } S[-] \quad \Gamma, \text{j isa } X, S[\text{j}] : S'}{\Gamma : S'}\ (relE) \quad , \ \text{j fresh}$$

$$(8.3.43)$$

The simplified derived E-rules are:

$$\cfrac{\Gamma : \text{j isa } X \text{ who } S[-]}{\Gamma : \text{j isa } X}\ (rel\hat{E})_1 \qquad \cfrac{\Gamma : \text{j isa } X \text{ who } S[-]}{\Gamma : S[\text{j}]}\ (rel\hat{E})_2 \qquad (8.3.44)$$

As an example of a derivation in this fragment, consider

$$\vdash_{N_1^+} \text{some girl who smiles sings} : \text{some girl sings} \qquad (8.3.45)$$

exhibiting the model-theoretical *upward monotonicity* of some in its first argument (a more comprehensive discussion on monotonicity of determiners is presented in Section 10.7).

$$\cfrac{\text{some girl who smiles sings} \quad \cfrac{\cfrac{[\text{r isa girl who smiles}]_1}{\text{r isa girl}}\ (rel\hat{E})_1 \quad [\text{r sings}]_2}{\text{some girl sings}}\ (sE^{1,2})}{\text{some girl sings}} \quad (sI)$$

$$(8.3.46)$$

Similarly, the following witness of the downward monotonicity of 'every' (in its first argument) can be derived.

$$\vdash_{N_1^+} \text{every girl sings} : \text{every girl who smiles sings} \qquad (8.3.47)$$

$$\cfrac{\cfrac{\text{every girl sings} \quad \cfrac{[\text{j isa girl who smiles}]_1}{\text{j isa girl}}\ (rel\hat{E}_1)}{\text{j sings}}\ (e\hat{E})}{\text{every girl who smiles sings}}\ (eI^1)$$

$$(8.3.48)$$

$$\overline{\Gamma, S : S} \ (ID)$$

$$\frac{\Gamma, \mathbf{j} \text{ isa } X, S[(\text{every } X)_{r(S[\mathbf{j}])+1}], S[\mathbf{j}] : S'}{\Gamma, \mathbf{j} \text{ isa } X, S[(\text{every } X)_{r(S[\mathbf{j}])+1}] : S'} \ (Le) \qquad \frac{\Gamma, \mathbf{j} \text{ isa } X : S[\mathbf{j}]}{\Gamma : S[(\text{every } X)_{r(S[\mathbf{j}])+1}]} \ (Re)$$

$$\frac{\Gamma, \mathbf{j} \text{ isa } X, S[\mathbf{j}] : S'}{\Gamma, S[(\text{some } X)_{r(S[\mathbf{j}])+1}] : S'} \ (Ls) \qquad \frac{\Gamma : \mathbf{j} \text{ isa } X \quad \Gamma : S[\mathbf{j}]}{\Gamma : S[(\text{some } X)_{r(S[\mathbf{j}])+1}]} \ (Rs)$$

where $\mathbf{j}$ is fresh in Re and Ls.

Figure 8.2: A sequent-calculus SC_0^+ for L_0^+

8.3.6   Properties of the proof-systems

I now present some of the properties of the N_i^+ ND-systems, qualifying them for being meaning conferring.

Proposition 8.3.12 (closure under derivation composition). *All the N_i^+ systems are closed under derivation composition.*

The proof is by a simple inductive argument on derivations and is omitted.

8.3.6.1   Decidability of N_0^+ derivability

I first attend to the issue of *decidability* of derivability in N_0^+. The positive result provided here makes PTS-based meaning effective for L_0^+. Figure 8.2 displays a sequent-calculus SC_0^+ for L_0^+, easily shown equivalent to N_0^+ (in having the same provable sequents). The rules are arranged in the usual way of L-rules (introduction in the antecedent) and R-rules (introduction in the succedent). The following claims are routinely established for SC_0^+.

- The structural rules of *weakening* (W) and *contraction* (C) are primitive.

- (Cut) is admissible.

The existence of a terminating proof-search procedure follows. The essence of the proof is as follows. First, observe that for all rules except (Le), the premise is simpler than the conclusion. Secondly, for (Le), even though $S[(\text{every } X)_{r(S[\mathbf{j}])+1}]$ is retained in the premise (causing non-simplification), the rule is applicable only with $\mathbf{j}$ isa X already in the context Γ. So, this rule is applicable only finitely often, as Γ is finite, and every rule that may contribute $\mathbf{j}$ isa X to the context is itself only finitely often applicable.

The extension to the sequent calculus SC_0^+ corresponding to the addition of proper names consists of the following rules.

$$\frac{\Gamma, \mathbf{j} \text{ is } N, S[\mathbf{j}] \; : \; S'}{\Gamma, S[N] \; : \; S'} \; (Ln) \qquad \frac{\Gamma \; : \; \mathbf{j} \text{ is } N \quad \Gamma \; : \; S[\mathbf{j}]}{\Gamma \; : \; S[N]} \; (Rn) \qquad (8.3.49)$$

The extension to a sequent-calculus $SC_{0,adj}^+$ for $L_{0,adj}^+$ for adding adjectives is:

$$\frac{\Gamma, \mathbf{j} \text{ is } A, \mathbf{j} \text{ isa } X : S'}{\Gamma, \mathbf{j} \text{ isa } A \, X : S'} \; (Ladj) \qquad \frac{\Gamma : \mathbf{j} \text{ is } A \quad \Gamma : \mathbf{j} \text{ isa } X}{\Gamma : \mathbf{j} \text{ isa } A \, X} \; (Radj) \qquad (8.3.50)$$

Finally, decidability of derivability with added relative clauses is shown by means of the following additional sequent-calculus rules, added to SC_0^+, to form SC_1^+.

$$\frac{\Gamma, \mathbf{j} \text{ isa } X, S[\mathbf{j}] : S'}{\Gamma, \mathbf{j} \text{ isa } X \text{ who } S[-] \; : \; S'} \; (Lrel) \qquad \frac{\Gamma : \mathbf{j} \text{ isa } X \quad \Gamma : S[\mathbf{j}]}{\Gamma : \mathbf{j} \text{ isa } X \text{ who } S[-]} \; (Rrel) \qquad (8.3.51)$$

8.3.6.2 Harmony and Stability

Obviously, the E-rules are form-harmonious w.r.t. the I-rules, as they are in the GE-form, harmoniously induced by the I-rules.

I show local-soundness for N_0^+, even though this follows from the form of the rules, as shown in Section 3.6.3. I do, however, omit showing the reductions/expansions for the extensions of the fragment presented below. For simplicity, the scope levels are omitted.

every: **local-soundness:**

$$\frac{\dfrac{\dfrac{[\mathbf{j} \text{ isa } X]_i}{\mathcal{D}_1}}{\dfrac{S[\mathbf{j}]}{S[(\text{every } X)]} \; (eI^i)} \qquad \dfrac{\mathcal{D}_2}{\mathbf{k} \text{ isa } X} \qquad \dfrac{\dfrac{[S[\mathbf{k}]]_j}{\mathcal{D}_3}}{S'}}{S'} \; (eE^j) \qquad \leadsto_r$$

$$\mathcal{D}_3[S[\mathbf{k}] := \dfrac{\mathcal{D}_1[\mathbf{j} \text{ isa } X := \dfrac{\mathcal{D}_2}{\mathbf{k} \text{ isa } X}, \mathbf{j} := \mathbf{k}]}{S[\mathbf{j}]} \;] \qquad\qquad (8.3.52)$$

Since $\mathbf{j}$ is fresh for the assumptions on which $\mathcal{D}_1$ depends, the replacement of $\mathbf{j}$ by $\mathbf{k}$ is permissible.

local-completeness:

$$\frac{\mathcal{D}}{S[(\text{every } X)]} \quad \rightsquigarrow_e \qquad \frac{\dfrac{\dfrac{\mathcal{D}}{S[(\text{every } X)]} \quad [\mathbf{j} \text{ isa } X]_1 \quad [S[\mathbf{j}]]_2}{S[\mathbf{j}]} \, (eE^2)}{S[(\text{every } X)]} \, (eI^1)$$

$$(8.3.53)$$

some: local-soundness:

$$\frac{\dfrac{\dfrac{\mathcal{D}_1}{\mathbf{j} \text{ isa } X} \quad \dfrac{\mathcal{D}_2}{S[\mathbf{j}]}}{S[(\text{some } X)]} \, (sI) \qquad \dfrac{[\mathbf{k} \text{ isa } X]_1 \quad [S[\mathbf{k}]]_2}{\dfrac{\mathcal{D}_3}{S'}}}{S'} \, (sE^{1,2}) \quad \rightsquigarrow_r$$

$$\frac{\mathcal{D}_3[\mathbf{k} \text{ isa } X := \mathbf{j} \text{ isa } X, S[\mathbf{k}] := S[\mathbf{j}]]}{S'}^{\mathcal{D}_1 \qquad\qquad \mathcal{D}_2}$$

$$(8.3.54)$$

Again, a fresh **k** has been replaced.

Local-completeness:

$$\frac{\mathcal{D}}{S[(\text{some } X)]} \quad \rightsquigarrow_e \qquad \frac{\dfrac{\mathcal{D}}{S[(\text{some } X)]} \quad \dfrac{[\mathbf{j} \text{ isa } X]_1 \quad [S[\mathbf{j}]]_2}{S[(\text{some } X)]} \, (sI)}{S[(\text{some } X)]} \, (sE^{1,2}) \qquad (8.3.55)$$

8.3.6.3 Closed derivations

Recall that a derivation of $\Gamma : S$ is *closed* iff $\Gamma = \varnothing$. In logic, as already mentioned above, closed derivations are a central topic, determining the *(formal) theorems* of the logic (recall Schroeder-Heister's criticism of the priority of the categorical over the hypothetical in Section 6.2). In particular, for bivalent logics, they induce the (syntactic) notions *tautology* and *contradiction*.

In L_0^+, in the absence of negation and *negative determiners* (like *no*), there is no natural notion of a contradiction. Furthermore, the only "positive" closed derivation in N_0^+ is for sentences of the form every X isa X. The closed derivation is shown below.

$$\frac{\mathbf{j} \text{ isa } X : \mathbf{j} \text{ isa } X}{: \text{every } X \text{ isa } X} \, (eI)$$

$$(8.3.56)$$

In particular, note that $\nvdash:$ some X isa X. However, once negative determiners are introduced into the fragment (in Section 10.4), new contradictions emerge, like no girl isa girl, a contradiction.

8.4 Passivisation

I start with the incorporation into the extensional fragment of passivisation as sentences headed by a transitive verb (TV), which is a simple task, not raising any new proof-theoretic problems or insights. The fragment is extended with sentences like

$$\text{Rachel/ some/every girl is kissed by Jacob/every/some girl} \qquad (8.4.57)$$

In the proof-language, I will use R^{-1} as the passive form of the TV R. I take passivisation to take place at the level of ground (atomic) sentences. The I/E-rules for passivisation are the following.

$$\frac{\text{j } R \text{ k}}{\text{k is } R^{-1} \text{ by j}} \; (passI) \qquad \frac{\text{k is } R^{-1} \text{ by j}}{\text{j } R \text{ k}} \; (passE) \qquad (8.4.58)$$

For example,

$$\frac{\text{j kisses k}}{\text{k is kissed by j}} \; (passI)$$

The passive counterpart $(spassI)$ of (sI) is

$$\frac{\text{k is } R^{-1} \text{ by j} \quad \text{k isa } X}{\text{some } X \text{ is } R^{-1} \text{ by j}} \; (spassI) \qquad (8.4.59)$$

From $(spassI)$, we easily derive the following derived rule $(spass\hat{I})$, lifting passivisation to the DP-level.

$$\frac{\text{j } R \text{ some } X}{\text{some } X \text{ is } R^{-1} \text{ by j}} \; (spass\hat{I}) \qquad (8.4.60)$$

For example,

$$\frac{\text{j kisses some girl}}{\text{some girl is kissed by j}} \; (spassI)$$

The derivation of $(s_n\hat{I})$ is (with Γ omitted)

$$\cfrac{\text{j } R \text{ some } X \quad \cfrac{\cfrac{[\text{j } R \text{ k}]_1}{\text{k is } R^{-1} \text{ by j}} \; (spassi) \qquad [\text{k isa } X]_2}{\text{some } X \text{ is } R^{-1} \text{ by j}} \; (spassI)}{\text{some } X \text{ is } R^{-1} \text{ by j}} \; (sE^{1,2})$$

Similar rules can be formulated for **every** and other (positive) determiners.

8.5 The sentential proof-theoretic meaning

Similarly to the case of PTS for logics, I take here the reified PTS-meaning of an E_i^+ sentence S, and also of an L_i^+ non-ground pseudo-sentence S, to be the function from contexts Γ returning the collection of all the *I-canonical* derivations in N_i^+ of S from Γ. Thus, the *I*-view of PTS is employed here.

Below are two examples of canonical derivations in N_1^+, exemplifying the two clauses in the definition of canonicity.

$$\dfrac{\text{every girl smiles} \quad \dfrac{\dfrac{[\text{j isa beautiful girl}]_1}{\text{j isa girl}}\,(Adj\hat{E})}{\text{j smiles}}\,(e\hat{E})}{\text{every beautiful girl smiles}}\,(eI^1) \tag{8.5.61}$$

$$\dfrac{\text{some beutiful girl smiles} \quad \dfrac{\dfrac{[\text{j isa beautiful girl}]_1}{\text{j isa girl}}\,(Adj\hat{E}) \quad [\text{j smiles}]_2}{\text{some girl smiles}}\,(sI)}{\text{some girl smiles}}\,(sE^{1,2}) \tag{8.5.62}$$

To see once more the significance of employing canonical derivations in defining meaning, this time in a natural language context, consider the following pseudo-sentence in L_1^+:

$$\text{j isa beautiful girl} \tag{8.5.63}$$

An indirect, non-canonical way of deriving (8.5.63) from some assumptions Γ is

$$\dfrac{\Gamma : \text{j isa girl} \quad \Gamma : \text{every girl isa beautiful girl}}{\Gamma : \text{j isa beautiful girl}}\,(eE)$$

This is not part of the meaning of adjectival modification. A canonical way, according to the meaning of adjectival modification, will end with applying $(adjI)$.

$$\dfrac{\Gamma : \text{j isa girl} \quad \Gamma : \text{j is beautiful}}{\Gamma : \text{j isa beautiful girl}}\,(adjI)$$

Recall that for a ground L_i^+ pseudo-sentence S, its meaning is assumed *given*, and the meaning of E_i^+ sentences, as well as non-ground L_i^+ pseudo-sentences, is defined *relative to the given meanings of ground pseudo-sentences*. Here too I employ the same convention of *atomic canonicity* for ground pseudo-sentences, to overcome the indeterminacy of meaning of atomic sentences when given externally.

In accordance with many views in the philosophy of language, every derivation in the meaning of a sentence S can be viewed as providing $GA[\![S]\!]$, *grounds for assertion* of S (recall that ground pseudo-sentences are not used for making any assertion, as they are not part of the natural language, only of the extension to a language for defining meanings by derivations). Semantic equivalence of sentences is based on equality of meaning (and *not* inter-derivability). In addition, a weaker semantic equivalence is based on equality of grounds for assertion.

Definition 8.5.63. (PTS-meaning, equivalence, grounds):

1. *For $S \in L_i^+$:*

$$[[S]]_{L_i^+}^{PTS} =^{df.} \lambda\Gamma.[[S]]_\Gamma^{Ic} \qquad GA[[S]] =^{df.} \{\Gamma \mid \vdash^{Ic}\Gamma : S\} \qquad (8.5.64)$$

   where:

   (a) *For S a sentence in E_i^+, Γ consists of E_i^+-sentences only. Parameters are not "observable" in grounds for assertion. In Chapter 12, pseudo-sentences in Γ are used to express contexts for meaning variation.*

   (b) *For S a pseudo-sentence in L_i^+, Γ may also contain pseudo-sentences with parameters.*

2. *For S_1, S_2 in L_i^+,*

   (a) $S_1 \equiv_m S_2$ *iff* $[[S_1]]_{L_i^+}^{PTS} = [[S_2]]_{L_i^+}^{PTS}$.

   (b) $S_1 \equiv_g S_2$ *iff* $GA_{L_i^+}^{PTS}[[S_1]] = GA_{L_i^+}^{PTS}[[S_2]]$.

When it is clear which language is meant, the subscript L_i^+ will be omitted, as well as the PTS superscript.

I do not dwell further here on the induced equivalences. As for the grounds for assertion, a member $\Gamma \in GA[[S]]$ can be seen, in Dummett's terms, as a *warrant* for the assertion of S (by a speaker). Being in possession of Γ, and of a I-canonical derivation of S from Γ, are a justification of the proper assertion of S. There are various ways of viewing "possession" of Γ. It may reflect the knowledge (or belief) of the speaker, or some non-linguistic (e.g., visual) observation.

The main formal property of meanings (under this definition) is the overcoming (at least for the fragment considered) of the manifestation argument against MTS: asserting a sentence S is based on (algorithmically) decidable grounds (see Section 8.3.6.1). A speaker in possession of Γ *can decide* whether $\vdash^{Ic}\Gamma : S$. Some properties of meanings of specific sentences are discussed below (in particular, see Section 10.2.3), showing that the proposed rules of N_i^+ do fit our pre-theoretic concept of the use of the E_i^+ sentences.

As mentioned in the introduction chapter, the main "visible" semantic property of (affirmative) sentences is that of *entailment*. Recall that in MTS for NL, entailment is defined in terms of truth conditions in models. A sentence S_1 entails a sentence S_2 iff for every (suitable) model $\mathcal{M}$, if $\mathcal{M} \models S_1$ then $\mathcal{M} \models S_2$. Below, this model-theoretic definition is replaced by a definition of *proof-theoretic consequence*, an immediate adaptation of Definition 5.2.39 for logic, suitable for NL, again based on the *propagation of grounds (for assertion)*. This definition is heavily relied upon in Chapter 10, where properties of determiners like conservativity and monotonicity (of determiners) are explored in detail, and is shown to leading to stronger results (regarding conservativity) than the model-theoretic counterpart.

Definition 8.5.64 (proof-theoretic consequence). *Let* $S_1, S_2 \in L_1^+$. *S_1 is a proof-theoretic consequence* of S_2, *denoted* $S_2 \Vdash S_1$, *iff* $GA[\![S_2]\!] \subseteq GA[\![S_1]\!]$.

If S_1 is a *proof-theoretic consequence* of S_2, I will also say that S_2 (proof-theoretically) *entails* S_1. Throughout, I abbreviate 'proof-theoretic consequence' to merely 'consequence'. Note that consequence is reflexive and transitive due to $\subseteq$ satisfying the same properties. When a bilateral proof-system is taken as meaning-conferring (like in the case of negative determiners in Section 10.4), the definition of proof-theoretic consequence is refined accordingly (cf. Definition 10.5.74).

Example 8.5.59. *To show an example exemplifying that proof-theoretic entailment conforms to the pre-theoretic notion of NL-entailment I show that* (some X)$_2$ R (every Y)$_1$ $\Vdash$ (some X)$_1$ R (every Y)$_2$. *I have to show the following: For every* Γ: *if* $\vdash^{I_c} \Gamma$: (some X)$_2$ R (every Y)$_1$ *then* $\vdash^{I_c} \Gamma$: (some X)$_1$ R (every Y)$_2$.

Assume that $\Gamma \vdash^c$ (some X)$_2$ R (every Y)$_1$. *So there is a canonical derivation of* (some X)$_2$ R (every Y)$_1$ *from* Γ, *whose last step is an application of* (sI^+). *It follows that both* $\Gamma \vdash$ **j** isa X *and* $\Gamma \vdash$ **j** R (every Y)$_1$. *The following derivation shows that* $\Gamma \vdash^{I_c}$ (some X)$_1$ R (every Y)$_2$.

$$
\cfrac{
  \cfrac{
    \cfrac{\Gamma : \mathbf{j}\, R\, (\text{every } Y)_1 \quad \Gamma, \mathbf{k} \text{ isa } Y : \mathbf{k} \text{ isa } Y}{\Gamma, [\mathbf{k} \text{ isa } Y]_i : \mathbf{j}\, R\, \mathbf{k}} \ (e\hat{E})
    \quad \Gamma : \mathbf{j} \text{ isa } X
  }{\Gamma, \mathbf{k} \text{ isa } Y : (\text{some } X)_1\, R\, \mathbf{k}} \ (sI)
}{\Gamma : (\text{some } X)_1\, R\, (\text{every } Y)_2} \ (eI)
$$

8.5.1   Further advantages of sentential meanings

In this section I delineate several additional advantages of proof-theoretic sentential meanings.

Unity of the proposition: This problem, which has its origin in antiquity (see http://en.wikipedia.org/wiki/Unity-of-the-proposition for a description and further references) can be stated as follows:
(q) *what distinguishes a sentence from a mere list of words?*
In model-theoretic semantics, when the question is posed, say, regarding Mary smiles, or Mary loves John, the question is what "glues together" the lexical meanings of the words, denotations of certain kinds, to produce a truth-value. The words Mary, loves and John are stipulated by MTS to have the denotations they have independently of any state-of-affairs (fact).

According to the PTS view, the answer to (q) is: sentences, in contrast to lists of words, *have proofs (derivations from other sentences)*! Sentences do not derive their meanings from meanings of the words of which they consist – rather, from their canonical derivations from the meaning-conferring the ND-system of rules. It is words that have meanings derived from their contribution to sentences in which they occur. This is further elaborated upon, including the technicalities, in Chapter 9. Those canonical derivations form the "glue" endowing the proposition its unity of meaning. This view answers also the question what distinguishes any two true sentences: having different meanings, their assertability is established on different grounds. This observation may prove important to a more comprehensive study of intensionality.

Granularity: PTS provides a more adequate granularity of meanings, not identifying logically equivalent propositions. In particular, not all logically true propositions have identical meaning. For example,

$$[[\text{every girl isa girl}]]^{PTS} \neq [[\text{every boy isa boy}]]^{PTS} \qquad (8.5.65)$$

No logical form: By this way of defining sentential meanings, I do not allude to any "logical form" of the sentence, differing from its surface form.

8.6 Conclusions

In this chapter, I have set the stage for the application of PTS to a natural language fragment. After the introduction of the core fragment and some extensions thereof, a meaning-conferring natural-deduction proof-system was introduced, determining the sentential meanings for sentences in the fragment. Various properties of the proof-system, and in particular its harmony and stability that qualify it a meaning conferring, were shown.

In the coming chapters of the second part of the book the viability of the application of PTS to NL is shown by tackling some more complicated problems, some of which are notoriously hard to treat in MTS.

Chapter 9

Proof-Theoretic Meanings of Sub-sentential Phrases

9.1 Introduction

In this chapter, I employ the method used in Chapter 5 in order to derive proof-theoretic semantic values for the logical constants, to derive proof-theoretic semantic values for sub-sentential phrases in the studied fragments. In particular, I derive semantic values for single *words*, based on their contribution to sentential meanings as obtained in Section 8.5. I again appeal to Frege's context principle for identifying those contributions, based on TLGs for the fragments.

9.1.1 Proof-Theoretic Type-Interpretation

In order to express the required abstractions over sentential meanings, I again propose a *proof-theoretic interpretation* of types, replacing the usual Montagovian model-theoretic interpretation of those types ((full) Henkin models), using *proof-theoretic domains*, comprised of functions from contexts to collections of N_0^+-derivations or functions therein. In contrast to the type used for logic in Section 5.3.2, here the collection of types actually used is much richer, as are their domains of interpretation. I start with types interpreted relative to N_0^+.

Definition 9.1.65 (types, proof-theoretic type-interpretation). *Let T_0^+ be the following system of types, ranged over by α, β:*

- *t is a (basic) type, with $D_t = \{[\![S]\!] \mid S \in L_0^+\}$. This is the main departure from*

263

Montague's model-theoretic type-interpretation, whereby t denotes the collection of sentential meanings instead of truth values.

- *p is a (basic) type of (individual) parameters, with $D_p = P$. This is a syntactic domain, replacing the model-dependent sets of objects (entities) in MTS.*

- *If α, β are types, then (α, β) is a functional type, with $D_{(\alpha,\beta)} = D_\beta^{D_\alpha}$, the collection of all functions from D_α to D_β.*

Note that for $z \in D_t$, $\rho(z)$ (the retrieval function) is a singleton. Specifically, if $z = \lambda\Gamma.[[S]]_\Gamma^{Ic}$, then $\rho(z) = \{S\}$. I also assume that $\rho([[S]]) = S$ for a *ground* S, to simplify the notation. Recall the appeal to Frege's distinction between assertoric content and ingredient sense (in Chapter 5). This facilitates the following definition, of a function mapping the meaning of S to its contributing semantic value:

$$\mathbf{ex} =^{df.} \lambda z^t \lambda\Gamma.[[\rho(z)]]_\Gamma^*$$

Here again $\mathbf{ex}([[S]])$ is the *exportation* of (the meaning of) S. Again, the role of the exportation is to take a sentential meaning (contributed semantic value), comprised of I-canonical derivations only, and convert it to the collection of all derivations of the sentence (the root of the meaning), constituting the contributing semantic value. See its use below in the I-functions. This function facilitates "burying" the difference between the two semantic values, the source of non-direct-compositionality. Thus, in the TLG-lexicon, only meanings need to be provided.

I again introduce a means for forming certain *subtypes*, for some of the more frequently used functional types, where the argument parameter has to occupy some position in a pseudo-sentence type (i.e., preventing constant functions). The sub typing is what is known as *subsumptive*.

- t_p is a subtype of (p, t) s.t. $D_{t_p} = \{\lambda\mathbf{j}.[[S[\mathbf{j}]]] \mid S[\mathbf{j}] \in L_0^+\}$.

- $t_{p,p}$ is a subtype of $(p, (p, t))$ s.t. $D_{t_{p,p}} = \{\lambda\mathbf{k}\lambda\mathbf{j}.[[S[\mathbf{j}, \mathbf{k}]]] \mid S[\mathbf{j}, \mathbf{k}] \in L_0^+\}$.

- n is a subtype of (p, t) s.t. $D_n = \{\lambda\mathbf{j}.[[\mathbf{j} \text{ isa } X]] \mid X \text{ a noun}\}$.
  I let $\nu(\lambda\mathbf{j}.[[\mathbf{j} \text{ isa } X]]) =^{df.} X$, recovering the noun from an element of D_n.

- a is a subtype of (p, t) s.t. $D_a = \{\lambda\mathbf{j}.[[\mathbf{j} \text{ is } A]] \mid A \text{ an adjective}\}$. I let $\alpha(\lambda\mathbf{j}.[[\mathbf{j} \text{ is } A]]) =^{df.} A$, recovering the adjective from an element of D_a.

Note the deviation here from the traditional Montegovian approach which equates the types of nouns and intransitive verbs (as well as general verb-phrases), viewing them all as *predicates* of type (e, t). More on the motivation for distinguishing[1] the types of the meaning of nouns and verb-phrases can be found in Section 10.2.3.

For expressing meanings I use PT_0^+-typed variables z_1, z_2.

---

[1]For another approach that distinguishes between the meanings of nouns and verb-phrases, where the former are not predicates, see [108]. The framework is modern type-theory (MTT), and there each specific noun is a type of its own, in contrast to the unique type of nouns employed here.

$$I_e = \lambda z_1^n \lambda z_2^{t_p} \lambda \mathbf{j} \lambda \Gamma . \{\overline{\Gamma : \rho(\mathcal{D})[(\text{every } \nu(z_1))_{r(\rho(\mathcal{D}))+1}/\mathbf{j}]}^{\mathcal{D}} \stackrel{(eI^i)}{:} \\ \mathcal{D} \epsilon \mathbf{ex}(z_2(\mathbf{j}))(\Gamma, [\rho(z_1(\mathbf{j}))]_i \& \mathbf{j} \text{ fresh}\} \qquad (9.1.1)$$

Note that the *value* of z_1 does not contribute to the result, only its *noun* X is used (retrieved via ν), by augmenting Γ with $\mathbf{j}$ isa X. This reflects the role of $\mathbf{j}$ isa X as a discharged assumption in the N_0^+-derivation of $S[\text{every } X]$ from Γ.

$$I_s = \lambda z_1^n \lambda z_2^{t_p} \lambda \mathbf{j} \lambda \Gamma . \{\overline{\Gamma : \rho(\mathcal{D}_2)[(\text{some } \nu(z_1))_{r(\rho(\mathcal{D}_2))+1}/\mathbf{j}]}^{\mathcal{D}_1 \ \mathcal{D}_2} \stackrel{(sI)}{:} \\ \mathcal{D}_1 \epsilon \mathbf{ex}(z_1(\mathbf{j}))(\Gamma), \ \mathcal{D}_2 \epsilon \mathbf{ex}(z_2(\mathbf{j}))(\Gamma)\} \qquad (9.1.2)$$

Note that here the value of the argument z_2 *is* used.

Figure 9.1: I-functions for the core of N_0^+

9.1.2 I-functions

The I-functions based on the I-rules of N_0^+ are presented in Figure 9.1. Note that when I_e is applied to a non-fresh $\mathbf{j}$, the resulting set of derivations is empty. Note also the use of exportation within the I-functions, the latter forming the interface between contributed and contributing semantic values.

9.2 Extracting PTS sub-sentential meanings

In this section, I demonstrate the extraction of PT-meanings for the main E_0^+ sub-sentential phrases: determiner-phrases (*dp*s) and determiners. Thus, I have a nesting of sub-sentential phrases to cope with, where a *dp* is a sub-phrase of a sentence, and the determiner itself is a sub-phrase of a *dp*. The syntactic lexicon driving the process is again based on the (associative) Lambek calculus **L** (see the Appendix). The basic categories are n (noun[2]), dp (determiner-phrase) and s (sentence). Recall that *directed arrows* are used for forming functional categories. Abbreviate the *raised* categories[3] of dp, namely $(s \leftarrow (dp \rightarrow s))$ and $((s \leftarrow dp) \rightarrow s)$ as $dp_n \uparrow$ and $dp_a \uparrow$, respectively.

---

[2]Note that I use n both as a category and as a type; as categories and types appear in different contexts, no confusion should occur.

[3]This distinction between *nominative* and *accusative* categories is morphologically realized in the form of the determiner itself in many languages.

Following is a typical syntactic lexicon[4] for (the core of) E_0^+.

$$
\begin{array}{c|c}
nouns & n \\
adjectives & a \\
 & n \leftarrow n \\
determiners & (dp_n \uparrow\!\!\leftarrow n) \\
 & (dp_a \uparrow\!\!\leftarrow n) \\
intransitive\ verbs & (dp\!\rightarrow\!s) \\
transitive\ verbs & ((dp\!\rightarrow\!s) \leftarrow dp)
\end{array}
\qquad (9.2.3)
$$

9.2.1 Determiner-phrases and determiners

I start by noting that by applying decompositions δ (as described in (5.3.46)), I extract semantic values for subsentential phrases that serve as functions, by abstracting over the argument (shown below in detail). However, some expressions are "pure arguments" in their contribution to sentential meaning. For such expression, their meaning is obtained via (the given) meanings of ground pseudo-sentences. I have three such cases.

Nouns: Nouns are of a basic category, and can only contribute arguments. I take their meaning (of type n, a subtype of (p, t)) to originate from the corresponding (*given!*) meaning of the ground pseudo-sentence: $[\![X]\!] = \lambda \mathbf{j}.[\![\mathbf{j}\ \text{isa}\ X]\!]$.

Intransitive verbs: If P is an intransitive verb, its meaning (of type t_p) originates from the (*given*) meaning of a ground pseudo-sentence headed by it: $[\![P]\!] = \lambda \mathbf{j}.[\![\mathbf{j}\ P]\!]$.

Transitive verbs: If R is a transitive verb, its meaning (of type $t_{p,p}$) originates from the (given) meaning of a ground pseudo-sentence headed by it:

$$[\![R]\!] = \lambda \mathbf{k}\lambda \mathbf{j}.[\![\mathbf{j}\ R\ \mathbf{k}]\!].$$

The point of departure is a schematic sentence, with a quantified dp in its subject position, say of the form

$$S = \text{every}\ X\ VP \qquad (9.2.4)$$

Here VP, a verb-phrase, can be either an intransitive verb, or a transitive verb with its object dp with a quantifier, already incorporated, or with an object being an individual parameter. Consider first the first two possibilities, behaving identically. The (sentential) meaning of S is given by

$$[\![S]\!] = \lambda \Gamma.\cup_{\mathbf{j}} I_e([\![X]\!])([\![VP]\!])(\mathbf{j})(\Gamma) \qquad (9.2.5)$$

---

[4]The convention here is that each word-class abbreviates the collection of all words of that class, as sharing the same category. For example, *noun* abbreviates girl, boy, etc., and similarly for the other classes.

where the type of $[[VP]]$ is t_p. By the syntactic categorization, S has the following derivation in the **L**-based TLG.

$$\frac{\begin{array}{cc} \text{every } X & VP \\ \overline{dp_n \uparrow : Q_s} & \overline{(dp \rightarrow s) : V} \end{array}}{s : (Q_s V)} \ (\leftarrow E) \tag{9.2.6}$$

Here Q_s is a variable of type (t_p, t), and V of type t_p. From (9.2.6) we get

$$[[S]] = [[\text{every } X]]([[VP]]) \tag{9.2.7}$$

By combining (9.2.5) and (9.2.7), introducing a variable z_2 of type t_p, and abstracting over it, we obtain the following meaning (contributed semantic value) for the subject dp.

$$[[\text{every } X]] = \lambda z_2^{t_p} \lambda \Gamma . \cup_j I_e([[X]])(z_2)(\mathbf{j})(\Gamma) \tag{9.2.8}$$

Note that in the second case, I_e introduces every X with $sl = 2$ in case the value of z_2 has already an np with a quantifier incorporated.

Next, I have from the categorization and the inner derivation of every X in the grammar that

$$[[\text{every } X]] = [[\text{every}]]([[X]]) \tag{9.2.9}$$

Combining (9.2.9) with (9.2.8), we can introduce another variable z_1, of type n, and abstract over it, to obtain a meaning (contributed semantic value) for the determiner itself.

$$[[\text{every}]] = \lambda z_1^n \lambda z_2^{t_p} \lambda \Gamma . \cup_j I_e(z_1)(z_2)(\mathbf{j})(\Gamma) \tag{9.2.10}$$

By a similar analysis of, say, $S = $ some $X \, VP$, I obtain

$$[[\text{some } X]] = \lambda z_2^{t_p} \lambda \Gamma . \cup_j I_s([[X]])(z_2)(\mathbf{j})(\Gamma) \tag{9.2.11}$$

and

$$[[\text{some}]] = \lambda z_1^n \lambda z_2^{t_p} \lambda \Gamma . \cup_j I_s(z_1)(z_2)(\mathbf{j})(\Gamma) \tag{9.2.12}$$

Next, I consider the remaining case of the VP. If its object is an individual parameter, say, $\mathbf{k}$, like in every X R $\mathbf{k}$, then $[[R]]$ is of type $t_{p,p}$. The function I_e will cause the introduction of every X with index $sl = 1$, bearing a low scope to the object quantifier, once introduced too. Examples of determiners in object position are shown below, in the discussion of reconstructing quantifier scope ambiguity.

Below is a sample derivation (after β-reductions), with actual noun and (intransitive) verb. This derivation exemplifies how, during parsing, the lexical proof-theoretic meanings are combined, to reconstruct the pre-existing sentential meaning. To save space, I ignore the unions $\cup_j$, assuming $\mathbf{j}$ is fresh.

$$\cfrac{\cfrac{\text{every}}{\begin{array}{c}(dp_n \uparrow \leftarrow n):\\ \lambda z_1^n \lambda z_2^{t_p}\lambda\Gamma.I_e(z_1)(z_2)(\mathbf{j})(\Gamma)\end{array}} \quad \cfrac{\text{girl}}{\begin{array}{c}n:\\ [[\text{girl}]]\end{array}}}{\begin{array}{c}dp_n \uparrow:\\ \lambda z_2^{t_p}\lambda\Gamma.I_e([[\text{girl}]])(z_2)(\mathbf{j})(\Gamma)\end{array}}(\leftarrow E) \quad \cfrac{\cfrac{\text{smiles}}{\begin{array}{c}(dp \rightarrow s):\\ [[\text{smiles}]]\end{array}}}{}$$

$$\cfrac{}{\begin{array}{c}s:\\ \lambda\Gamma.I_e([[\text{girl}]])([[\text{smiles}]])(\mathbf{j})(\Gamma) = [[\text{every girl smiles}]]\end{array}}(\leftarrow E)$$

$$\text{(9.2.13)}$$

I end the discussion of dp meanings by showing how the quantifier-scope ambiguity is generated. Note that in the following derivation, the power of $\mathbf{L}$ to discharge assumptions is relied upon.

Our point of departure will be a schematic sentence, headed by a transitive verb, with two determiners:

$$S = \text{every } X \; R \text{ some } Y$$

Recall that such an S has quantifier-scope ambiguity, reflected in N_0^+ by the order of introduction into its two dps. By the syntactic categorization, S has two derivations in $\mathbf{L}$, one for each scopal relation, as shown below. I note here that in displaying the derivation, I deviate somewhat from the standard TLG presentation (as explained in the appendix). The reason is that, in standard TLG, meanings are themselves represented as λ-terms, that can be substituted for the respective free variables of the proof-term of the derivations, bound variables left intact. Here, meanings are proof-theoretic objects, written in a pseudo λ-notation in the meta-language. Hence, I use meanings in the derivation "on-the-fly". In particular, bound variables are those of L_0^+, over which our proof-theoretic objects are defined. Still, to keep readability, I do use free meta-variables for the proof-theoretic objects, substituting for them at the end of the derivation, and keep only bound variables as bound parameters.

$$\cfrac{\cfrac{\text{every } X}{\begin{array}{c}dp_n\uparrow:\\ Q_s\end{array}} \quad \cfrac{\cfrac{\cfrac{\cfrac{\cfrac{R}{\begin{array}{c}((dp\rightarrow s)\leftarrow dp):\\ R\end{array}} \quad [dp:\mathbf{k}]_1}{\begin{array}{c}(dp\rightarrow s):\\ R(\mathbf{k})\end{array}}(\leftarrow E) \quad [dp:\mathbf{j}]_2}{\begin{array}{c}s:\\ R(\mathbf{k})(\mathbf{j})\end{array}}(\rightarrow E)}{\begin{array}{c}(s\leftarrow dp):\\ \lambda\mathbf{k}.R(\mathbf{k})(\mathbf{j})\end{array}}(\leftarrow I_1) \quad \cfrac{\text{some } Y}{\begin{array}{c}dp_a\uparrow:\\ Q_o\end{array}}}{\begin{array}{c}s:\\ Q_o(\lambda\mathbf{k}.R(\mathbf{k})(\mathbf{j}))\end{array}}(\rightarrow E)}{\begin{array}{c}(dp\rightarrow s):\\ \lambda\mathbf{j}.Q_o(\lambda\mathbf{k}.R(\mathbf{k})(\mathbf{j}))\end{array}}(\rightarrow I_2)}{\begin{array}{c}s:\\ Q_s(\lambda\mathbf{j}.Q_o(\lambda\mathbf{k}.R(\mathbf{k})(\mathbf{j})))\end{array}}(\leftarrow E)$$

$$\text{(9.2.14)}$$

$$
\cfrac{
\cfrac{
\cfrac{
\cfrac{R}{((dp{\to}s)\leftarrow dp):}}{R} \quad [dp:\mathbf{k}]_1
}{(dp{\to}s):\;R(y)}\;(\leftarrow E)
}{\;}
$$

$$
\cfrac{
\cfrac{
[dp:\mathbf{j}]_2 \qquad \cfrac{(dp{\to}s):\;R(y)}{s:\;R(\mathbf{k})(\mathbf{j})}\;(\to E)
}{(dp{\to}s):\;\lambda\mathbf{j}.R(\mathbf{k})(\mathbf{j})}\;(\to I_2)
}{\;}
$$

$$
\cfrac{
\cfrac{\text{every }X}{\cfrac{dp_n\uparrow:}{Q_s}} \qquad
\cfrac{s:\;Q_s(\lambda\mathbf{j}.Q_o(\lambda\mathbf{k}.R(\mathbf{k})(\mathbf{j})))}{(s\leftarrow dp):\;\lambda\mathbf{k}.Q_s(\lambda\mathbf{j}.R(\mathbf{k})(\mathbf{j}))}\;(\leftarrow I_1)
\qquad
\cfrac{\text{some }Y}{\cfrac{dp_a\uparrow:}{Q_o}}
}{s:\;Q_o(\lambda\mathbf{k}.Q_s(\lambda\mathbf{j}.R(\mathbf{k})(\mathbf{j})))}\;(\to E)
$$

$$(9.2.15)$$

By substituting the dp meanings for Q_s and Q_o, namely

$$
Q_s = \lambda z_2^{t_p}\lambda\Gamma.\cup_{\mathbf{j}} I_e([\![X]\!])(z_2)(\mathbf{j})(\Gamma)
$$

$$
Q_o = \lambda z_2^{t_p}\lambda\Gamma.\cup_{\mathbf{k}} I_s([\![Y]\!])(z_2)(\mathbf{k})(\Gamma)
$$

we obtain the following two readings for the S above.

$$
[\![(\text{every }X)_2\; R\; (\text{some }Y)_1]\!] =
$$
$$
\lambda\Gamma.\cup_{\mathbf{j}} I_e([\![X]\!])(\lambda\mathbf{j}'\lambda\Gamma'.\cup_{\mathbf{k}}.I_s([\![Y]\!])(\lambda\mathbf{k}'.[\![\mathbf{j}'\; R\; \mathbf{k}']\!])])(\mathbf{k})(\Gamma'))(\mathbf{j})(\Gamma)
$$

$$(9.2.16)$$

$$
[\![(\text{every }X)_1\; R\; (\text{some }Y)_2]\!] =
$$
$$
\lambda\Gamma.\cup_{\mathbf{k}} I_s([\![Y]\!])(\lambda\mathbf{k}'\lambda\Gamma'.\cup_{\mathbf{j}} I_e([\![X]\!])(\lambda\mathbf{j}'.[\![\mathbf{j}'\; R\; \mathbf{k}']\!])(\mathbf{j})(\Gamma'))(\mathbf{k})(\Gamma)
$$

$$(9.2.17)$$

I show the first derivation. Substituting the value of Q_s into (9.2.14), renaming $\mathbf{j}$ and omitting the type of z_2, we obtain

$$
\lambda z_2\lambda\Gamma.\cup_{\mathbf{j}} I_e([\![X]\!])(z_2)(\mathbf{j})(\Gamma)(\lambda\mathbf{j}'.Q_o(\lambda\mathbf{k}.R(\mathbf{k})(\mathbf{j}')))
$$

which reduces to

$$
\lambda\Gamma.\cup_{\mathbf{j}} I_e([\![X]\!])(\lambda\mathbf{j}'.Q_o(\lambda\mathbf{k}.R(\mathbf{k})(\mathbf{j}'))(\mathbf{j})(\Gamma)
$$

Substituting the value of Q_o, renaming $\mathbf{k}$ and Γ, yields

$$
\lambda\Gamma.\cup_{\mathbf{j}} I_e([\![X]\!])(\lambda\mathbf{j}'\lambda z_2\lambda\Gamma'.\cup_{\mathbf{k}} I_s([\![Y]\!])(z_2)(\mathbf{k})(\Gamma'))(\lambda\mathbf{k}'.R(\mathbf{k}')(\mathbf{j}'))(\mathbf{j})(\Gamma)
$$

which reduces to

$$
\lambda\Gamma.\cup_{\mathbf{j}} I_e([\![X]\!])(\lambda\mathbf{j}'\lambda\Gamma'\cup_{\mathbf{k}} I_s([\![Y]\!])(\lambda\mathbf{k}'.R(\mathbf{k}')(\mathbf{j}'))(\mathbf{k})(\Gamma'))(\mathbf{j})(\Gamma)
$$

and after substituting for R its lexical meaning $[\![\mathbf{j}'\; R\; \mathbf{k}']\!]$ the result follows.

9.2.2 E_1^+: Adding Relative Clauses

I next attend the N_1^+ system. The syntactic categories of the relative pronouns are

$$\left|\begin{array}{l|l} \text{who} & ((n \leftarrow n) \leftarrow (dp{\rightarrow}s)) \\ \text{whom} & ((n{\rightarrow}n) \leftarrow (s \leftarrow dp)) \end{array}\right| \qquad (9.2.18)$$

forming a compound noun from a simpler noun and a sentence missing its object. The L-derivations of compound nouns are given below.

$$\begin{array}{c} \dfrac{X}{n} \quad \dfrac{\dfrac{\text{who}}{((n{\rightarrow}n) \leftarrow (dp{\rightarrow}s))} \quad \dfrac{S[-]}{(dp{\rightarrow}s)}}{(n{\rightarrow}n)} \; (\leftarrow E) \\ \hline n \end{array} \; ({\rightarrow}E) \qquad (9.2.19)$$

$$\begin{array}{c} \dfrac{X}{n} \quad \dfrac{\dfrac{\text{whom}}{((n{\rightarrow}n) \leftarrow (s \leftarrow dp))} \quad \dfrac{S[-]}{(s \leftarrow dp)}}{(n{\rightarrow}n)} \; (\leftarrow E) \\ \hline n \end{array} \; ({\rightarrow}E) \qquad (9.2.20)$$

9.2.3   Extracting meanings for relative clauses and relative pronouns

As before, I introduce an I-function for (rel).

$$I_r = \lambda z_1^n \lambda z_2^{t_p} \lambda \mathbf{j} \lambda \Gamma. \{ \dfrac{\mathcal{D}_1 \quad \mathcal{D}_2}{\Gamma \vdash \mathbf{j} \text{ isa } \nu(z_1) \text{ who } \rho(\mathcal{D}_2)[-]} \; (relI) \; : \; \begin{matrix} \mathcal{D}_1 \epsilon \mathbf{ex}(z_1)(\mathbf{j})(\Gamma) \\ \& \mathcal{D}_2 \epsilon \mathbf{ex}(z_2)(\mathbf{j})(\Gamma) \end{matrix} \}$$
$$(9.2.21)$$

The meaning of compound nouns resembles that of lexical nouns.

$$[\![X \text{ who}(\text{m}) \, S[-]]\!] = \lambda \mathbf{j}.[\![\mathbf{j} \text{ isa } X \text{ who}(\text{m}) \, S[-]]\!] = \lambda \mathbf{j} \lambda \Gamma. I_r([\![X]\!])([\![S[\mathbf{j}]]\!])(\mathbf{j})(\Gamma) \qquad (9.2.22)$$

By (9.2.19, 9.2.20),

$$[\![X \text{ who}(\text{m}) \, S[-]]\!] = [\![\text{who}(\text{m})]\!]([\![S[-]]\!]([\![X]\!])) \qquad (9.2.23)$$

Introducing a variable z_1^n and abstracting over it in (9.2.22), by $\mathbf{F}$, yields

$$[\![\text{who}(\text{m}) \, S[-]]\!] = \lambda z_1^n \lambda \mathbf{j} \lambda \Gamma. I_r(z_1)([\![S[\mathbf{j}]]\!])(\mathbf{j})(\Gamma) \qquad (9.2.24)$$

and by another variable introduction and abstraction over it,

$$[\![\text{who}(\text{m})]\!] = \lambda z_2^{t_p} \lambda z_1^n \lambda \mathbf{j} \lambda \Gamma. I_r(z_1)(z_2)(\mathbf{j})(\Gamma) \qquad (9.2.25)$$

9.2.4 Extracting meaning for (intersective) adjectives

Finally, I apply the meaning extraction procedure to obtain meanings for intersective adjectives. I introduce an I-function for (adj).

$$I_{adj} = \lambda z_1^n \lambda z_2^a \lambda \mathbf{j} \lambda \Gamma . \{\Gamma \vdash \mathbf{j} \text{ isa } \alpha(z_1) \nu(z_2) \; \dfrac{\mathcal{D}_1 \quad \mathcal{D}_2}{} \; (adjI) \; : \; \begin{matrix} \mathcal{D}_1 \epsilon \mathbf{ex}(z_1)(\mathbf{j})(\Gamma) \\ \& \mathcal{D}_2 \epsilon \mathbf{ex}(z_2)(\mathbf{j})(\Gamma) \end{matrix} \; \} \quad (9.2.26)$$

By abstraction, we obtain the following meaning for an adjective in its attributive role.

$$[[A]] = \lambda z_2^a \lambda z_1^n \lambda \mathbf{j} \lambda \Gamma . I_{adj}(z_1)(z_2)(\mathbf{j})(\Gamma) \quad (9.2.27)$$

9.3 The resulting E_0^+-TLG

I summarize the full grammar of E_1^+ (i.e., including PT-meanings) in the following lexicon. To fit it into the figure, I again omit the (standard) categories.

	type	meaning
nouns X	n	$\lambda \mathbf{j}.[[\mathbf{j} \; isa \; X]]$
adjectives A	a	$\lambda \mathbf{j}.[[\mathbf{j} \; is \; A]]$
	$(n,(a,n))$	$\lambda z_2^a \lambda z_1^n \lambda \mathbf{j} \lambda \Gamma . I_{adj}(z_1)(z_2)(\mathbf{j})(\Gamma)$
determiner : every	$(n,(t_p,t))$	$\lambda z_1^n \lambda z_2^{t_p} \lambda \Gamma .$ $\cup_{\mathbf{j}} I_e(z_1)(z_2)(\mathbf{j})(\Gamma)$
determiner : some	$(n,(t_p,t))$	$\lambda z_1^n \lambda z_2^{t_p} \lambda \Gamma .$ $\cup_{\mathbf{j}} I_s(z_1)(z_2)(\mathbf{j})(\Gamma)$
intransitive verbs V	t_p	$\lambda \mathbf{j}.[[\mathbf{j} \; V]]$
transitive verbs R	$t_{p,p}$	$\lambda \mathbf{k} \lambda \mathbf{j}.[[\mathbf{j} \; R \; \mathbf{k}]]$
relative pronoun	$(n,(t_p,n))$	$\lambda z_2^{t_p} \lambda z_1^n \lambda \mathbf{j} \lambda \Gamma . I_r(z_1)(z_2)(\mathbf{j})(\Gamma)$
copula : isa	(n,n)	$\lambda z^n.z$
copula : is	(a,a)	$\lambda z^a.z$

9.4   Summary

In this chapter, I have presented the procedure for extracting the meaning of subsentential phrases, down to single lexical items (words), from the sentential meanings of sentences in which the words appear. Note again the direction of flow, from sentential meanings *to* word meanings – the reverse direction employed in Montague's MTS. The extraction takes placed based on the *contributions* of the words to the sentential meanings, according to Frege's principle. Contributions are taken from the syntax given by means of TLG, where δ, the basic equality, is used for *decomposition*, not for composition. The resulting subsentential meanings are also reified (explicitly defined).

Chapter 10

A Study of Determiners' Meanings

10.1   Introduction

A cornerstone of model-theoretic semantics (MTS) is the theory of *generalized quanti-fiers (GQs),* viewed as originating from relations between sets of entities (in a model), some such relations taken standardly as the denotations of determiners; see, for example, Barwise and Cooper ([7], Peters and Westerståhl [142]. The aim of this chapter is the utilisation of determiners' meanings within *proof-theoretic semantics (PTS),* for the investigation of one of their most important properties: *conservativity,* also *defined here proof-theoretically,* without any appeal to models and entities, in contrast to their traditional study as part of the GQ-theory. I do not develop here a *full* proof-theoretical theory of determiners – just enough of such a theory to prove its feasibility and advantage over traditional GQ-theory as underlying determiner (and *dp*) meanings. In the PTS for subsentential phrases (cf. Chapter 9), a special attention is paid to the meanings of determiners. In fact, determiners are the only lexical items in the core of the fragment considered here with *extracted,* as opposed to *given,* meanings (see Chapter 9). I assume here that determiners are *regular,* in that they combine with a noun and a verb-phrase to yield a sentence. An example of non-regular determiners (in Salish languages) can be found in Matthewson [112].

The main result in this chapter is a theorem stating that every (regular) determiner is conservative in at least one of its arguments. This result provides an additional, strong argument in favor of the PTS theory of meaning, beyond the claims shown in Chapter 8 that PTS escapes the usual criticism of MTS as a theory of meaning (for NL) by various philosophers of language.

I follow here Keenan's distinction in [88] between first-argument conservativity and second-argument conservativity (explained in Section 10.3, see Definition 10.3.68). Ever since Barwise and Cooper [7], this property of determiners was *stipulated* to consist in a semantic universal. There were attempts in the literature (e.g., Keenan and Stavi [89] and Keenan [87]) to characterize GQs as closures of certain classes of basic determiners' meanings, the latter found conservative by inspection. However, this is not a full answer to the question why only conservative determiners are possible in natural language. I am able to prove the conservativity of all NL determiners, based on their proof-theoretic definitions that exclude as inexpressible non-conservative determiners. See also van Fintel and Matthewson [228] for a discussion of this universal and related ones.

In order to be able to define conservativity proof-theoretically, there is a need to appeal to a suitable proof-theoretic notion of *consequence*, to replace the model-theoretic consequence employed by GQ-theory, and some form of a proof-theoretic *conjunction*, to replace the model-theoretic intersection in boolean models employed by MTS. The definition of a proof-theoretic notion of *consequence* I employ here is the one in Definition 8.5.64 (in Chapter 8). Recall that this definition of consequence is based on *propagating grounds (for assertion)* (see Definition 8.5.63): every Γ that derives canonically S_2, already derives canonically S_1.

The proof-theoretic conjunction suitable for the NL-fragment is defined in terms of ND-rules for *relative clauses*, present in E_1^+ (in 8.1.2).

Regarding only, there is a disagreement in the literature whether it is a determiner at all. For those who consider it to be so, it sometimes is considered exceptional, as not being conservative, since it is not conservative in its first argument. However, it was observed that it *is* conservative in its second argument[1].   I adhere to this approach, and the definition of conservativity does cover only (and related determiners). This point is clarified in Section 10.3.

For comprehensibility, the chapter is structured as having two parts. The first part deals with *positive* determiners only, using the unilateral PTS defined in Chapter 8. The second part, then, presents the extension with negative determiners, appealing, as mentioned above, to bilateralism (cf. 4.4). I propose the following definition for the distinction between positive and negative determiners.

Definition 10.1.66 (determiners classification). *A determiner D is positive iff for every noun X and verb-phrase V (see later for the notation) the sentence D X V is* compatible *with the sentence* every X V. *Otherwise, D is negative.*

Compatibility here is taken *proof-theoretically* as the coherence (see Section 4.4.1.3) of accepting both sentences. Incompatibility, thus, means that accepting one sentences enforces, by coherence, rejecting the other. Note that for a negative determiner D to be negative (hence incompatible with every), j V has to be *denied* for at least one j satisfying j isa X, leading naturally to bilateralism. Thus, at least two is positive.

---

[1] A property attributed to focus effects. References are given in Section 10.3

$$\frac{\Gamma : \mathbf{j} \text{ isa } X \quad \Gamma : S[\mathbf{j}]}{\Gamma : S[(\text{donk } X)_{r(S[\mathbf{j}])+1}]} \; (donkI)$$

$$\frac{\Gamma : S[(\text{donk } X)_{r(S[\mathbf{j}])+1}] \quad \Gamma : \mathbf{j} \text{ isa } X \quad \Gamma, S[\mathbf{j}] : S'}{\Gamma : S'} \; (donkE)$$

Figure 10.1: The disharmonious determiner *donk*

For example, at least two girls smile is compatible with every girl smiles. On the other hand, at most two is negative, as at most two girls smile is not, in general, compatible with every girl smiles.

For another way to see the need of harmony, I introduce a determiner *donk*, with an *I*-rule of some and an *E*-rule of every (see Figure 10.1), that obviously is disharmonious. Since the rules here are not on the sentential level, the effect of the disharmony of *donk* is not as devastating as that of the connective *tonk*, the latter trivializing the whole deducibility relation '⊢'. Still, its effect is bad enough, as will be shown below. Similarly to the case of every, $(donkE)$ gives rise to the following special case as a derived rule $(donk\hat{E})$.

$$\frac{\Gamma : S[(\text{donk } X)_{r(S[\mathbf{k}])+1}] \quad \Gamma : \mathbf{k} \text{ isa } X}{\Gamma : S[\mathbf{k}]} \; (donk\hat{E})$$

Now, consider the following derivation, introducing an irreducible maximal formula containing *donk*, failing local-soundness. For simplicity, It is presented without contexts and with scope-levels ignored.

$$\frac{\dfrac{\mathbf{j} \text{ isa } X \quad S[\mathbf{j}]}{S[\text{donk } X]} \; (donkI) \quad \mathbf{k} \text{ isa } X}{S[\mathbf{k}]} \; (donk\hat{E})$$

The effect of disharmony here is the following trivialization of predication: if *something* which is an X satisfies S, then everything which is an X satisfies S – an all or nothing effect.

10.2 The proof-theoretic meaning of determiners

I now turn to a discussion of the general structure of the proof-theoretic meaning of determiners, as reflected by their proof-theoretic type interpretation (in Section 9.1.1).

10.2.1   The general structure of (DI) and (DE)

What is a possible form of an I-rule for a (regular) determiner D? The regularity implies that the determiner combines with a noun, say X, and then with a verb-phrase, say $S[-]$, to form a sentence S of the form $D \, X \, S[-]$. Hence, (DI) may use premises of the form $\mathbf{j}$ isa X, and $S[\mathbf{j}]$ (possibly for more than one $\mathbf{j}$, like $(\geq mI)$). Any such premise can be discharged (like in (eI) or (oI)), or not, like in (sI) or $(\geq mI)$. This structure is reflected in the general form of a determiner meaning, as presented in Section 10.2.2.

From this general form of (DI) and the harmony requirement, (and based on the explicit construction in Section 3.6.3), the following observation regarding the general form of the corresponding (DE) rules can be made, and will be used below.

Observation:

1. If an I-rule discharges an assumption S, then the corresponding GE-rule has a premise $\Gamma : S$.

2. If an I-rule has a premise $\Gamma : S$, then the corresponding GE-rule has a premise $\Gamma, S : S'$, for the S' which forms the arbitrary conclusion of the rule, discharging the assumption S.

I distinguish between *basic* (regular) determiners, like the ones in N_1^+ (and negative ones like no and at most n, added later (in Section 10.6)), and *compound* (regular) determiners, like *possessives* (like every/some girl's) and *coordinated* ones (like at least n and/or at most m). Compound determiners are deferred to Section 10.3.2.

10.2.2   Meanings of basic determiners

Unlike nouns and verbs, the meaning of a basic determiner is *extracted*. The process by which the meaning is extracted is described in details in Chapter 9 for the determiners every and some.   However, this extraction depends only on the syntactic category of determiners (in an underlying type-logical grammar for E_1^+), a category reflecting the regularity assumption, and hence is readily extensible for any regular determiner.

In general, it is convenient for the definition of the meaning of a basic determiner D to use an *I-function* I_D, induced by the respective I-rule(s). These functions are presented in Figure 10.2 (cf. 10.2.1). To avoid notational clutter, I abuse the notation by using z_2 instead of $\rho(z_2)$.

The following notes may be helpful for the understanding of the definitions of these I-functions:

$$I_e = \lambda z_1^n \lambda z_2^{t_p} \lambda \mathbf{j} \lambda \Gamma . \left\{ \frac{\mathcal{D}}{\Gamma : z_2(\mathbf{j})[(\text{every } \nu(z_1))_{r(z_2(\mathbf{j}))+1}/\mathbf{j}]} (eI^i) \right\} : \begin{array}{c} \mathcal{D} \in [\![z_2(\mathbf{j})]\!]^*_{\Gamma,[z_1(\mathbf{j})]_i} \text{ \&} \\ \mathbf{j} \text{ fresh for } \Gamma \end{array}$$

$$I_s = \lambda z_1^n \lambda z_2^{t_p} \lambda \mathbf{j} \lambda \Gamma . \left\{ \frac{\mathcal{D}_1 \quad \mathcal{D}_2}{\Gamma : z_2(\mathbf{j})[(\text{some } \nu(z_1))_{r(z_2(\mathbf{j}))+1}/\mathbf{j}]} (sI) \right\} : \begin{array}{c} \mathcal{D}_1 \in [\![z_1(\mathbf{j})]\!]^*_{\Gamma} \text{ \&} \\ \mathcal{D}_2 \in [\![z_2(\mathbf{j})]\!]^*_{\Gamma} \end{array}$$

$$I_o = \lambda z_1^n \lambda z_2^{t_p} \lambda \mathbf{j} \lambda \Gamma . \left\{ \frac{\mathcal{D}}{\Gamma : z_2(\mathbf{j})[(\text{only } \nu(z_1))_{r(z_2(\mathbf{j}))+1}/\mathbf{j}]} (oI^i) \right\} : \begin{array}{c} \mathcal{D} \in [\![z_1(\mathbf{j})]\!]^*_{\Gamma,[z_2(\mathbf{j})]_i} \text{ \&} \\ \mathbf{j} \text{ fresh for } \Gamma \end{array}$$

$$I_{\geq m} = \lambda z_1^n \lambda z_2^{t_p} \lambda \mathbf{j}_1 \cdots \lambda \mathbf{j}_m \lambda \Gamma . \left\{ \frac{\mathcal{D}_1^1 \cdots \mathcal{D}_m^1 \quad \mathcal{D}_1^2 \cdots \mathcal{D}_m^2}{\Gamma : z_2(\mathbf{j}_1)[(\text{at least } m \; \nu(z_1))_{r(z_2(\mathbf{j}_1))+1}/\mathbf{j}_1]} (\geq mI) : \mathcal{D}_i^k \in [\![z_k(\mathbf{j}_i)]\!]^*_{\Gamma} \right\}$$

Figure 10.2: I-functions for N_1^+

- If, for instance, $z_1 = \lambda \mathbf{k}.[\![\mathbf{k} \text{ isa } X]\!]$ and $z_2 = \lambda \mathbf{k}.[\![S[\mathbf{k}]]\!]$, then $z_1(\mathbf{j}) = \mathbf{j}$ isa X, $z_2(\mathbf{j}) = S[\mathbf{j}]$.

- Note the interface between contributed and contributing semantic values, manifested by the use of sets of *all* derivations (e.g., $[\![z_1(\mathbf{j})]\!]^*_{\Gamma}$).

- In I_e, the *value* of z_1 does not contribute to the result, only its *noun* X is used (retrieved via ν), by augmenting Γ with $\mathbf{j}$ isa X. This reflects the role of $\mathbf{j}$ isa X as a discharged assumption in the N_1^+-derivation of e.g., $S[(\text{every } X)]$ from Γ.

- In I_s, on the other hand, the value of the argument z_1 *is* used.

- When I_e or I_o are applied to a non-fresh $\mathbf{j}$, the resulting set of derivations is empty.

The general form of the (extracted) meaning of a basic determiner D is given by:

$$[\![D]\!] = \lambda z_1^n \lambda z_2^{t_p} \lambda \Gamma . \bigcup_{\mathbf{j}_1,\ldots,\mathbf{j}_m \in \mathcal{P}} I_D(z_1)(z_2)(\mathbf{j}_1) \cdots (\mathbf{j}_m)(\Gamma) \tag{10.2.1}$$

where I_D is the I-function for D. I refer to z_1^n as the *nominal argument*, and to $z_2^{t_p}$ as the *verbal argument*. The retrieval function associated with determiners' meanings is 'δ', given by

$$\delta([\![D]\!]) = D \tag{10.2.2}$$

For the determiners **every**, **some** and **only**, $m = 1$. However, for **at least** m, m may be greater than 1. Note that, while the type of an I-function is $(n, (t_p, (p^m, t)))$, where m is the number of parameters, the type of a determiner's meaning is $(n, (t_p, t))$. Thus, the meaning of a determiner does not apply directly to parameters. I abbreviate $(\mathbf{j}_1) \cdots (\mathbf{j}_m)$ to $(\bar{\mathbf{j}})$. I stress the fact that the nominal argument and verbal argument have *different types* in the type system employed here (n and t_p, respectively), which play different roles in the I-rules. This is in strong contrast to standard MTS semantic typing, where there is only one type of predicate, namely (e, t), and both arguments of a determiner are of the same type. Some observations related to **every**, based on this difference in typing, are presented in Section 10.2.3.

I note here that for *negative* determiners, like no and at most n, there is a certain change in the details (mainly, the type proof-theoretic interpretation), but the results do not change, and the conservativity theorem proved below holds for negative (regular) determiners too, as shown in Section 10.6. From the type of the proof-theoretic basic determiners' meanings it can be seen, that these meanings are taken from a much more restricted domain than the one GQs, which are taken as underlying determiners' meanings by MTS, are taken from. GQs are definable by means of arbitrary properties of sets (subsets of the universe of a model), including reference to their cardinalities, and thereby give rise easily to non-conservative GQs. Thus, GQ-based determiners like the following, which are non-conservative in the MTS definition, have no counterpart in the realm of determiners' proof-theoretic meanings. Let A and B be arbitrary subsets of the domain E of any model.

$$G_1(A)(B) \Leftrightarrow |A| > |B|, \quad G_2(A)(B) \Leftrightarrow |A| = |B| \quad G_3(A)(B) \Leftrightarrow (E - A) \subseteq B$$
$$(10.2.3)$$

Another discrepancy of determiners cannot arise: dependency of their MT-denotation on the cardinality of the domain. For example, a definition like

$$[[D]] = \{ \begin{array}{ll} [[\text{every}]] & |E| \geq 100 \\ [[\text{some}]] & |E| < 100 \end{array} \qquad (10.2.4)$$

In MTS, such denotations are excluded from being a determiner's meaning by requiring (ext), the invariance of the denotation under the extension of the universe of the model (cf. Peters and Westerståhl [142], p. 138).

I emphasize again that the I_D function has access only to its two arguments, though each argument can be applied to any number of individual parameters. This generates a fairly restricted class of possible determiner meanings. I classify the I_D functions according to the discharging policy of their corresponding I-rules.

non-discharging: The I-rule does not discharge any of its premises. Such rules have a certain symmetry, further discussed below. The I-rules for some and at least m are non-discharging.

nominal discharge: The I-rule discharges its nominal argument. every is of this form.

verbal discharge: The I-rule discharges its verbal argument. only is of this form.

10.2.3   Digression: the meaning of every and of nouns

There is an important consequence of the every I-rule (reflected in its type as shown above) regarding a difference between PTS (as proposed here) and (current proposals for) MTS. In standard MTS, the meanings of the two complements of every (one a noun, the other a verb-phrase) have *the same semantic type*; namely, they are both[2]

---

[2]In addition, this is also the type of adjectives when used attributively.

predicates (of type (e, t)), having arbitrary subsets of the domain of individual elements as their extensions in models. In the truth-conditions assigned to $S[(\text{every } X)]$, both complements have similar roles. Thus, if $\forall x. P(x) \rightarrow Q(x)$ is semantically well-formed, so is $\forall x. Q(x) \rightarrow P(x)$. This does not predict the asymmetry of the roles of the nominal predicate and the verbal predicate (noted already in Barwise and Cooper [7]), and the *semantic* anomaly of (*) every smiles girl, or (*) every loves Rachel smiles, leaving them to be ruled out by the syntax.

On the other hand, our (eI) I-rule[3] is *asymmetric* w.r.t. the types of its two arguments. The discharged assumption is that of a nominal predicate (j isa X), and its role in the rule *cannot* be filled by a predicate originating in an intransitive verb (or a verb-phrase)! Thus the semantic abnormality of (*) every smiles girl is predicted by the rule. So, our PTS makes finer type-distinctions among predicates, according to their semantic contribution to sentential meaning. This does not imply that no finer MTS can be proposed, incorporating such a distinction; however, it is not obvious on what to base such a refinement within a model-theoretic denotation-based type-system.

The role of a noun X in $S[(\text{every } X)]$ is *not* that of a predicate. Rather, it *determines (or specifies) the domain of quantification*! For a recent persuasive argument to that effect, see Ben-Yami [10]. See also Francez [45] for a more general discussion of related matters. The notion of a universal domain of quantification, consisting of "property-less" objects, may well be an artefact of MTS (following the use of FOL in mathematics), not really needed[4] for conferring meaning on natural language sentences with localized quantification. In that, the view here of universal quantification in NL, as *evidence-transforming*, is different *both* from the model-theoretic view as ranging over some external domain of objects, and from some view in logic, by means of ranging over all substitution instances.

I mention in passing that in Hebrew, for example, since present tense forms of verbs are analogous to nouns (or adjectives), the analogue of (*) every smiles girl *is* semantically well-formed[5]; thus, different rules should be used for a PTS of a similar fragment of Hebrew. This fact also demonstrates that the Montagovian semantic type system is too coarse.

---

[3] The (sI) rule is also, in a less apparent way, asymmetric.

[4] Even when a *dp* like 'everything' is considered, unrestricted quantification is not used, as languages provide nouns such as 'thing', whose extension in MTS should be the whole domain of quantification.

[5] The proper rendering every smiles girl into English is by means of a relative clause, e.g., every thing that smiles is a girl., or by every smiler is a girl.

10.3   Provable conservativity of determiners' meanings

10.3.1   Simple determiners

First-argument conservativity is manifested in NL by *equivalences* (mutual entailment) like that between (10.3.5a) and (10.3.5b)

$a.$ every student smiles
$b.$ every student is a student who smiles.   (10.3.5)

Apparently, only does not satisfy the analog equivalence between (10.3.6a) and (10.3.6b).

$a.$ Only boys smile.
$b.$ Only boys are boys who smile.   (10.3.6)

and was considered by some as an exception to the universal that all NL determiners are conservative.

However, as has been noted by several authors (e.g., Herburger [72], de Mey [25]), for some determiners related to *focus*, notably[6] only, the equivalence is between[7] (10.3.7a) and (10.3.7b), exhibiting second-argument conservativity.

$a.$ only students smile
$b.$ only smilers are smilers who are students.   (10.3.7)

This is based on viewing (10.3.7a) equivalent to every smiler is a student (note the change of the argument that is relativized).

In [240], Zuber presents a general definition of the *inverse* D^{-1} of a determiner D, defined by

$$D^{-1}(A)(B) = D(B)(A) \qquad (10.3.8)$$

(for $A, B \subseteq E$), whereby only^{-1} = every. Zuber brings such determiners from Polish, and considers them, like only, to be non-conservative. I use a proof-theoretic variant of the definition of the inverse determiner, and by the definition of conservativity (see below), that requires conservativity in at least one of the arguments, they do not fall out of the class of conservative determiners.

Apparently, in the definition of conservativity, there is some form of *conjunction* of the noun's meaning with that of the VP meaning. In MTS, this conjunction is defined directly as set intersection. In our PTS setting, however, these two meanings are not necessarily in the same domain, so it does not make sense to define conjunction as set intersection.

Alternatively, I define a form of proof-theoretic conjunction, based on the rules for *relativization* in Section 8.3.5. Examples for expressions conjoin able by means of this

---

[6]Still, some consider only not to be a determiner at all.
[7]I ignore here the issue that only requires its arguments to be plural form.

$$I_r = \lambda z_1^n \lambda z_2^{t_p} \lambda \mathbf{j} \lambda \Gamma . \left\{ \frac{\mathcal{D}_1 \qquad \mathcal{D}_2}{\Gamma : z_1(\mathbf{j} \text{ who } z_2(\mathbf{j})[-/\mathbf{j}])} \; rI : \begin{array}{l} \mathcal{D}_1 \in [[z_1(\mathbf{j})]]_\Gamma^* \; \& \\ \mathcal{D}_2 \in [[z_2(\mathbf{j})]]_\Gamma^* \end{array} \right\}$$

Figure 10.3: I-function for relativization in N_1^+

proof-theoretic conjunction are (10.3.9a), in which a parameter is removed from the subject position in **j loves every girl**, and (10.3.9b), in which a parameter is removed from the object position in **every girl loves k**.

$$\begin{array}{ll} a. \text{ loves every girl} \\ b. \text{ every girl loves} \end{array} \qquad (10.3.9)$$

The I-function for relativization is defined in Figure 10.3, and is of type $(n, (t_p, (p, t_p)))$. Using this I-function, I define *proof-theoretic conjunction* as an operator '$\sqcap$' of type $(n, (t_p, (p, t_p)))$, taking the meaning of a noun and the meaning of sentence missing a subject or an object, and returning the meaning of the compound noun that is composed of the two arguments with a relative clause introduction.

$$\sqcap =^{df.} \lambda z_1^n \lambda z_2^{t_p} \lambda \mathbf{j} \lambda \Gamma . I_r(z_1)(z_2)(\mathbf{j})(\Gamma) \qquad (10.3.10)$$

In order to be able to express the change of the relativized argument brought about by second-argument conservative determiners like **only**, I assume a 1-1 *nominalization mapping* **n**, mapping verb-phrases to nouns. For example,

$$\mathbf{n}(\text{smile}) = \text{smiler}, \quad \mathbf{n}(\text{love Schubert}) = \text{Schubert lover} \qquad (10.3.11)$$

This mapping satisfies, for every **j** and V,

$$[[\mathbf{j} \; V]] = [[\mathbf{j} \text{ isa } \mathbf{n}(V)]] \qquad (10.3.12)$$

For example, $[[\mathbf{j} \text{ smiles}]] = [[\mathbf{j} \text{ isa smiler}]]$. I postulate the following closure property of our fragment.

Nominalization closure: The fragment E_1^+ is closed under **n**.

Apparently, full English satisfies this closure under some mild idealization. In Hebrew, this property holds with **n** being the identity function, since every verb (in present tense) is also a noun. I also assume that the externally given meaning of ground sentences, relative to which our PTS is defined, respects **n**, so that $[[P]] = [[\mathbf{n}(P)]]$.

In addition, I need a second conversion function, **v**, that maps nouns to a verb-phrases, using copular predication. For example,

$$\mathbf{v}(\text{girl}) = \text{isa girl} \qquad (10.3.13)$$

Again, the externally given meanings for ground pseudo-sentences are assumed to satisfy $[\![X]\!] = [\![\mathbf{v}(X)]\!]$. It is easily seen that the fragment E_1^+ is closed under $\mathbf{v}$. By abuse of notation, I apply the functions $\mathbf{n}$ and $\mathbf{v}$ also to *meanings* of the appropriate type. Thus, $\mathbf{n}([\![\text{smile}]\!])$ means $[\![\mathbf{n}(\text{smile})]\!]$.

I now define the inverse of a determiner in this framework.

Definition 10.3.67 (inverse determiner).

$$D^{-1}(a)(b) =^{\text{df.}} D(\mathbf{n}(b))(\mathbf{v}(a)) \qquad\qquad (10.3.14)$$

The type of D^{-1} is the same as that of D. Note that for symmetric determiners like some, $D = D^{-1}$.

Definition 10.3.68. *(Conservativity)*

- *A determiner D is 1-conservative iff for all $a \in D_n$ and $b \in D_{t_p}$:*

  $$[\![D]\!](a)(b) \Vdash [\![D]\!](a)(\sqcap(a)(b)) \text{ and } [\![D]\!](a)(\sqcap(a)(b)) \Vdash [\![D]\!](a)(b)$$

- *A determiner D is 2-conservative iff D^{-1} is 1-conservative, namely, for all $a \in D_n$ and $b \in D_{t_p}$:*

  $$[\![D]\!](a)(b) \Vdash [\![D]\!](\mathbf{n}(b))(\sqcap(\mathbf{n}(b))(\mathbf{v}(a)))$$

  and

  $$[\![D]\!](\mathbf{n}(b))(\sqcap(\mathbf{n}(b))(\mathbf{v}(a))) \Vdash [\![D]\!](a)(b)$$

- *A determiner D is* conservative *iff it is either 1-conservative or 2-conservative.*

The following lemma follows directly from the definitions.

Lemma 10.3.1. *If $a \in D_n$ and $b \in D_{t_p}$, then there are a noun X and a pseudo-sentence $S[\mathbf{j}]$, such that $a = \lambda\mathbf{j}.[\![\mathbf{j} \text{ isa } X]\!]$ and $b = \lambda\mathbf{j}.[\![S[\mathbf{j}]]\!]$, and*

- $a(\mathbf{j}) = [\![\mathbf{j} \text{ isa } X]\!]$

- $b(\mathbf{j}) = [\![S[\mathbf{j}]]\!]$

- $(\sqcap(a)(b))(\mathbf{j}) = [\![\mathbf{j} \text{ isa } X \text{ who } S[-]]\!]$

- $[\![D]\!](a)(b) = [\![D \ X \ S[-]]\!]$

- $[\![D]\!](a)(\sqcap(a)(b)) = [\![D \ X \text{ isa } X \text{ who } S[-]]\!]$

Theorem 10.3.11 (conservativity). *Every basic determiner is conservative.*

Proof: I prove the result here only for positive basic (regular) determiners. However, as shown in Section 10.6.4, the results hold for the negative ones too. Let D be any positive (regular) determiner. To simplify the notation, assume w.l.o.g. that only one individual parameter $\mathbf{j}$ is involved in the application of I_D (otherwise, a vectored notation $\bar{\mathbf{j}}$ needs to be used). Consider two cases.

I_D **discharges no assumption.** In this case, I claim that D is both 1-conservative and 2-conservative. To establish 1-conservativity, I have to show that $[\![D]\!](a)(b) \Vdash D(a)(\sqcap(a)(b))$ and
$[\![D]\!](a)(\sqcap(a)(b)) \Vdash [\![D]\!](a)(b)$.

1. By Lemma 10.3.1, there are a noun X and a verb-phrase $S[-]$ s.t. $[\![D]\!](a)(b)$ is $[\![D\ X\ S[-]]\!]$. Suppose that for some Γ it holds that $\vdash^c \Gamma : D\ X\ S[-]$. By the non-discharge assumption, there is a canonical derivation

$$\frac{\Gamma : \mathbf{j}\ \text{isa}\ X \quad \Gamma : S[\mathbf{j}]}{\Gamma : D\ X\ S[-]}\ (DI)$$

So, by Observation 2 above, (DE) has as its discharged assumptions $\mathbf{j}$ isa X and $S[\mathbf{j}]$. But in this case, there is also a canonical derivation $\Gamma \vdash^c D\ X$ isa X who $S[-]$ (displayed in stages with contexts omitted to fit he page). Let

$$\mathcal{D}_1 : \frac{D\ X\ S[-] \quad \dfrac{[\mathbf{j}\ \text{isa}\ X]_1, [S[\mathbf{j}]]_2}{\mathbf{j}\ \text{isa}\ X}\ (Ax)}{\mathbf{j}\ \text{isa}\ X}\ (DE^{1,2})$$

$$\mathcal{D}_2 : \frac{D\ X\ S[-] \quad \dfrac{[\mathbf{j}\ \text{isa}\ X]_3, [S[\mathbf{j}]]_4}{S[\mathbf{j}]}\ (Ax)}{S[\mathbf{j}]}\ (DE^{3,4})$$

Then, the main derivation is

$$\frac{\dfrac{\mathcal{D}_1 \quad \mathcal{D}_2}{\dfrac{\mathbf{j}\ \text{isa}\ X \quad S[\mathbf{j}]}{\mathbf{j}\ \text{isa}\ X\ \text{who}\ S[-]}\ (rI)} \quad \dfrac{\mathcal{D}_2}{S[\mathbf{j}]}}{D\ X\ \text{isa}\ X\ \text{who}\ S[-]}\ (DI)$$

This establishes $[\![D]\!](a)(b) \Vdash [\![D]\!](a)(\sqcap(a)(b))$.

2. By Lemma 10.3.1, there are a noun X and a verb-phrase $S[-]$ s.t. $[\![D]\!](a)(\sqcap(a)(b))$ is $[\![D\ X\ \text{isa}\ X\ \text{who}\ S[-]]\!]$. Suppose that for some Γ it holds that $\vdash^c \Gamma : D\ X$ isa X who $S[-]$. Therefore, there is a canonical derivation

$$\frac{\Gamma : \mathbf{j}\ \text{isa}\ X\ \text{who}\ S[-] \quad \Gamma : S[\mathbf{j}]}{\Gamma : D\ X\ \text{isa}\ X\ \text{who}\ S[-]}\ (DI)$$

So, by Observation 2 above, (DE) has as its discharged assumptions $\mathbf{j}$ isa X who $S[-]$ and $S[\mathbf{j}]$. But then, also $\vdash^c \Gamma : D\ X\ S[-]$, as shown below. Let

$$\mathcal{D}_1: \quad \dfrac{D\ X \text{ isa } X \text{ who } S[-] \qquad \dfrac{\dfrac{[\mathbf{j} \text{ isa } X \text{ who } S[-]]_1, [S[\mathbf{j}]]_2}{S[\mathbf{j}]}\ (Ax)}{S[\mathbf{j}]}\ (DE^{1,2})}{S[\mathbf{j}]}$$

$$\mathcal{D}_2: \quad \dfrac{\dfrac{D\ X \text{ isa } X \text{ who } S[-] \qquad \dfrac{[\mathbf{j} \text{ isa } X \text{ who } S[-]]_3, [S[\mathbf{j}]]_4}{\mathbf{j} \text{ isa } X \text{ who } S[-]}\ (Ax)}{\mathbf{j} \text{ isa } X \text{ who } S[-]}\ (DE^{3,4})}{\mathbf{j} \text{ isa } X}\ (r\hat{E})$$

Then the main derivation is

$$\dfrac{\begin{array}{cc}\mathcal{D}_1 & \mathcal{D}_2\\ S[\mathbf{j}] & \mathbf{j} \text{ isa } X\end{array}}{D\ X\ S[-]}\ (DI)$$

This establishes $[[D]](a)(\sqcap(a)(b)) \Vdash [[D]](a)(b)$. To show 2-conservativity, the argument is similar. Actually, since in this case $D = D^{-1}$, one can repeat the above argument for D^{-1}.

I_D **discharges an assumption.** This case is split in two, depending on which is the discharged assumption.

The discharged assumption is the nominal argument: In this case D is 1-conservative. Again, I have to show that
$[[D]](a)(b) \Vdash [[D]](a)(\sqcap(a)(b))$ and $[[D]](a)(\sqcap(a)(b)) \Vdash [[D]](a)(b)$.

1. Once again, by Lemma 10.3.1, there are a noun X and a verb-phrase $S[-]$ s.t. $[[D]](a)(b)$ is $[[D\ X\ S[-]]]$. Suppose that for some Γ it holds that $\vdash^c \Gamma : D\ X\ S[-]$. By the discharge assumption, there is a derivation

$$\dfrac{\Gamma, \mathbf{j} \text{ isa } X : S[\mathbf{j}]}{\Gamma : D\ X\ S[-]}\ (DI)$$

By Observation 1, (DE) has a premise $\mathbf{j}$ isa X, and by Observation 2, it has a premise with a discharged assumption $S[\mathbf{j}]$. Hence, there is also a derivation

$$\dfrac{\dfrac{D\ X\ S[-] \qquad \dfrac{[\mathbf{j} \text{ isa } X]_1 \quad [S[\mathbf{j}]]_2}{\mathbf{j} \text{ isa } X \text{ who } S[-]}\ (rI)}{\mathbf{j} \text{ isa } X \text{ who } S[-]}\ (DE^2)}{D\ X \text{ isa } X \text{ who } S[-]}\ (DI^1)$$

which establishes $[[D]](a)(b)\mathbf{j} \Vdash [[D]](a)(\sqcap(a)(b))$.

2. By Lemma 10.3.1, there are a noun X and a verb-phrase $S[-]$ s.t. $[[D]](a)(\sqcap(a)(b))$ is $[[D\ X \text{ isa } X \text{ who } S[-]]]$. Suppose that for some Γ it holds that $\vdash^c \Gamma : D\ X \text{ isa } X \text{ who } S[-]$. Therefore, using

Observations 1 and 2, the required canonical derivation is

$$\cfrac{D\ X \text{ isa } X \text{ who } S[-] \quad [\mathbf{j} \text{ isa } X]_1 \quad \cfrac{\cfrac{[\mathbf{j} \text{ isa } X \text{ who } S[-]]_2}{S[\mathbf{j}]} (r\hat{E})}{S[\mathbf{j}]} (DE^2)}{\cfrac{S[\mathbf{j}]}{D\ X\ S[-]} (DI^1)}$$

Establishing $[\![D]\!](a)(\sqcap(a)(b)) \Vdash [\![D]\!](a)(b)$.

The discharged assumption is the verbal argument: In this case, D is 2-conservative. This argument is similar, actually arguing the same about D^{-1}.

Below, I show typical examples for the various clauses of the proof.

- A typical case of I_D not discharging any assumption is some. We have the following canonical derivations.

$$\mathcal{D}_1: \cfrac{\text{some girl smiles} \quad \cfrac{[\mathbf{j} \text{ isa girl}]_1, [\mathbf{j} \text{ smiles}]_2}{\mathbf{j} \text{ isa girl}} (Ax)}{\mathbf{j} \text{ isa girl}} (sE^{1,2})$$

$$\mathcal{D}_2: \cfrac{\text{some girl smiles} \quad \cfrac{[\mathbf{j} \text{ isa girl}]_3, [\mathbf{j} \text{ smiles}]_4}{\mathbf{j} \text{ smiles}} (Ax)}{\mathbf{j} \text{ smiles}} (sE^{3,4})$$

Then, the main derivation is

$$\cfrac{\cfrac{\mathcal{D}_1 \qquad \mathcal{D}_2}{\cfrac{\mathbf{j} \text{ isa girl} \quad \mathbf{j} \text{ smiles}}{\mathbf{j} \text{ isa girl who smiles}} (rI)} \qquad \cfrac{\mathcal{D}_2}{\mathbf{j} \text{ smiles}}}{\text{some girl isa girl who smiles}} (sI)$$

and

$$\mathcal{D}_1: \cfrac{\text{some girl isa girl who smiles} \quad \cfrac{[\mathbf{j} \text{ isa girl who smiles}]_1, [\mathbf{j} \text{ smiles}]_2}{\mathbf{j} \text{ smiles}} (Ax)}{\mathbf{j} \text{ smiles}} (sE^{1,2})$$

$$\mathcal{D}_2: \cfrac{\text{some girl isa girl who smiles} \quad \cfrac{\cfrac{[\mathbf{j} \text{ isa girl who smiles}]_3, [S[\mathbf{j}]]_4}{\mathbf{j} \text{ isa girl who smiles}} (Ax)}{} (sE^{3,4})}{\cfrac{\mathbf{j} \text{ isa girl who smiles}}{\mathbf{j} \text{ isa girl}} (r\hat{E})}$$

The main derivation is

$$\cfrac{\cfrac{\mathcal{D}_1 \qquad \mathcal{D}_2}{\mathbf{j} \text{ smiles} \quad \mathbf{j} \text{ isa girl}}}{\text{some girl smiles}} (sI)$$

- A typical case for I_D discharging its nominal argument assumption is every. We have the following derivations.

$$\cfrac{\text{every girl smiles} \quad \cfrac{[\mathbf{j}\text{ isa girl}_1 \quad [\mathbf{j}\text{ smiles}]_2}{\mathbf{j}\text{ isa girl who smiles}}\ (rI)}{\cfrac{\mathbf{j}\text{ isa girl who smiles}}{\text{every girl isa girl who smiles}}\ (eI^1)}\ (eE^2)$$

and

$$\cfrac{\text{every girl isa girl who smiles} \quad [\mathbf{j}\text{ isa girl}]_1 \quad \cfrac{[\mathbf{j}\text{ isa girl who smiles}]_2}{\mathbf{j}\text{ smiles}}\ (r\hat{E})}{\cfrac{\mathbf{j}\text{ smiles}}{\text{every girl smiles}}\ (eI^1)}\ (eE^2)$$

- A typical case of I_D discharging its verbal argument assumption is only. We have the following derivations.

$$\cfrac{\text{only girls smile} \quad \cfrac{[\mathbf{j}\text{ isa girl}]_2 \quad [\mathbf{j}\text{ smiles}]_1}{\mathbf{j}\text{ isa smiler who isa girl}}\ (rI)}{\cfrac{\mathbf{j}\text{ isa smiler who isa girl}}{\text{only smilers are smilers who are girls}}\ (oI^1)}\ (oE^2)$$

and

$$\cfrac{\text{only smilers are smilers who are girls} \quad [\mathbf{j}\text{ smiles}]_1 \quad [\mathbf{j}\text{ isa smilerl who isa girl}]_2}{\cfrac{\cfrac{\mathbf{j}\text{ isa smiler who isa girl}}{\mathbf{j}\text{ isa girl}}\ (r\hat{E})}{\text{only girls smile}}\ (oI^1)}\ (oE^2)$$

10.3.2   Compound determiners

I now attend to the conservativity compound determiners.

10.3.2.1   Possessives

I first add to the fragment *possessive* determiners. To avoid a certain complication involved in considering proper names (cf. Section 8.3.3), I do not treat here possessives like Jacob's sheep, and focus on *quantified possessives* like

$$\text{every/some girl's dog barked} \tag{10.3.15}$$

I focus here on positive, *pre-nominal* possessives like in (10.3.15). The extension to negative possessives, like no girl's dog barked is added later (in Section 10.6.3.1). To simplify the technical presentation, I ignore here *partitive* possessives. Below I list some of the pre-theoretic attributes of possessives' meanings.

1. The quantification is over *possessors*. Thus, in (10.3.15), the quantification is over girls, not over dogs.

2. As observed by Barker [5], the quantification in (10.3.15) is not over *all* girls – only over girls who own dogs; otherwise, we may have incorrect inferences. This phenomenon is referred to as (primary) *narrowing*.

3. There is a second, implicit, quantifier involved, over the second argument, namely over the possessed entities. Thus, (10.3.15) can be paraphrased in (at least) three ways.

   (a) Definite: Every/some girl who owns a dog owns exactly one dog, and that dog barked.

   (b) Universal: Every/some girl who owns a dog owns any number of dogs, and all of them barked.

   (c) Existential: Every/some girl who owns a dog owns any number of dogs, and some of them barked.

   Here 'them' is the e-type pronoun meanings 'the dogs owned by the girl'. It is sometimes claimed (wrongly!) that only the definite reading obtains. More often, the implicit quantification is considered as determined by context. Note that this quantification is also over a *narrowing*, but of the second argument, called secondary narrowing: in (10.3.15), the second, implicit quantification is over dogs owned by the girl bound by the main quantification. A detailed MTS detailing this issue can be found in Francez [41].

4. The actual *full* lexical meaning of the possession relation is left *free*, not further specified, as it can vary considerably, not always in a systematic way.

5. The scope of the quantifier within the possessive is always higher than the possession relation (between possessor and possessed) in the matrix clause. Thus, in (10.3.15) (in the universal case), it is not meant that a dog possessed by every girl barked ... However, the usual quantifier-scope ambiguity between *dp*s is present with possessives too. Thus, in (10.3.16) below

   $$\text{every girl's dog chased some cat} \qquad (10.3.16)$$

   the usual two readings are present. One by which there is one (specific) cat chased by every girl's dog, and another by which the chased cats vary with the girl.

Note that the MTS of possessives faces difficulties in attaining compositionality, leading to the employment of unorthodox semantic operations (see Francez [41]).

The first step in devising the PTS is to augment the proof-language L_1^+ with a ground pseudo-sentence $j\mathcal{R}k$, representing the possession of $\mathbf{k}$ by $\mathbf{j}$. This augmentation takes care of the freedom of the possession relation by $\mathcal{R}$ left further unspecified. Similarly to the MTS treatment of the possession marker $'s$, it is defined as a function from a pair of a determiner's meaning, say D, and a noun's meaning, say X, returning a determiner's meaning. This view deviates from simple compositionality, which would require $'s$ to have *one* argument, the result of applying D to X. This deviation is necessary because of the primary narrowing, that modifies the argument X to X who $\mathcal{R}$ **k**, where **k** satisfies the argument Y of $\delta X's Y$. In the example above (in the universal case), every is changed from applying to girl to applying to girl who owns some dog.

The way I treat $'s$ in the proof-theoretic setting is to define the appropriate I-function $I_{(D\,X's\,Y)}$ in terms of I_D and I_Q, where Q is the implicit quantifier related to above. For simplicity, I present[8] the definition with a single parameter **j**, avoiding vectorized notation.

$$[\![{}'s]\!] = \begin{array}{l} \lambda u_0^{(n,(t_p,(p,t)))} \lambda u_1^n \lambda z_1^n \lambda z_2^{t_p} \lambda \mathbf{j} \lambda \Gamma. \\ I_{\delta(u_0)}(I_r(u_1)(I_s(z_1)([\![\mathcal{R}]\!])(\mathbf{j})(\Gamma))(\mathbf{j})(\Gamma)) \\ (I_Q(I_r(z_1)([\![\mathcal{R}]\!])(\mathbf{j})(\Gamma))(\mathbf{j})(\Gamma)(z_2))(\mathbf{j})(\Gamma) \end{array} \tag{10.3.17}$$

For a determiner D and noun X, we get

$$\begin{array}{l} [\![D\,X's]\!] = [\![{}'s]\!]([\![D]\!])([\![X]\!]) = \lambda z_1^n \lambda z_2^{t_p} \lambda \Gamma. \\ \cup_{\mathbf{j}\in\mathcal{P}}[\![D]\!](I_r([\![X]\!])(I_s(z_1)([\![\mathcal{R}]\!])(\mathbf{j})(\Gamma))(\mathbf{j})(\Gamma)) \\ (I_Q(I_r(z_1)([\![\mathcal{R}]\!])(\mathbf{j})(\Gamma))(\mathbf{j})(\Gamma)(z_2))(\mathbf{j})(\Gamma) \end{array} \tag{10.3.18}$$

For a determiner D and nouns X, Y, this generates the following I-rule for $D\,X's\,Y$ (recall that dps are introduced, not determiners).

$$\begin{array}{ll} I_{(D\,X's\,Y)} & = [\![{}'s]\!]([\![D]\!])([\![X]\!])([\![Y]\!]) \\ & = \lambda z_2^{t_p} \lambda \mathbf{j} \lambda \Gamma. \\ & I_D(I_r([\![X]\!])(I_s([\![Y]\!])([\![\mathcal{R}]\!])(\mathbf{j})(\Gamma))(\mathbf{j})(\Gamma)) \\ & (I_e(I_r([\![Y]\!])([\![\mathcal{R}]\!])(\mathbf{j})(\Gamma))(z_2)(\mathbf{j})(\Gamma))(\mathbf{j})(\Gamma) \end{array} \tag{10.3.19}$$

The corresponding E-rule is the one harmoniously-induced by the above I-rule, depending on D.

To understand how this definition works, consider the application of $'s$ to every, then to girl and finally to dog, choosing Q also as every.

$$\begin{array}{ll} I_{(\text{every girl's dog})} & = [\![{}'s]\!]([\![\text{every}]\!])([\![\text{girl}]\!])([\![\text{dog}]\!]) \\ & = \lambda z_2^{t_p} \lambda \mathbf{j} \lambda \Gamma. \\ & I_e(I_r([\![\text{girl}]\!])(I_s([\![\text{dog}]\!])([\![\mathcal{R}]\!])(\mathbf{j})(\Gamma))(\mathbf{j})(\Gamma)) \\ & (I_e(I_r([\![\text{dog}]\!])([\![\mathcal{R}]\!])(\mathbf{j})(\Gamma))(z_2)(\mathbf{j})(\Gamma))(\mathbf{j})(\Gamma) \end{array} \tag{10.3.20}$$

Let us see the meanings of some of the subexpressions.

- $I_s([\![\text{dog}]\!])([\![\mathcal{R}]\!])(\mathbf{j})(\Gamma)$: **j** $\mathcal{R}$ some dog (namely, **j** possesses some dog).

---

[8]Note that functions here are Curried. A binary function applied to its two arguments is displayed as $f(a)(b)$, not as $f(a,b)$.

$$\frac{\Gamma, \mathbf{j} \text{ isa } X, \mathbf{d} \text{ isa } Y, \mathbf{j}\,\mathcal{R}\,\mathbf{d} : S[\mathbf{d}]}{\Gamma : S[(\text{every } X's\ Y)]} \ ((e\ X's\ Y)I) \qquad \frac{\Gamma : \mathbf{j} \text{ isa } X \quad \Gamma : \mathbf{d} \text{ isa } Y \quad \Gamma : \mathbf{j}\,\mathcal{R}\,\mathbf{d} \quad \Gamma : S[\mathbf{d}]}{\Gamma : S[(\text{some } X's\ Y)]} \ ((s\ X's\ Y)I)$$

$$\frac{\Gamma : S[(\text{every } X's\ Y)] \quad \Gamma : \mathbf{j} \text{ isa } X \quad \Gamma : \mathbf{d} \text{ isa } Y \quad \Gamma : \mathbf{j}\mathcal{R}\mathbf{d} \quad \Gamma, S[\mathbf{d}] : S'}{\Gamma \vdash S'} \ ((e\ X's\ Y)E)$$

$$\frac{\Gamma : S[(\text{some } X's\ Y)] \quad \Gamma, \mathbf{j} \text{ isa } X, \mathbf{d} \text{ isa } Y, \mathbf{j}\mathcal{R}\mathbf{d} : S'}{\Gamma : S'} \ ((s\ X's\ Y)E)$$

where **j**, **d** are fresh for Γ in $((e\ X's\ Y)I)$, $((s\ X's\ Y)E)$.

Figure 10.4: I/E-rules for **every/some** X's Y

- $I_e(I_r([\![\text{girl}]\!])(I_s([\![\text{dog}]\!])([\![\mathcal{R}]\!])(\mathbf{j})(\Gamma))(\mathbf{j})(\Gamma))(\mathbf{j})(\Gamma)$:   every girl who possesses some dog. This shows the effect of primary narrowing.

- $I_e(I_r([\![\text{dog}]\!])([\![\mathcal{R}]\!])(\mathbf{j})(\Gamma))(z_2)(\mathbf{j})(\Gamma)$ (after substituting $[\![\text{barked}]\!]$ for z_2): every dog possessed by **j** barked' exhibiting secondary narrowing and universal implicit quantification on dogs possessed by girls.

Figure 10.4 displays the I/E-rules for **every/some** X's Y induced by $[\![\ 's]\!]$. Scope levels are not shown for brevity. The E-rules are the harmoniously-induced GE-rules.

For example, when X, Y are instantiated to **girl** and **dog**, respectively, we get the following I-rule for **every girl's dog**

$$\frac{\Gamma, \mathbf{j} \text{ isa girl}, \mathbf{d} \text{ isa dog}, \mathbf{j}\,\mathcal{R}\,\mathbf{d} : S[\mathbf{d}]}{\Gamma : S[(\text{every girl's dog})]} \ ((e\ \text{girl's dog})I) \tag{10.3.21}$$

Next, I attend to the conservativity of quantified possessive. Their 1-conservativity is exhibited by equivalences like the one between (10.3.22 a) and (10.3.22 b) below.

$a.$ every girl's dog barked
$b.$ every girl's dog is a dog that barked
$\tag{10.3.22}$

However, since our meanings include narrowing and implicit quantification on the possessed, what one has to show is actually the equivalence between the following[9] two explications of (10.3.22 a) and (10.3.22 b):

$a.$ every girl who possesses some dog is such that Q dogs she possesses barked
$b.$ every girl who possesses some dog is such that Q dogs she possesses are dogs that barked

$\tag{10.3.23}$

I now prove a theorem to the effect of the propagation of conservativity from the quantifier within the possessive to the possessive determiner itself.

Theorem 10.3.12 (conservativity of possessives). *For every determiner D and noun X, D X's is conservative in every argument for which D is conservative.*

---

[9]I am not familiar with any alluding to this observation in the literature.

Proof: Suppose D is 1-conservative. Let $a \in D_n$ and $b \in D_{t_p}$. $[[D\ X's]](a)(b) \iff$ $[[D\ X's]](a)(\sqcap(a,b))$ follows immediately, using (10.3.18), from the conservativity of D, the topmost operator in (10.3.18), allowing to conjoin its second argument with the first.

The case of 2-conservativity is similar and omitted.

Note that this step, defining an I-rule in terms of other I-rules, is innovative in proof-theory, where usually I-rules are given directly, not in terms of other I-rules.

10.3.2.2   Coordinated determiners

Another way in which compound determiners can be formed is by means of *coordination*, using coordinating words like and/or/... Examples are

$$a. \text{ at least } n \text{ and at most } m \text{ girls smiled}$$
$$b. \text{ all and only girls smiled} \tag{10.3.24}$$

(equating here all with every).

The proof-theoretic framework provides a convenient and uniform way to define the meaning of coordinated determiners. It mimics the traditional way of forming *sentential coordination* (recall that the latter is not assumed in the fragment under consideration). I exemplify this way for and and or. Let D_1, D_2 be two determiners. Then, $(D_1 \text{ and } D_2)I$ and $(D_1 \text{ or } D_2)I$ are given below, where r abbreviates $r(S[\mathbf{j}]) + 1$.

$$\frac{\Gamma : S[(D_1 X)_r] \quad \Gamma : S[(D_2 X)_r]}{\Gamma : S[(D_1 \text{ and } D_2\ X)_r]} \ ((D_1 \text{ and } D_2)\ I)$$

$$\frac{\Gamma : S[(D_1 X)_r]}{\Gamma : S[(D_1 \text{ or } D_2\ X)_r]} \ ((D_1 \text{ or } D_2)_1\ I) \tag{10.3.25}$$

$$\frac{\Gamma : S[(D_1 X)_r]}{\Gamma : S[(D_2 \text{ or } D_2\ X)_r]} \ ((D_1 \text{ or } D_2)_2\ I)$$

Note that the coordinated determiner phrase is introduced at the same scope as the separate dps have in the premises, the latter implicitly assumed to relate to the *same* dp position within S. Note also that the rule for conjoining determiners is different than the rule for conjoining nps, which is

$$\frac{\Gamma : S[(D_1 X)_r] \quad \Gamma : S[(D_2 Y)_r]}{\Gamma : S[(D_1\ X \text{ and } D_2\ Y)_r]} \ ((D_1\ X \text{ and } D_2\ Y)\ I)$$

The corresponding harmonious GE-rules are shown below.

$$\frac{\Gamma : S[(D_1 \text{ and } D_2\ X)_r] \quad \Gamma, S[(D_1\ X)_r], S[(D_2\ X)_r] : S'}{\Gamma : S'} \quad ((D_1 \text{ and } D_2)E)$$

$$\frac{\Gamma : S[(D_1 \text{ or } D_2\ X)_r] \quad \Gamma, S[(D_1\ X)_r] : S' \quad \Gamma, S[(D_2\ X)_r] : S'}{\Gamma : S'} \quad ((D_1 \text{ or } D_2)E)$$

$$(10.3.26)$$

Theorem 10.3.13 (conservativity of coordinated determiners). *The coordinated determiners D_1 and D_2 and D_1 or D_2 are conservative in any argument in which both D_1 and D_2 are conservative.*

Proof: I show 1-conservativity of D_1 and D_2; the other proofs are similar and omitted. Assume both D_1 and D_2 are 1-conservative, and suppose $(*)[[(D_1 \text{ and } D_2)]](a)(b)$ holds. I have to show that $(**)[[(D_1 \text{ and } D_2)]](a)(\sqcap(a,b))$ holds too.

From $(*)$ and Lemma 10.3.1, there are a noun X and a verb-phrase $S[-]$ s.t. $[[(D_1 \text{ and } D_2)]](a)(b)$ is $[[(D_1 \text{ and } D_2)\ X\ S[-]]]$. Suppose that for some Γ it holds that

$$\Gamma \vdash (D_1 \text{ and } D_2)\ X\ S[-] \tag{10.3.27}$$

We thus have (again used the derived simple E-rules)

$$\frac{\Gamma : (D_1 \text{ and } D_2)\ X\ S[-]}{\Gamma : D_1\ X\ S[-]} \quad ((D_1 \text{ and } D_2)\hat{E}) \qquad \frac{\Gamma : (D_1 \text{ and } D_2)\ X\ S[-]}{\Gamma : D_2\ X\ S[-]} \quad ((D_1 \text{ and } D_2)\hat{E})$$

$$(10.3.28)$$

This means that both $[[D_1]](a)(b)$ and $[[D_2]](a)(b)$ hold. By the assumed 1-conservativity of both D_1 and D_2, both $[[D_1]](a)(\sqcap(a,b))$ and $[[D_2]](a)(\sqcap(a,b))$ hold. Hence,

$$\Gamma : D_1\ X \text{ isa } X \text{ that } S[-], \quad \Gamma : D_2\ X \text{ isa } X \text{ that } S[-] \tag{10.3.29}$$

Hence,

$$\frac{\Gamma : D_1\ X \text{ isa } X \text{ that } S[-] \quad \Gamma : D_2\ X \text{ isa } X \text{ that } S[-]}{\Gamma : (D_1 \text{ and } D_2)\ X \text{ isa(are) } X \text{ that } S[-]} \quad ((D_1 \text{ and } D_2)I) \quad (10.3.30)$$

establishing $(**)$, as required.

10.4 Introducing negative determiners

In this part of the chapter, I extend the PTS also for *negative* determiners like no and at most n, using *bilateralism* (cf. Section 4.4), the latter claimed to be the most suitable approach for the former task.

Recall that Bilateralism considers *denial* as a primitive speech act, on par with *assertion*, the latter underlying my previously proposed PTS. Let the resulting fragment be E_1, and let its extension for formulating the proof system be L_1 (omitting the '+' superscript indicating positivity of the fragment).

I show here how negative determiners can be treated in an extension of the PTS framework discussed above, while avoiding the use of the controversial sentence level negation. To this end, I move to a *bilateral ND-proof system* for the fragment (explained below), and again mark sentences in the ND proof-system with a *force marker*, indicating *assertion* and *denial*.

After reconsidering the general form of the meaning of determiners, I extend the results in the first part of the chapter about conservativity to negative determiners too, using their bilateral meanings. In order to achieve this extension, I need to adapt this property to the bilateral setup, in particular revising the notion of proof-theoretic consequence. I demonstrate the application of the revised definition of conservativity on some determiners in the extended fragment, both positive and negative.

10.5   A Bilateral PTS for Sentences

10.5.1   Bilateralism

I adopt again the notation of Rumfitt [179] and add a force marker, '+' or '−', in front of each sentence of L_1 (cf. Section 4.4).

A positively marked sentence of the form '$+S$' is to be understood as indicating an *assertion*, referred also as acceptance, of S, while a negatively marked sentence '$-S$' is to be understood as indicating *denial*, referred to also as rejection, of S. A context Γ is extended to contain both positively and negatively marked sentences. Thus, a sequent like $\Gamma : +S$ (respectively, $\Gamma : -S$) indicates that accepting all sentences in Γ that are marked with '+' and rejecting all sentences in Γ that are marked with '−', implies accepting (respectively, rejecting) S. I prefer the use of the terms *acceptance* and *rejection* here, over *assertion* and *denial*, because I need to apply these notions not only to E_1 sentences, but also to pseudo-sentences. And it sounds odd to say that a pseudo-sentence is *asserted* or *denied*. Saying that a pseudo-sentence is *accepted* or *rejected* sounds (at least a little bit) better.

Meta-variables α range again over force-marked sentences, combinations of a force marker and a content (like $+S$ or $-S$). I let $att(\alpha)$ retrieve the force of α, and $cnt(\alpha)$ retrieve its content. Thus, $att(+S) = +$, $att(-S) = -$ and $cnt(+S) = cnt(-S) = S$.

Note again that $-S$ should not be confused with an operator expressing the sentential negation of S, the latter absent from our fragments. In particular, negation can be nested, while $+ + S$, $- + S$ and the like are not well formed. Also, sub-sentential phrases, like sub-formulas in logics, can be negated, while only full sentences can be marked with '+' or '−'.

10.5.2   The Sentential Bilateral ND Proof System

There are two main new questions to be answered:

1. How to define assertion and denial rules for negative determiners?

2. How to define denial rules for the positive determiners.

Let us reexamine the I-rules for every and some from Figure 8.1. They can be seen as saying that accepting j isa X, and also accepting j P, implies accepting some X P.[10] And whenever accepting j isa X (for a fresh j) leads to accepting j P, accepting every X P follows.

Under what circumstances should no X P be accepted? A possible answer, adopted here, is that accepting j isa X (for a fresh j) leads to *rejecting* j P. Extending this idea leads to the bilateral sentential PTS presented below.

The I-rules and E-rules for sentences marked with '+' (acceptance) are given in Figure 10.5. The rules for every and some are similar to the old ones from Figure 8.1. I again refer to such rules, obtained from unilateral rules by marking '+' all the premises and the conclusion, as the *natural adaptation* of unilateral rules to bilateral ones. The new rules are for no and not every. The motivation for the I-rule for no has already been described above. The idea behind the I-rule for not every is that accepting j isa X but rejecting j P implies accepting not every X P.

Since I have now positively marked sentences as well as negatively marked ones, I also define I-rules (and E-rules) for the latter. These are shown in Figure 10.6.

Note that accepting (respectively, rejecting) every X P is "equivalent" to rejecting (respectively, accepting) not every X P. Similarly, accepting (respectively, rejecting) some X P is "equivalent" to rejecting (respectively, accepting) no X P. This is an outcome of the horizontal harmony of N_1 (see Section 4.4.1.7).

The rules in Figure 10.5 together with the rules in Figure 10.6, and Figure 10.7 for relativization, separated for convenience, comprise the proof-system N_1.

---

[10]For the purpose of the presentation, I occasionally use simple sentences of the form det X P instead of the general, but more cumbersome, form $S[(\det X)]$.

$$\frac{}{\Gamma, +S : +S} \ (Ax^+)$$

$$\frac{\Gamma, +(\mathbf{j} \text{ isa } X) : +S[\mathbf{j}]}{\Gamma : +S[(\text{every } X)_{r(S[\mathbf{j}])+1}]} \ (eI^+) \qquad \frac{\Gamma : +(\mathbf{j} \text{ isa } X) \qquad \Gamma : +S[\mathbf{j}]}{\Gamma : +S[(\text{some } X)_{r(S[\mathbf{j}])+1}]} \ (sI^+)$$

$$\frac{\Gamma, +(\mathbf{j} \text{ isa } X) : -S[\mathbf{j}]}{\Gamma : +S[(\text{no } X)_{r(S[\mathbf{j}])+1}]} \ (nI^+) \qquad \frac{\Gamma : +(\mathbf{j} \text{ isa } X) \qquad \Gamma : -S[\mathbf{j}]}{\Gamma : +S[(\text{not every } X)_{r(S[\mathbf{j}])+1}]} \ (neI^+)$$

$$\frac{\Gamma : +S[(\text{every } X)_{r(S[\mathbf{j}])+1}] \quad \Gamma : +(\mathbf{j} \text{ isa } X) \quad \Gamma, +S[\mathbf{j}] : \alpha}{\Gamma : \alpha} \ (eE^+)$$

$$\frac{\Gamma : +S[(\text{some } X)_{r(S[\mathbf{j}])+1}] \quad \Gamma, +(\mathbf{j} \text{ isa } X), [+S[\mathbf{j}]]_i : \alpha}{\Gamma : \alpha} \ (sE^+)$$

$$\frac{\Gamma : +S[(\text{no } X)_{r(S[\mathbf{j}])+1}] \quad \Gamma : +(\mathbf{j} \text{ isa } X) \quad \Gamma, -S[\mathbf{j}] : \alpha}{\Gamma : \alpha} \ (nE^+)$$

$$\frac{\Gamma : +S[(\text{not every } X)_{r(S[\mathbf{j}])+1}] \quad \Gamma, +(\mathbf{j} \text{ isa } X), -S[\mathbf{j}] : \alpha}{\Gamma : \alpha} \ (neE^+)$$

where $\mathbf{j}$ is fresh for Γ in (eI^+) and (nI^+), and for Γ, α in (sE^+) and (neE^+).

Figure 10.5: Rules for acceptance

$$\frac{}{\Gamma, -S : -S} \ (Ax\ )$$

$$\frac{\Gamma : +(\mathbf{j}\ \mathsf{isa}\ X) \qquad \Gamma : -S[\mathbf{j}]}{\Gamma : -S[(\mathsf{every}\ X)_{r(S[\mathbf{j}])+1}]} \ (eI^-) \qquad\qquad \frac{\Gamma, +(\mathbf{j}\ \mathsf{isa}X) : -S[\mathbf{j}]}{\Gamma : -S[(\mathsf{some}\ X)_{r(S[\mathbf{j}])+1}]} \ (sI^-)$$

$$\frac{\Gamma : +(\mathbf{j}\ \mathsf{isa}\ X) \qquad \Gamma : +S[\mathbf{j}]}{\Gamma : -S[(\mathsf{no}\ X)_{r(S[\mathbf{j}])+1}]} \ (nI^-) \qquad\qquad \frac{\Gamma, +(\mathbf{j}\ \mathsf{isa}X) : +S[\mathbf{j}]}{\Gamma : -S[(\mathsf{not\ every}\ X)_{r(S[\mathbf{j}])+1}]} \ (neI^-)$$

$$\frac{\Gamma : -S[(\mathsf{every}\ X)_{r(S[\mathbf{j}])+1}] \quad \Gamma, +(\mathbf{j}\ \mathsf{isa}\ X), -S[\mathbf{j}] : \alpha}{\Gamma : \alpha} \ (eE^-)$$

$$\frac{\Gamma : -S[(\mathsf{some}\ X)_{r(S[\mathbf{j}])+1}] \quad \Gamma : +(\mathbf{j}\ \mathsf{isa}\ X) \quad \Gamma, -S[\mathbf{j}] : \alpha}{\Gamma : \alpha} \ (sE^-)$$

$$\frac{\Gamma : -S[(\mathsf{no}\ X)_{r(S[\mathbf{j}])+1}] \quad \Gamma, +(\mathbf{j}\ \mathsf{isa}\ X), +S[\mathbf{j}] : \alpha}{\Gamma : \alpha} \ (nE^-)$$

$$\frac{\Gamma : -S[(\mathsf{not\ every}\ X)_{r(S[\mathbf{j}])+1}] \quad \Gamma : +(\mathbf{j}\ \mathsf{isa}\ X) \quad \Gamma, +S[\mathbf{j}] : \alpha}{\Gamma : \alpha} \ (neE^-)$$

where **j** is fresh for Γ in (sI^-) and (neI^-), and for Γ, α in (eE^-) and (nE^-).

Figure 10.6: Rules for rejection

$$\frac{\Gamma : +(\mathbf{j}\ \mathsf{isa}\ X) \qquad \Gamma : +S[\mathbf{j}]}{\Gamma : +(\mathbf{j}\ \mathsf{isa}\ X\ \mathsf{who}\ S[-])} \ (rI^+)$$

$$\frac{\Gamma : +(\mathbf{j}\ \mathsf{isa}\ X\ \mathsf{who}\ S[-]) \qquad \Gamma, +(\mathbf{j}\ \mathsf{isa}\ X), +S[\mathbf{j}] : \alpha}{\Gamma : \alpha} \ (rE^+)$$

$$\frac{\Gamma : -(\mathbf{j}\ \mathsf{isa}\ X)}{\Gamma : -(\mathbf{j}\ \mathsf{isa}\ X\ \mathsf{who}\ S[-])} \ (rI^-)_1$$

$$\frac{\Gamma : -S[\mathbf{j}]}{\Gamma : -(\mathbf{j}\ \mathsf{isa}\ X\ \mathsf{who}\ S[-])} \ (rI^-)_2$$

$$\frac{\Gamma : -(\mathbf{j}\ \mathsf{isa}\ X\ \mathsf{who}\ S[-]) \qquad \Gamma, -(\mathbf{j}\ \mathsf{isa}\ X) : \alpha \qquad \Gamma, -S[\mathbf{j}] : \alpha}{\Gamma : \alpha} \ (rE^-)$$

Figure 10.7: Rules for relativization

It is useful to also have the following derived E-rules for every and no, easily derivable using the respective E-rules in Figure 10.5.

$$\frac{\Gamma:+S[(\text{every }X)_{r(S[j])+1}] \quad \Gamma:+(\textbf{j isa }X)}{\Gamma:+S[\textbf{j}]} \quad (e\hat{E}^+)$$

$$\frac{\Gamma:+S[(\text{no }X)_{r(S[j])+1}] \quad \Gamma:+(\textbf{j isa }X)}{\Gamma:-S[\textbf{j}]} \quad (ne\hat{E}^+)$$

Similarly, the following derived E-rules for not every and some are derivable using
the respective E-rules in Figure 10.6.

$$\frac{\Gamma:-S[(\text{not every }X)_{r(S[j])+1}] \quad \Gamma:+(\textbf{j isa }X)}{\Gamma:-S[\textbf{j}]} \quad (ne\hat{E}^-)$$

$$\frac{\Gamma:-S[(\text{some }X)_{r(S[j])+1}] \quad \Gamma:+(\textbf{j isa }X)}{\Gamma:-S[\textbf{j}]} \quad (s\hat{E}^-)$$

Rules for some additional determiners, both positive and negative, are presented in
Figure 10.8. An inspection reveals that the rules for at least m generalize those for
some, where the latter is a special case where $m = 1$. Similarly, the rules for at most
m generalizes those for no, where the latter is a special case where $m = 0$. Below is
an example of a derivation using both acceptance and rejection rules.

Example 10.5.60. *I show that*

$$\vdash + (\text{every editor})_2 \text{ read (at least two manuscripts)}_1 :$$
$$+(\text{no editor})_2 \text{ read (at most one manuscript)}_1$$

and

$$\vdash + (\text{no editor})_2 \text{ read (at most one manuscript)}_1 :$$
$$+(\text{every editor})_2 \text{ read (at least two manuscripts)}_1$$

I use some self-evident abbreviation of predicate names, and atm, atl for at most
and at least, *respectively.*

$$\frac{\cfrac{[+\textbf{j isa ed}]_1 \quad +(\text{every ed})_2\, R\,(\text{atl 2 ms})_1}{+\textbf{j}\, R\,\text{atl 2 ms}}\,(e\hat{E}^+) \quad \cfrac{[+\textbf{k isa ms}]_2 \quad [+\textbf{m isa ms}]_3 \quad [+\textbf{j}\, R\, \textbf{k}]_4 \quad [+\textbf{j}\, R\, \textbf{m}]_5}{-\textbf{j}\, R\, \text{atm 1 ms}}\,(\geq 2E^-_{2,3,4,5})}{\cfrac{-\textbf{j}\, R\, \text{atm 1 ms}}{+(\text{no ed})_2\, R\,(\text{atm 1 ms})_1}\,(nI^+_1)}\,(\leq 1I^-)$$

$$\frac{\cfrac{[+\textbf{j isa ed}]_1 \quad +\text{no ed}\, R\,\text{atm 1 ms}}{-\textbf{j}\, R\, \text{atm 1 ms}}\,(n\hat{E}^+) \quad \cfrac{[\textbf{k isa ed}]_2 \quad [\textbf{m isa ed}]_3 \quad [\textbf{j}\, R\, \textbf{k}]_4 \quad [\textbf{j}\, R\, \textbf{m}]_5}{+\textbf{j}\, R\, \text{atl 2 ms}}\,(\geq eI^+)}{\cfrac{+\textbf{j}\, R\, \text{atl 2 ms}}{(\text{every ed})_2\, R\,(\text{atl 2 ms})_1}\,(eI^+_1)}\,(\leq 1E^-_{2,3,4,5})$$

The definition of canonical derivations is carried over from N_1^+ to N_1, so that now
a derivation is *canonical* if it essentially ends with any kind of I-rule (positive or
negative).

Definition 10.5.69 (bilaterally canonical derivation). *A derivation is* bilaterally canonical *iff it essentially ends with an application of an I-rule (either I^+ or I^-).*

$$\frac{\{\Gamma : +\mathbf{j}_i \text{ isa } X\}_{1\le i\le m} \qquad \{\Gamma : +S[\mathbf{j}_i]\}_{1\le i\le m}}{\Gamma : +S[(\text{at least m } X)]_{r(S[\mathbf{j}_1])+1}} \ (\ge mI^+)$$

$$\frac{\Gamma, \{+S[\mathbf{j}_i]\}_{1\le i\le m}, \{+\mathbf{j}_i \text{ isa } X\}_{1\le i\le m+1} : -S[\mathbf{j}_{m+1}]}{\Gamma : +S[(\text{at most m } X)]_{r(S[\mathbf{j}_1])+1}} \ (\le mI^+)$$

$$\frac{\Gamma, \{+S[\mathbf{j}_i]\}_{1\le i\le m}, \{+\mathbf{j}_i \text{ isa } X\}_{1\le i\le m} : -S[\mathbf{j}_m]}{\Gamma : -S[(\text{at least m } X)]_{r(S[\mathbf{j}_1])+1}} \ (\ge mI^-)$$

$$\frac{\{\Gamma : +\mathbf{j}_i \text{ isa } X\}_{1\le i\le m+1} \qquad \{\Gamma : +S[\mathbf{j}_i]\}_{1\le i\le m+1}}{\Gamma : -S[(\text{at most m } X)]_{r(S[\mathbf{j}_1])+1}} \ (\le mI^-)$$

$$\frac{\Gamma : +S[(\text{at least m } X)]_{r(S[\mathbf{j}_1])+1} \qquad \Gamma, \{+\mathbf{j}_i \text{ isa } X\}_{1\le i\le m}, \{+S[\mathbf{j}_i]\}_{1\le i\le m} : \alpha}{\Gamma : \alpha} \ (\ge mE^+)$$

$$\frac{\Gamma : +S[(\text{at most m } X)]_{r(S[\mathbf{j}_1])+1} \quad \{\Gamma : +S[\mathbf{j}_i]\}_{1\le i\le m} \quad \{\Gamma : +\mathbf{j}_i \text{ isa } X\}_{1\le i\le m+1} \quad \Gamma, -S[\mathbf{j}_{m+1}] : \alpha}{\Gamma : \alpha} \ (\le mE^+)$$

$$\frac{\Gamma : -S[(\text{at least m } X)]_{r(S[\mathbf{j}_1])+1} \quad \{\Gamma : +S[\mathbf{j}_i]\}_{1\le i\le m-1} \quad \{\Gamma : +\mathbf{j}_i \text{ isa } X\}_{1\le i\le m} \quad \Gamma, -S[\mathbf{j}_m] : \alpha}{\Gamma : \alpha} \ (\ge mE^-)$$

$$\frac{\Gamma : -S[(\text{at most m } X)]_{r(S[\mathbf{j}_1])+1} \qquad \Gamma, \{+\mathbf{j}_i \text{ isa } X\}_{1\le i\le m+1}, \{+S[\mathbf{j}_i]\}_{2\le i\le m+2} : \alpha}{\Gamma : \alpha} \ (\le mE^-)$$

In each rule the $\mathbf{j}_i$s are pairwise distinct (meaning e.g., $|\{\mathbf{j}_1,\ldots,\mathbf{j}_m\}| = m$.)
In $(\le mI^+)$, $(\ge mI^-)$, the $\mathbf{j}_i$s are fresh for Γ;
in $(\ge mE^+)$ and $(\le mE^-)$, the $\mathbf{j}_i$s are fresh for $\Gamma \cup \{\alpha\}$.

Figure 10.8: Rules for at least/at most m

10.5.3 Coherence (of contexts)

Recall that an important property of bilateral contexts is that of *coherence* (cf. Section 4.4.1.3), restated below in terms of E_1.

Definition 10.5.70 (NL-Coherent context). *A (bilateral) context Γ is coherent iff for every sentence S, at most one of $\vdash\Gamma : +S$ and $\vdash\Gamma : -S$ holds. Otherwise, Γ is incoherent.*

It is easy to see that, having the structural rule $(*)$ (see Section 10.5.4), everything is derivable from an incoherent context.

Example 10.5.61. *Some examples of incoherent contexts are listed below.*

- $\Gamma, +S, -S.$

- $+\text{some } X\ P, +\text{no } X\ P$, *since:*

$$\cfrac{+\text{some } X\ P \quad \cfrac{[+\mathbf{j} \text{ isa } X]_1 \quad [+\mathbf{j}\ P]_2}{-\text{no } X\ P}\,(nI^-)}{-\text{no } X\ P}\,(sE^+_{1,2})$$

- $+\text{no } X \text{ isa } X, +\mathbf{j} \text{ isa } X$, *since:*

$$\cfrac{+\text{no } X \text{ isa } X \qquad +\mathbf{j} \text{ isa } X}{-\mathbf{j} \text{ isa } X}\,(nE)$$

The following proposition follows immediately by using the structural rule $(*)$.

Proposition 10.5.13 (incoherence). *If $\vdash\Gamma : +S$ (respectively, $\vdash\Gamma : -S$) holds, then $\Gamma, -S$ (respectively, $\Gamma, +S$) is incoherent.*

10.5.4 More structural rules

Just as before, I still assume the structural rules of *weakening* (W), *contraction* (C) and *exchange* (E). Additionally, I also assume the following structural rule, reminiscent of the intuitionistic negation elimination (cf. Section 2.1.4).

$$\cfrac{\Gamma : +S \qquad \Gamma : -S}{\Gamma : \alpha}\,(*)$$

Note, however, that unlike intuitionistic negation elimination no connective is involved in this rule. That is why it is a structural rule.

From rules (W) and $(*)$ we can derive the following two related rules.

$$\frac{\Gamma:+S}{\Gamma,-S:\alpha}\,(*+) \qquad \frac{\Gamma:-S}{\Gamma,+S:\alpha}\,(*-)$$

Below is a derivation of $(*+)$. The derivation of $(*-)$ is similar.

$$\frac{\dfrac{\Gamma:+S}{\Gamma,-S:+S}\,(W) \qquad \dfrac{}{\Gamma,-S:-S}\,(Ax^+)}{\Gamma,-S:\alpha}\,(*)$$

10.5.5 Justifying the rules as meaning conferring

Recall that in Section 4.4.1.5, harmony is extended to bilateral ND-systems in two respects: vertical harmony and horizontal harmony. It is easy to see that both the positive and negative E-rules of N_1 have the right GE-form, so vertical harmony of the N_1 ND-system holds.

As for the horizontal harmony, it holds too, as can be verified by inspection of the rules. For example, the rules for **every**, (eI_i^+) and (eI^-), stand in a similar relationship to each other as the implication rules (cf. Section 4.4.102).

10.5.6 Bilateral sentential meanings

Like in Chapter 9, each sentence is assigned two semantic values: its *meaning (contributed semantic value)* and its *contributing* semantic value. The difference is that now I use *pairs* of collections of derivations, instead of a single collection. Compare the following definition with Definition 8.5.63.

Definition 10.5.71 (Bilateral sentential meaning and semantic value).

1. *The* meaning *of a non-ground (force unmarked) sentence $S \in L_1$ is given by*

$$[[S]] \stackrel{def}{=} \lambda\Gamma.\langle[[+S]]_\Gamma^c, [[-S]]_\Gamma^c\rangle \qquad (10.5.31)$$

where:

   (a) *For a non-pseudo sentence S, Γ consists of non-pseudo-sentences only. Parameters are not "observable" in meanings.*

   (b) *For a pseudo-sentence S, Γ may also contain pseudo-sentences.*

2. *For an arbitrary (force unmarked) sentence $S \in L_1$, its* contributing *semantic value is given by*

$$[[S]]^* \stackrel{def}{=} \lambda\Gamma.\langle[[+S]]_\Gamma^*, [[-S]]_\Gamma^*\rangle \qquad (10.5.32)$$

Again, the meaning of a ground sentence is assumed to be (externally) given. Note that what is given consists of *both* acceptance and rejection of ground sentences.

Next, I revise the notion of grounds for assertion, adding to it also its ground for denial counterpart (cf. Definition 8.5.63).

Definition 10.5.72 (bilateral grounds). *Let S be a sentence in E_1 (i.e., a non-pseudo-sentence). The grounds for assertion of S, $GA[\![S]\!]$, and the grounds for denial of S, $GD[\![S]\!]$ are defined as follows.*

$$GA[\![S]\!] = \{\Gamma \subseteq E_1 \mid \Gamma \text{ is coherent and } \vdash^c \Gamma : +S\}$$
$$GD[\![S]\!] = \{\Gamma \subseteq E_1 \mid \Gamma \text{ is coherent and } \vdash^c \Gamma : -S\} \tag{10.5.33}$$

Note that Γ is in $GA[\![S]\!]$ or $GD[\![S]\!]$ only if it consists of non-pseudo-sentences only. Also, because such Γ is by definition coherent, it follows that $GA[\![S]\!] \cap GD[\![S]\!] = \varnothing$.

For convenience only, I extend these notions to pseudo-sentences as well in the following definition.

Definition 10.5.73. *For a pseudo-sentence $S \in L_1$:*

$$GA[\![S]\!] = \{\Gamma \subseteq L_1 \mid \Gamma \text{ is coherent and } \vdash^c \Gamma : +S\} \tag{10.5.34}$$
$$GD[\![S]\!] = \{\Gamma \subseteq L_1 \mid \Gamma \text{ is coherent and } \vdash^c \Gamma : -S\}$$

Note that here Γ may contain not only non-pseudo, but also pseudo-sentences.

Based on these definitions of grounds, I revise the notion of proof-theoretic consequence to *bilateral (proof-theoretic) consequence* (cf. Definition 8.5.64).

Definition 10.5.74 (bilateral consequence, contributing consequence). *Let $S_1, S_2 \in L_1$.*

1. *S_1 is a (bilateral) proof-theoretic consequence of S_2 ($S_2 \Vdash S_1$) iff both $GA(S_2) \subseteq GA(S_1)$ and $GD(S_1) \subseteq GD(S_2)$.*

2. *S_1 is a (bilateral) proof-theoretic contributing consequence of S_2 ($S_2 \Vdash^* S_1$) iff, for every coherent Γ, both*

$$\vdash^c \Gamma : +S_2 \Rightarrow \vdash^c \Gamma : +S_1$$

and

$$\vdash^c \Gamma : -S_1 \Rightarrow \vdash^c \Gamma : -S_2$$

Example 10.5.62. *I show that* $(\text{some } X)_2 \, R \, (\text{every } Y)_1 \Vdash (\text{some } X)_1 \, R \, (\text{every } Y)_2$. *There are two facts to verify (where Γ is assumed coherent):*

1. *For every Γ: if $\vdash^c \Gamma : +(\text{some } X)_2 \, R \, (\text{every } Y)_1$ then $\vdash^c \Gamma : +(\text{some } X)_1 \, R$ $(\text{every } Y)_2$.*

2. *For every Γ: if $\vdash^c \Gamma : -(\text{some } X)_1 R (\text{every } Y)_2$ then $\vdash^c \Gamma : -(\text{some } X)_2 R$*
   $(\text{every } Y)_1$.

Let us start with the first, and assume that $\vdash^c \Gamma : +(\text{some } X)_2 R (\text{every } Y)_1$. So there
is a canonical derivation of $+(\text{some } X)_2 R (\text{every } Y)_1$ from Γ, whose last step is an
application of (sI^+). It follows that both $dr\Gamma : +\mathbf{j}$ isa X and $dr\Gamma : +\mathbf{j} R (\text{every } Y)_1$.
The following derivation shows that $\vdash^c \Gamma : +(\text{some } X)_1 R (\text{every } Y)_2$.

$$
\cfrac{
\cfrac{
\Gamma : +\mathbf{j} R (\text{every } Y)_1 \qquad
\cfrac{[+\mathbf{k} \text{ isa } Y]_i : +\mathbf{k} \text{ isa } Y}{}(Ax^+)
}{
\cfrac{\Gamma, [+\mathbf{k} \text{ isa } Y]_i \vdash +\mathbf{j} R \mathbf{k}}{\Gamma, [+\mathbf{k} \text{ isa } Y]_i : +(\text{some } X)_1 R \mathbf{k}}(eE)
} \qquad \Gamma : +\mathbf{j} \text{ isa } X
}{
\Gamma : +(\text{some } X)_1 R (\text{every } Y)_2
}(eI_i^+) \quad (sI^+)
$$

For the second, assume that $\vdash^c \Gamma : -(\text{some } X)_1 R (\text{every } Y)_2$. So there is a canonical
derivation of $-(\text{some } X)_1 R (\text{every } Y)_2$ from Γ whose last step is an application of
(eI^-), hence both $\vdash \Gamma : +\mathbf{k} \text{ isa } Y$ and $\vdash \Gamma : -(\text{some } X)_1 R \mathbf{k}$. The following
derivation shows that $\vdash^c \Gamma : -(\text{some } X)_2 R (\text{every } Y)_1$.

$$
\cfrac{
\cfrac{
\Gamma \vdash -(\text{some } X)_1 R \mathbf{k} \qquad
\cfrac{[+\mathbf{j} \text{ isa } X]_i : +\mathbf{j} \text{ isa } X}{}(Ax^+)
}{
\cfrac{\Gamma, [+\mathbf{j} \text{ isa } X]_i : -\mathbf{j} R \mathbf{k}}{\Gamma, [+\mathbf{j} \text{ isa } X]_i : -\mathbf{j} R (\text{every } Y)_1}(sE)
} \qquad \Gamma : +\mathbf{k} \text{ isa } Y
}{
\Gamma : -(\text{some } X)_2 R (\text{every } Y)_1
}(sI_i^-) \quad (eI^-)
$$

10.6   A bilateral PTS for sub-sentential phrases

10.6.1   Bilateral proof-theoretic type-Interpretation

In moving to bilateralism, I keep the same type-system, but modify type t to be interpreted as collections of *bilateral* sentential meanings, based on N_1 canonical derivations; this change is propagated to functional types too.

In the sequel, it will improve readability by using $+z$ and $-z$: whenever z is the
meaning of a sentence S, then $+z$ is the same as $+S$, and $-z$ is the same as $-S$. In
such cases, I further identify $cnt(z)$ with S.

10.6.2   Meanings of basic determiners

In the bilateral setting, we have I-functions I_D induced by the respective I-rule(s),
both positive and negative. There are four such functions for the core of N_1, one for

$I_e^+ = \lambda z_1^n \lambda z_2^{t_p} \lambda \mathbf{j} \lambda \Gamma.$

$$\left\{ \dfrac{\mathcal{D}}{\Gamma : +cnt(z_2(\mathbf{j}))[(\text{every } \nu(z_1))_{r(cnt(z_2(\mathbf{j})))+1}/\mathbf{j}]} \ (el_i^+): \ \begin{array}{l} \mathcal{D} \in [\![+z_2(\mathbf{j})]\!]^*_{\Gamma,[+z_1(\mathbf{j})]_i} \ \& \\ f(\Gamma)(\mathbf{j}) \end{array} \right\}$$

$I_s^+ = \lambda z_1^n \lambda z_2^{t_p} \lambda \mathbf{j} \lambda \Gamma.$

$$\left\{ \dfrac{\mathcal{D}_1 \qquad \mathcal{D}_2}{\Gamma : +cnt(z_2(\mathbf{j}))[(\text{some } \nu(z_1))_{r(cnt(z_2(\mathbf{j})))+1}/\mathbf{j}]} \ (sl^+): \ \begin{array}{l} \mathcal{D}_1 \in [\![+z_1(\mathbf{j})]\!]^*_{\Gamma} \ \& \\ \mathcal{D}_2 \in [\![+z_2(\mathbf{j})]\!]^*_{\Gamma} \end{array} \right\}$$

$I_n^+ = \lambda z_1^n \lambda z_2^{t_p} \lambda \mathbf{j} \lambda \Gamma.$

$$\left\{ \dfrac{\mathcal{D}}{\Gamma : +cnt(z_2(\mathbf{j}))[(\text{no } \nu(z_1))_{r(cnt(z_2(\mathbf{j})))+1}/\mathbf{j}]} \ (nl_i^+): \ \begin{array}{l} \mathcal{D} \in [\![-z_2(\mathbf{j})]\!]^*_{\Gamma,[+z_1(\mathbf{j})]_i} \ \& \\ f(\Gamma)(\mathbf{j}) \end{array} \right\}$$

$I_{ne}^+ = \lambda z_1^n \lambda z_2^{t_p} \lambda \mathbf{j} \lambda \Gamma.$

$$\left\{ \dfrac{\mathcal{D}_1 \qquad \mathcal{D}_2}{\Gamma : +cnt(z_2(\mathbf{j}))[(\text{not every } \nu(z_1))_{r(cnt(z_2(\mathbf{j})))+1}/\mathbf{j}]} \ (nel^+): \ \begin{array}{l} \mathcal{D}_1 \in [\![+z_1(\mathbf{j})]\!]^*_{\Gamma} \ \& \\ \mathcal{D}_2 \in [\![-z_2(\mathbf{j})]\!]^*_{\Gamma} \end{array} \right\}$$

$I_e^- = \lambda z_1^n \lambda z_2^{t_p} \lambda \mathbf{j} \lambda \Gamma.$

$$\left\{ \dfrac{\mathcal{D}_1 \qquad \mathcal{D}_2}{\Gamma : -cnt(z_2(\mathbf{j}))[(\text{every } \nu(z_1))_{r(cnt(z_2(\mathbf{j})))+1}/\mathbf{j}]} \ (sl^+): \ \begin{array}{l} \mathcal{D}_1 \in [\![+z_1(\mathbf{j})]\!]^*_{\Gamma} \ \& \\ \mathcal{D}_2 \in [\![-z_2(\mathbf{j})]\!]^*_{\Gamma} \end{array} \right\}$$

$I_s^- = \lambda z_1^n \lambda z_2^{t_p} \lambda \mathbf{j} \lambda \Gamma.$

$$\left\{ \dfrac{\mathcal{D}}{\Gamma : -cnt(z_2(\mathbf{j}))[(\text{some } \nu(z_1))_{r(cnt(z_2(\mathbf{j})))+1}/\mathbf{j}]} \ (el_i^+): \ \begin{array}{l} \mathcal{D} \in [\![-z_2(\mathbf{j})]\!]^*_{\Gamma,[+z_1(\mathbf{j})]_i} \ \& \\ f(\Gamma)(\mathbf{j}) \end{array} \right\}$$

$I_n^- = \lambda z_1^n \lambda z_2^{t_p} \lambda \mathbf{j} \lambda \Gamma.$

$$\left\{ \dfrac{\mathcal{D}_1 \qquad \mathcal{D}_2}{\Gamma : -cnt(z_2(\mathbf{j}))[(\text{no } \nu(z_1))_{r(cnt(z_2(\mathbf{j})))+1}/\mathbf{j}]} \ (sl^+): \ \begin{array}{l} \mathcal{D}_1 \in [\![+z_1(\mathbf{j})]\!]^*_{\Gamma} \ \& \\ \mathcal{D}_2 \in [\![+z_2(\mathbf{j})]\!]^*_{\Gamma} \end{array} \right\}$$

$I_{ne}^- = \lambda z_1^n \lambda z_2^{t_p} \lambda \mathbf{j} \lambda \Gamma.$

$$\left\{ \dfrac{\mathcal{D}}{\Gamma : -cnt(z_2(\mathbf{j}))[(\text{not every } \nu(z_1))_{r(cnt(z_2(\mathbf{j})))+1}/\mathbf{j}]} \ (el_i^+): \ \begin{array}{l} \mathcal{D} \in [\![+z_2(\mathbf{j})]\!]^*_{\Gamma,[+z_1(\mathbf{j})]_i} \ \& \\ f(\Gamma)(\mathbf{j}) \end{array} \right\}$$

Figure 10.9: I-functions for the core of N_1

each determiner. These functions are presented in Figure 10.9. The I-functions for the additional determiners are presented in Figure 10.11, and for relativization in Figure 10.10.

The general form of the bilateral meaning of a determiner D, naturally generalizing the unilateral meaning considered before, is given by:

$$[\![D]\!] = \lambda z_1^n \lambda z_2^{t_p} \lambda \Gamma.$$
$$\left\langle \bigcup_{\mathbf{j}_1,\dots,\mathbf{j}_m \in \mathcal{P}} I_D^+(z_1)(z_2)(\mathbf{j}_1)\cdots(\mathbf{j}_m)(\Gamma), \bigcup_{\mathbf{j}_1,\dots,\mathbf{j}_r \in \mathcal{P}} I_D^-(z_1)(z_2)(\mathbf{j}_1)\cdots(\mathbf{j}_r)(\Gamma) \right\rangle$$

where I_D^+ and I_D^- are the I-functions for D.

For example, For the determiners in the core of E_1, both $m = 1$ and $r = 1$. For the other determiners (in Figure 10.11), m and/or r may be grater than 1. It can be easily realized, that $[\![D]\!]$ is obtained by abstracting over the noun argument X and vp-argument $S[-]$ in the sentential meaning $[\![D\ X\ S[-]]\!]$. Thus, it is based on canonical derivations of asserting $D\ X\ S[-]$ and of denying it. This general form is again the main tool for establishing the conservativity of negative determiners.

$$I_r^+ = \lambda z_1^n \lambda z_2^{tp} \lambda \mathbf{j} \lambda \Gamma.$$

$$\left\{ \frac{\mathcal{D}_1 \qquad \mathcal{D}_2}{\Gamma : +(cnt(z_1(\mathbf{j})) \text{ who } cnt(z_2(\mathbf{j}))[-/\mathbf{j}])} (rI^+) : \begin{array}{l} \mathcal{D}_1 \in [[+z_1(\mathbf{j})]]_\Gamma^* \ \& \\ \mathcal{D}_2 \in [[+z_2(\mathbf{j})]]_\Gamma^* \end{array} \right\}$$

$$I_r^- = \lambda z_1^n \lambda z_2^{tp} \lambda \mathbf{j} \lambda \Gamma.$$

$$\left\{ \frac{\mathcal{D}}{\Gamma : -(cnt(z_1(\mathbf{j})) \text{ who } cnt(z_2(\mathbf{j}))[-/\mathbf{j}])} (rI^-)_1 : \mathcal{D} \in [[-z_1(\mathbf{j})]]_\Gamma^* \right\}$$

$$\cup \left\{ \frac{\mathcal{D}}{\Gamma : -(cnt(z_1(\mathbf{j})) \text{ who } cnt(z_2(\mathbf{j}))[-/\mathbf{j}])} (rI^-)_2 : \mathcal{D} \in [[-z_2(\mathbf{j})]]_\Gamma^* \right\}$$

Figure 10.10: I-functions for relativization N_1

10.6.3 Meanings of compound determiners

As mentioned above, I consider two forms of compound determiners: possessives, dealt with in Section 10.6.3.1, and coordinated determiners, dealt with in Section 10.6.3.2.

10.6.3.1 Negative possessives

I now have in the fragment also negative possessive determiners, like

$$\text{no girl's dog barked} \tag{10.6.37}$$

I thus extend the unilateral meanings assigned above to possessives with positive quantifiers only.

As mentioned above, an important observation of Francez [41], relevant to the bilateral meaning, is that the variability of the implicit quantifier is restricted to possessives with a positive quantifier only. Thus, in a possessive with a negative (downward entailing) quantifier like (10.6.37) the implicit quantifier can only be existential[11]. This observation is extended here to general denial of sentences with quantified possessive. In denying

$$\text{every girl's dog barked} \tag{10.6.38}$$

the implicit quantifier must be existential too.

The way I revise 's in the bilateral setting is to define the appropriate I-functions $I_{(D\ X)'s}^{+,-}$ in terms of $I_D^{+,-}$ and I_Q^+, where Q is the implicit quantifier related to above. For simplicity, I henceforth assume that Q is always existential, and present the defi-

---

[11] In Francez [41], this phenomenon is connected with a similar one related to "donkey pronouns".

$$I^+_{\geq m} = \lambda z_1^n \lambda z_2^{t_p} \lambda j_1 \cdots \lambda j_m \lambda \Gamma.$$

$$\left\{ \frac{D_1^1 \cdots D_m^1 \quad D_1^2 \cdots D_m^2}{\Gamma : +cnt(z_2(j_1))[(\text{at least m } \nu(z_1))_{r(cnt(z_2(j_1)))+1}/j_1]} \quad (\geq mI^+) : D_i^k \in [\![+z_k(j_i)]\!]_\Gamma^* \right\}$$

$$I^+_{\leq m} = \lambda z_1^n \lambda z_2^{t_p} \lambda j_1 \cdots \lambda j_{m+1} \lambda \Gamma.$$

$$\left\{ \frac{D}{\Gamma : +cnt(z_2(j_1))[(\text{at most m } \nu(z_1))_{r(cnt(z_2(j_1)))+1}/j_1]} \quad (\leq mI^+) : \begin{array}{l} D \in [\![-z_2(j_1)]\!]_{\Gamma,[+z_1(j_1)],\{[-j_1 \text{ is } j_i]\}_{2\leq i\leq m+1}}^* \\ \& f(\Gamma)(j_i) \end{array} \right.$$

$$I^-_{\geq m} = \lambda z_1^n \lambda z_2^{t_p} \lambda j_1 \cdots \lambda j_m \lambda \Gamma.$$

$$\left. \frac{D}{\Gamma : -cnt(z_2(j_1))[(\text{at least m } \nu(z_1))_{r(cnt(z_2(j_1)))+1}/j_1]} \quad (\geq mI^-) : D \in [\![-z_2(j_1)]\!]_{\Gamma,[+z_1(j_1)],\{[-j_1 \text{ is } j_i]\}_{2\leq i\leq m}}^* \right\}$$

$$I^-_{\leq m} = \lambda z_1^n \lambda z_2^{t_p} \lambda j_1 \cdots \lambda j_{m+1} \lambda \Gamma.$$

$$\left\{ \frac{D_1^1 \cdots D_{m+1}^1 \quad D_1^2 \cdots D_{m+1}^2}{\Gamma : -cnt(z_2(j_1))[(\text{at most m } \nu(z_1))_{r(cnt(z_2(j_1)))+1}/j_1]} \quad (\leq mI^-) : D_i^k \in [\![+z_k(j_i)]\!]_\Gamma^* \right\}$$

Figure 10.11: I-functions for at least/at most m

$$[\![\text{at least m}]\!] = \lambda z_1^n \lambda z_2^{t_p} \lambda \Gamma. \left\langle \bigcup_{\mathbf{j}_1,\ldots,\mathbf{j}_m \in \mathcal{P}} I_{\geq m}^+ (z_1)(z_2)(\mathbf{j}_1)\cdots(\mathbf{j}_m)(\Gamma), \bigcup_{\mathbf{j}_1,\ldots,\mathbf{j}_m \in \mathcal{P}} I_{\geq m}^- (z_1)(z_2)(\mathbf{j}_1)\cdots(\mathbf{j}_m)(\Gamma) \right\rangle$$

(10.6.35)

$$[\![\text{at most m}]\!] = \lambda z_1^n \lambda z_2^{t_p} \lambda \Gamma. \left\langle \bigcup_{\mathbf{j}_1,\ldots,\mathbf{j}_{m+1} \in \mathcal{P}} I_{\leq m}^+ (z_1)(z_2)(\mathbf{j}_1)\cdots(\mathbf{j}_{m+1})(\Gamma), \bigcup_{\mathbf{j}_1,\ldots,\mathbf{j}_{m+1} \in \mathcal{P}} I_{\leq m}^- (z_1)(z_2)(\mathbf{j}_1)\cdots(\mathbf{j}_{m+1})(\Gamma) \right\rangle$$

(10.6.36)

nition with a single parameter **j**, avoiding vectorized notation.

$$[['s]] = \begin{aligned} &\lambda u_0^{(n,(t_p,(p,t)))} \lambda u_1^n \lambda z_1^n \lambda z_2^{t_p} \lambda \mathbf{j} \lambda \Gamma. \\ &\langle I_{\delta(u_0)}^+ (I_r^+(u_1)(I_s^+(z_1)([[\mathcal{R}]])(\mathbf{j})(\Gamma))(\mathbf{j})(\Gamma)) \\ &\quad (I_s^+(I_r^+(z_1)([[\mathcal{R}]])(\mathbf{j})(\Gamma))(\mathbf{j})(\Gamma)(z_2))(\mathbf{j})(\Gamma), \\ &I_{\delta(u_0)}^- (I_r^+(u_1)(I_s^-(z_1)([[\mathcal{R}]])(\mathbf{j})(\Gamma))(\mathbf{j})(\Gamma)) \\ &\quad (I_s^+(I_r^+(z_1)([[\mathcal{R}]])(\mathbf{j})(\Gamma))(\mathbf{j})(\Gamma)(z_2))(\mathbf{j})(\Gamma), \rangle \end{aligned} \qquad (10.6.39)$$

For a determiner D and noun X, we get

$$[[D\ X's]] = [['s]]([[D]])([[X]]) = \begin{aligned} &\lambda z_1^n \lambda z_2^{t_p} \lambda \Gamma. \\ &\textstyle\bigcup_{\mathbf{j} \in \mathcal{P}} \\ &\langle I_D^+(I_r^+([[X]])(I_s^+(z_1)([[\mathcal{R}]])(\mathbf{j})(\Gamma))(\mathbf{j})(\Gamma)) \\ &\quad (I_s^+(I_r^+(z_1)([[\mathcal{R}]])(\mathbf{j})(\Gamma))(\mathbf{j})(\Gamma)(z_2))(\mathbf{j})(\Gamma), \\ &I_D^-(I_r^+([[X]])(I_s^+(z_1)([[\mathcal{R}]])(\mathbf{j})(\Gamma))(\mathbf{j})(\Gamma)) \\ &\quad (I_s^+(I_r^+(z_1)([[\mathcal{R}]])(\mathbf{j})(\Gamma))(\mathbf{j})(\Gamma)(z_2))(\mathbf{j})(\Gamma) \rangle \end{aligned}$$
$$(10.6.40)$$

For a determiner D and nouns X, Y, this generates the following I-rules for $D\ X's\ Y$ (recall that dps are introduced, not determiners).

$$\begin{aligned} I_{(D\ X's\ Y)}^+ \quad &= [['s]]([[D]])([[X]])([[Y]]) \\ &= \lambda z_2^{t_p} \lambda \mathbf{j} \lambda \Gamma. \\ &\quad I_D^+(I_r^+([[X]])(I_s^+([[Y]])([[\mathcal{R}]])(\mathbf{j})(\Gamma))(\mathbf{j})(\Gamma)) \\ &\quad (I_s^+(I_r^+([[Y]])([[\mathcal{R}]])(\mathbf{j})(\Gamma))(z_2))(\mathbf{j})(\Gamma) \end{aligned} \qquad (10.6.41)$$

and

$$\begin{aligned} I_{(D\ X's\ Y)}^- \quad &= [['s]]([[D]])([[X]])([[Y]]) \\ &= \lambda z_2^{t_p} \lambda \mathbf{j} \lambda \Gamma. \\ &\quad I_D^-(I_r^+([[X]])(I_s^+([[Y]])([[\mathcal{R}]])(\mathbf{j})(\Gamma))(\mathbf{j})(\Gamma)) \\ &\quad (I_s^+(I_r^+([[Y]])([[\mathcal{R}]])(\mathbf{j})(\Gamma))(z_2))(\mathbf{j})(\Gamma) \end{aligned} \qquad (10.6.42)$$

The corresponding GE-rule is the one harmoniously induced by the above I-rules, depending on D.

To understand how this definition works, consider the application of $[['s]]$ to $[[every]]$, then to $[[girl]]$ and finally to $[[dog]]$, choosing Q as **some**. For simplicity, the projection on the first (positive) and second (negative) components are omitted.

$$\begin{aligned} I_{(every\ girl's\ dog)}^+ \quad &= [['s]]([[every]])([[girl]])([[dog]]) \\ &= \lambda z_2^{t_p} \lambda \mathbf{j} \lambda \Gamma. \\ &\quad I_e^+(I_r^+([[girl]])(I_s([[dog]])([[\mathcal{R}]])(\mathbf{j})(\Gamma))(\mathbf{j})(\Gamma)) \\ &\quad (I_s^+(I_r^+([[dog]])([[\mathcal{R}]])(\mathbf{j})(\Gamma))(z_2))(\mathbf{j})(\Gamma) \end{aligned} \qquad (10.6.43)$$

$$\begin{aligned} I_{(every\ girl's\ dog)}^- \quad &= [['s]]([[every]])([[girl]])([[dog]]) \\ &= \lambda z_2^{t_p} \lambda \mathbf{j} \lambda \Gamma. \\ &\quad I_e^-(I_r^+([[girl]])(I_s([[dog]])([[\mathcal{R}]])(\mathbf{j})(\Gamma))(\mathbf{j})(\Gamma)) \\ &\quad (I_s^+(I_r^+([[dog]])([[\mathcal{R}]])(\mathbf{j})(\Gamma))(z_2))(\mathbf{j})(\Gamma) \end{aligned} \qquad (10.6.44)$$

The denial of **every girl's dog barked** is thus inferred, by the I^-_{every} rule, from accepting **j isa girl who owns some dogs** and the denying that some of **j**'s dogs barked.

The positive rules for **every/some** X's Y are the natural adaptation of the rules in Figure 10.4. Figure 10.12 displays the I/E-rules for denial of **every/some** X's Y induced by $[\![\text{'s}]\!]$. Scope levels are not shown for brevity. The GE-rules are the harmoniously induced ones.

For example, when X, Y are instantiated to **girl** and **dog**, respectively, we get the following positive I-rule (with the freshness restriction applied as before) for **every girl's dog**

$$\frac{\Gamma, +\textbf{j isa girl}, +\textbf{d isa dog}, +\textbf{j} \mathcal{R} \textbf{d} : +S[\textbf{d}]}{\Gamma : +S[(\textbf{every girl's dog})]} \quad ((e \text{ girl's dog})I^+)$$

The rules for **no** X's Y are presented in Figure 10.13.

10.6.3.2 Coordinated determiners

I have to revise the proof-theoretic meaning of coordinated determiners by adding also denial rules for them. Let D_1, D_2 be two determiners. Then, $((D_1 \text{ and } D_2) I^+)$ and $((D_1 \text{ or } D_2) I^+)$ are given below.

$$\frac{\Gamma : +S[(D_1 \ X)_r] \quad \Gamma : +S[(D_2 \ X)_r]}{\Gamma : +S[(D_1 \text{ and } D_2 \ X)_r]} \quad ((D_1 \text{ and } D_2) I^+)$$

$$\frac{\Gamma : +S[(D_1 \ X)_r]}{\Gamma : +S[(D_1 \text{ or } D_2 \ X)_r]} \quad ((D_1 \text{ or } D_2)_1 \ I^+) \qquad (10.6.45)$$

$$\frac{\Gamma : +S[(D_1 X)_r]}{\Gamma : +S[(D_2 \text{ or } D_2 \ X)_r]} \quad ((D_1 \text{ or } D_2)_2 \ I^+)$$

The denial rules for coordinated determiners are the following.

$$\frac{\Gamma : -S[(D_1 \ X)_r]}{\Gamma : -S[(D_1 \text{ and } D_2 \ X)_r]} \quad ((D_1 \text{ and } D_2)_1 \ I^-)$$

$$\frac{\Gamma : -S[(D_2 \ X)_r]}{\Gamma : -S[(D_1 \text{ and } D_2 \ X)_r]} \quad ((D_1 \text{ and } D_2)_2 \ I^-) \qquad (10.6.46)$$

$$\frac{\Gamma : -S[(D_1 \ X)_r] \quad \Gamma : -S[(D_2 \ X)_r]}{\Gamma : -S[(D_1 \text{ or } D_2 \ X)_r]} \quad ((D_1 \text{ or } D_2) \ I^-)$$

The corresponding harmonious GE-rules are shown below. (the reductions triggered by those E-rules may depend on the the structural rules being present).

$$\frac{\Gamma : +\mathbf{j}\,\text{isa}\,X \quad \Gamma : +\mathbf{d}\,\text{isa}\,Y \quad \Gamma : +\mathbf{j}\,\mathcal{R}\,\mathbf{d} \quad \Gamma : -S[\mathbf{d}]}{\Gamma : -S[(\text{every}\,X's\,Y)]} \;((e\,X's\,Y)I^-)$$

$$\frac{\Gamma, +\mathbf{j}\,\text{isa}\,X, +\mathbf{d}\,\text{isa}\,Y, +\mathbf{j}\,\mathcal{R}\,\mathbf{d} : -S[\mathbf{d}]}{\Gamma : -S[(\text{some}\,X's\,Y)]} \;((s\,X's\,Y)I^-)$$

$$\frac{\Gamma : -S[(\text{every}\,X's\,Y)] \quad \Gamma, +\mathbf{j}\,\text{isa}\,X, +\mathbf{d}\,\text{isa}\,Y, +\mathbf{j}\mathcal{R}\mathbf{d}, -S[\mathbf{d}] : \alpha}{\Gamma : \alpha} \;((e\,X's\,Y)E^-)$$

$$\frac{\Gamma : -S[(\text{some}\,X's\,Y)] \quad \Gamma : +\mathbf{j}\,\text{isa}\,X \quad \Gamma : +\mathbf{d}\,\text{isa}\,Y \quad \Gamma : +\mathbf{j}\mathcal{R}\mathbf{d} \quad \Gamma : -S[\mathbf{d}] : \alpha}{\Gamma : \alpha} \;((s\,X's\,Y)E^-)$$

where **j**, **d** are fresh for $((e\,X's\,Y)E^-)$ and $((s\,X's\,Y)I^-)$.

Figure 10.12: I/E-rules for denial of every/some X's Y

$$\frac{\Gamma, +j\text{ isa } X, +d\text{ isa } Y, +j\mathcal{R}d : -S[d]}{\Gamma : +S[(\text{no } X's\,Y)]} \; ((n\,X's\,Y)\,I^+)$$

$$\frac{\Gamma : +S[(\text{no } X's\,Y)] \quad \Gamma : +j\text{ isa } X \quad \Gamma : +d\text{ isa } Y \quad \Gamma : +j\mathcal{R}d \quad \Gamma, -S[d] : \alpha}{\Gamma : \alpha} \; ((n\,X's\,Y)\,E^+)$$

$$\frac{\Gamma : +j\text{ isa } X \quad \Gamma : +d\text{ isa } Y \quad \Gamma : +j\mathcal{R}d \quad \Gamma : +S[d]}{\Gamma : -S[(\text{no } X's\,Y)]} \; ((n\,X's\,Y)\,I^-)$$

$$\frac{\Gamma : -S[(\text{no } X's\,Y)] \quad \Gamma, +j\text{ isa } X, +d\text{ isa } Y, +j\mathcal{R}d, -S[d] : \alpha}{\Gamma : \alpha} \; ((n\,X's\,Y)\,E^-)$$

Figure 10.13: The I/E rules for no X's Y

where **j**, **d** are fresh for $((n\,X's\,Y)\,E^+)$ and $((n\,X's\,Y)\,E^-)$.

$$\frac{\Gamma : +S[(D_1 \text{ and } D_2\ X)_r] \quad \Gamma, +S[(D_1\ X)_r], +S[(D_2\ X)_r] : \alpha}{\Gamma : \alpha} \quad ((D_1 \text{ and } D_2)E^+)$$

$$\frac{\Gamma : +S[(D_1 \text{ or } D_2\ X)_r] \quad \Gamma, [+S[(D_1\ X)_r]]_i : \alpha \quad \Gamma[+S[(D_2\ X)_r]]_j : \alpha}{\Gamma : \alpha} \quad ((D_1 \text{ or } D_2)E^{+,i,j})$$

$$\frac{\Gamma : -S[(D_1 \text{ and } D_2\ X)_r] \quad \Gamma, -S[(D_1\ X)_r] : \alpha}{\Gamma : \alpha} \quad ((D_1 \text{ and } D_2)_1 E^-)$$

$$\frac{\Gamma : -S[(D_1 \text{ and } D_2\ X)_r] \quad \Gamma, -S[(D_2\ X)_r] : \alpha}{\Gamma : \alpha} \quad ((D_1 \text{ and } D_2)_2 E^-)$$

$$\frac{\Gamma : -S[(D_1 \text{ or } D_2\ X)_r] \quad \Gamma, -S[(D_1\ X)_r], -S[(D_2\ X)_r] : \alpha}{\Gamma : \alpha} \quad ((D_1 \text{ or } D_2)E^-)$$

(10.6.47)

10.6.4 Conservativity under bilateral meanings

The first step is to revise the definition of '$\sqcap$' to reflect also denials.

$$\sqcap =^{df.} \lambda z_1^n \lambda z_2^{t_p} \lambda \mathbf{j} \lambda \Gamma. \langle I_r^+(z_1)(z_2)(\mathbf{j})(\Gamma), I_r^-(z_1)(z_2)(\mathbf{j})(\Gamma) \rangle \qquad (10.6.48)$$

This definition refines the definition of '$\sqcap$' in (10.3.10), which is based on a unilateral version of I_r.

A closer inspection shows that the proof of the theorem to the effect that *all* the determiners are conservatives (under their unilateral meanings) carries over to the bilateral meaning employed here. The proof employs an argument based on

- The general form of $[\![D]\!]$.

- The way I_D discharges assumptions.

- The form of I_r (relativization).

It is easy to see that the general forms above "hide" the use of negative rules. As for the discharge of assumption, all that is required is to consider both the discharges of I_D^+ and I_D^-.

Instead of presenting the revised proof in full detail, I do here the following.

- Show *directly* the (first-argument) conservativity of every, under its bilateral meaning. This gives a good clue to the structure of the proof in the general case.

- Show *directly* the (first-argument) conservativity of no, that has no unilateral counterpart meaning.

Note that lemma 10.3.1 holds here too.

10.6.4.1 Conservativity of every

In order to show that every is conservative under its bilateral meaning, I need to show that for arbitrary $a \in D_n$ and $b \in D_{t_p}$:

$$\begin{aligned} &1.\ [\![\text{every}]\!](a)(b) \Vdash [\![\text{every}]\!](a)(\sqcap(a)(b)) \\ &2.\ [\![\text{every}]\!](a)(\sqcap(a)(b)) \Vdash [\![\text{every}]\!](a)(b) \end{aligned} \qquad (10.6.49)$$

Now, if $a \in D_n$ and $b \in D_{t_p}$, then, by Lemma 10.3.1, there are a noun X and a pseudo-sentence $S[\mathbf{j}]$, such that $a = \lambda \mathbf{j}.[\![\mathbf{j} \text{ isa } X]\!]$ and $b = \lambda \mathbf{j}.[\![S[\mathbf{j}]]\!]$. Thus:

- $+a(\mathbf{j}) = +\mathbf{j}$ isa X and $-a(\mathbf{j}) = -\mathbf{j}$ isa X

- $+b(\mathbf{j}) = +S[\mathbf{j}]$ and $-b(\mathbf{j}) = -S[\mathbf{j}]$

- $+(\sqcap(a)(b))(\mathbf{j}) = +\mathbf{j}$ isa X who $S[-]$ and $-(\sqcap(a)(b))(\mathbf{j}) = -\mathbf{j}$ isa X who $S[-]$

Clearly, (10.6.49) holds iff, for an arbitrary Γ, both (10.6.50a.) is equivalent to (10.6.50b.) and (10.6.51a.) is equivalent to (10.6.51b.), for $\mathbf{j}$ fresh for Γ.

$$\begin{aligned} a.\ &: \Gamma, +\mathbf{j} \text{ isa } X : +S[\mathbf{j}] \\ b.\ &: \Gamma, +\mathbf{j} \text{ isa } X : +\mathbf{j} \text{ isa } X \text{ who } S[-]] \end{aligned} \qquad (10.6.50)$$

$$\begin{aligned} a.\ &: \Gamma : -S[\mathbf{j}] \\ b.\ &: \Gamma : -\mathbf{j} \text{ isa } X \text{ who } S[-]] \end{aligned} \qquad (10.6.51)$$

So there are four entailments here, and I verify each of them by a suitable derivation.

- (10.6.50a) $\Rightarrow$ (10.6.50b):

$$\cfrac{\cfrac{}{\Gamma, +(\mathbf{j} \text{ isa } X) : +(\mathbf{j} \text{ isa } X)}\ (Ax^+) \qquad \Gamma, +(\mathbf{j} \text{ isa } X) : +S[\mathbf{j}]}{\Gamma, +(\mathbf{j} \text{ isa } X) : +(\mathbf{j} \text{ isa } X \text{ who } S[-])}\ (rI^+)$$

- (10.6.50b) $\Rightarrow$ (10.6.50a):

$$\cfrac{\Gamma, +(\mathbf{j} \text{ isa } X) : +(\mathbf{j} \text{ isa } X \text{ who } S[-]) \qquad \cfrac{}{\Gamma, +(\mathbf{j} \text{ isa } X), +S[\mathbf{j}] : +(\mathbf{j} \text{ isa } X)}\ (Ax^+)}{\Gamma, +(\mathbf{j} \text{ isa } X) : +(\mathbf{j} \text{ isa } X)}\ (rE^+)$$

- (10.6.51a) $\Rightarrow$ (10.6.51b):

$$\cfrac{\Gamma : -S[\mathbf{j}]}{\Gamma : -(\mathbf{j} \text{ isa } X \text{ who } S[-])}\ (rI^-)_2$$

- (10.6.51b) $\Rightarrow$ (10.6.51a):

$$\cfrac{\Gamma : -(\mathbf{j} \text{ isa } X \text{ who } S[-]) \qquad \cfrac{\cfrac{\Gamma : +(\mathbf{j} \text{ isa } X)}{\Gamma, -(\mathbf{j} \text{ isa } X) : -S[\mathbf{j}]}\ (**) \qquad \cfrac{}{\Gamma, -S[\mathbf{j}] : -S[\mathbf{j}]}\ (Ax^+)}{}\ (rE^-)}{\Gamma : -S[\mathbf{j}]}$$

10.6.4.2   Conservativity of no

In order to show that no is conservative, I need to show that for arbitrary $a \in D_n$ and $b \in D_{t_p}$:

$$\begin{aligned} &1.\ [\![no]\!](a)(b) \Vdash [\![no]\!](a)(\sqcap(a)(b)) \\ &2.\ [\![no]\!](a)(\sqcap(a)(b)) \Vdash [\![no]\!](a)(b) \end{aligned} \qquad (10.6.52)$$

Again, by Lemma 10.3.1, there are a noun X and a pseudo-sentence $S[\mathbf{j}]$, such that $a = \lambda\mathbf{j}.[\![\mathbf{j} \text{ isa } X]\!]$ and $b = \lambda\mathbf{j}.[\![S[\mathbf{j}]]\!]$. Thus:

- $+a(\mathbf{j}) = +\mathbf{j}$ isa X and $-a(\mathbf{j}) = -\mathbf{j}$ isa X

- $+b(\mathbf{j}) = +S[\mathbf{j}]$ and $-b(\mathbf{j}) = -S[\mathbf{j}]$

- $+(\sqcap(a)(b))(\mathbf{j}) = +\mathbf{j}$ isa X who $S[-]$ and $-(\sqcap(a)(b))(\mathbf{j}) = -\mathbf{j}$ isa X who $S[-]$

no is conservative iff both (10.6.53a.) is equivalent to (10.6.53b.) and (10.6.54a.) is equivalent to (10.6.54b), for $\mathbf{j}$ fresh for Γ.

$$
\begin{aligned}
&a.\ \vdash \Gamma, +\mathbf{j}\ \text{isa}\ X : -S[\mathbf{j}] \\
&b.\ \vdash \Gamma, +\mathbf{j}\ \text{isa}\ X : -\mathbf{j}\ \text{isa}\ X\ \text{who}\ S[-]
\end{aligned}
\qquad (10.6.53)
$$

$$
\begin{aligned}
&a.\ \vdash \Gamma : +S[\mathbf{j}]] \\
&b.\ \vdash \Gamma : +\mathbf{j}\ \text{isa}\ X\ \text{who}\ S[-]]
\end{aligned}
\qquad (10.6.54)
$$

So again there are four entailments here, and I verify each of them by a suitable derivation.

- (10.6.53a) $\Rightarrow$ (10.6.53b):

$$
\frac{\Gamma, +\mathbf{j}\ \text{isa}\ X : -S[\mathbf{j}]}{\Gamma, +\mathbf{j}\ \text{isa}\ X : -(\mathbf{j}\ \text{isa}\ X\ \text{who}\ S[-])}\ (rI^-)_2
$$

- (10.6.53b) $\Rightarrow$ (10.6.53a):

$$
\frac{\Gamma : +(\mathbf{j}\ \text{isa}\ X\ \text{who}\ S[-]) \qquad \dfrac{}{\Gamma, +(\mathbf{j}\ \text{isa}\ X), +S[\mathbf{j}] : +S[\mathbf{j}]}\ (Ax^+)}{\Gamma : +S[\mathbf{j}]}\ (rE^+)
$$

- (10.6.54a) $\Rightarrow$ (10.6.54b):

$$
\frac{\Gamma : +(\mathbf{j}\ \text{isa}\ X) \qquad \Gamma : +S[\mathbf{j}]}{\Gamma : +(\mathbf{j}\ \text{isa}\ X\ \text{who}\ S[-])}\ (rI^+)
$$

The other derivation is similar and omitted.

- (10.6.54b) $\Rightarrow$ (10.6.54a) (for the derivation to fit the page's width, I let $\Gamma' = \Gamma, +(\mathbf{j}\ \text{isa}\ X)$):

$$
\frac{\Gamma' : -(\mathbf{j}\ \text{isa}\ X\ \text{who}\ S[-]) \qquad \dfrac{\dfrac{}{\Gamma' : +(\mathbf{j}\ \text{isa}\ X)}\ (Ax^+)}{\Gamma', -(\mathbf{j}\ \text{isa}\ X) : -S[\mathbf{j}]}\ (**) \qquad \dfrac{}{\Gamma', -S[\mathbf{j}] : -S[\mathbf{j}]}\ \substack{(Ax^+)\\(rE^-)}}{\Gamma' : -S[\mathbf{j}]}
$$

10.7   Monotonicity of determiners

Monotonicity is a property of many NL determiners, extensively studied under MTS. To test if a determiner det is monotone in any of its arguments, one can use the sentences in (10.7.55) below.

a. det dogs barked
b. det dogs that ran barked   (10.7.55)
c. det dogs barked loudly

A determiner can be either *upward* or *downward* monotone in its first (nominal) argument: if $(10.7.55a)$ ⊩ $(10.7.55b)$, det is *left downward monotone* (example: every); if, on the other hand, $(10.7.55b)$ ⊩ $(10.7.55a)$, det is *left upward monotone* (examples are some and at least n). A determiner can also be either *upward* or *downward* monotone in its second (verbal) argument: if $(10.7.55a)$ ⊩ $(10.7.55c)$, det is *right downward monotone* (example: only); if, on the other hand, $(10.7.55c)$ ⊩ $(10.7.55a)$, det is *right upward monotone* (example: some).

In MTS, monotonicity (in either argument of a determiner) is defined in terms of preserving a partial order based on inclusion between subsets of the universe of the model – extensions of the predicates denoted, in that model, by the arguments of the determiner. Recall that in PTS there are no models, no universes with entities and no extensions. We need to rely on the available resource for an alternative definition.

There are at least two possible sources of partial order between nouns and between verb-phrases, to be denoted by '$\sqsubseteq$', based on our proof-theoretic definition of the determiner's meaning. Recall the division of determiners into positive determiners and negative ones. The distinction is "hidden" within the structure of the domain D_t, and either unilateral or bilateral will be employed, as the case requires.

1. *Externally imposed partial orders:* To obtain them, we have to assume that the externally given lexical meanings (possibly appealing also to world knowledge) based on ground pseudo-sentences provide them, in the form of *non-logical* axioms. For example, we might have

   $[\![\mathrm{dog}]\!] \sqsubseteq [\![\mathrm{animal}]\!]$ if ⊢j isa dog : j isa animal

   and check whether, say

   det dogs barked ⊩ det animals barked

   Such an extension of the fragments needs to be dealt with very carefully, since, as is well-known, such non-logical axioms interfere with cut elimination in the equivalent sequent-calculus used for establishing decidability of used in Chapter 8. I will avoid them here.

2. *Internally imposed partial-orders:* The partial-order is based on relativization, by letting, for example, for every X and S,

   $[\![X \text{ who } S[-]]\!] \sqsubseteq [\![X]\!]$

In other words, for a, b of type n, t_p, respectively, let $\sqcap(a)(b) \sqsubseteq a$ (recall that $\sqcap$ was defined in Definition 10.3.10). To strengthen this approach, I add also *intersective adjectives* (see Section 8.1.3.2) as a means for noun modification contributing to the partial order.

For typing adjectives, I introduce another subtype of (p, t), called d, with

$$D_d = \{\lambda \mathbf{j}.[\![\mathbf{j} \text{ is } A]\!] \mid A \text{ an (intersective) adjective}\} \tag{10.7.56}$$

In addition, I extend '$\sqsubseteq$' also to adjectives, by adding a second clause to the unilateral definition of $\sqcap$.

$$\sqcap =^{\mathrm{df.}} \lambda z_1^d \lambda z_2^n \lambda \Gamma \lambda \mathbf{j}. I_{adj}(z_1)(z_2)(\Gamma)(\mathbf{j}) \tag{10.7.57}$$

For verb-phrases, the partial-order is defined via their nominalization.

However, in view of the classification of determiners to positive or negative, a clause for the bilateral meaning is needed to. I overload the symbol $\sqcap$, leaving to context to determine whether the unilateral or bilateral definition is used. Thus, we end up with

$$\sqcap =^{\mathrm{df.}} \begin{cases} \begin{cases} \lambda z_1^n \lambda z_2^{t_p} \lambda \mathbf{j} \lambda \Gamma. I_r(z_1)(z_2)(\mathbf{j})(\Gamma) & \text{relativizitaion} \\ \text{or} & \\ \lambda z_1^d \lambda z_2^n \lambda \Gamma \lambda \mathbf{j}. I_{adj}(z_1)(z_2)(\Gamma)(\mathbf{j}) & \text{adjectival modification} \end{cases} & \text{unilateral} \\ \begin{cases} \lambda z_1^n \lambda z_2^{t_p} \lambda \mathbf{j} \lambda \Gamma. \langle I_r^+(z_1)(z_2)(\mathbf{j})(\Gamma), I_r^-(z_1)(z_2)(\mathbf{j})(\Gamma) \rangle & \text{relativization} \\ \text{or} & \\ \lambda z_1^d \lambda z_2^{t_p} \lambda \mathbf{j} \lambda \Gamma. \langle I_{adj}^+(z_1)(z_2)(\mathbf{j})(\Gamma), I_{adj}^-(z_1)(z_2)(\mathbf{j})(\Gamma) \rangle & \text{adjectival modification} \end{cases} & \text{bilateral} \end{cases}$$
$$\tag{10.7.58}$$

Definition 10.7.75 (the pre partial-order '$\sqsubseteq$'). *Let '$\sqsubseteq$' be the least pre partial-order over $D_n \cup D_d \cup D_{t_p}$ satisfying:*

- *For $a \in D_n$ and $b \in D_{t_p}$, $\sqcap(a)(b) \sqsubseteq a$.*

- *For $a \in D_d$ and $b \in D_n$, $\sqcap(a)(b) \sqsubseteq b$.*

- *For $a, b \in D_{t_p}$, $a \sqsubseteq b$ if $\mathbf{n}(a) \sqsubseteq \mathbf{n}(b)$.*

The following lemma, augmenting Lemma 10.3.1, follows directly from the definitions.

Lemma 10.7.2. *If $a \in D_d$ and $b \in D_n$, then there is and adjective A and a noun X, such that*

- $a(\mathbf{j}) = [\![\mathbf{j} \text{ is } A]\!]$

- $b(\mathbf{j}) = [\![\mathbf{j} \text{ isa } X]\!]$

- $(\sqcap(a)(b))(\mathbf{j}) = [\![\mathbf{j} \text{ isa } A\, X]\!]$

Similarly, the proof-theoretic consequence symbol $\Vdash$ is overloaded for both the unilateral and bilateral cases, leaving to context the determination of the applied case.

$$S_2 \Vdash S_1 \text{iff} \begin{cases} G[\![S_2]\!] \subseteq G[\![S_1]\!] & \text{unilateral} \\ \text{or} \\ GA(S_2) \subseteq GA(S_1) \text{ and } GD(S_1) \subseteq GD(S_2) & \text{bilateral} \end{cases} \qquad (10.7.59)$$

Definition 10.7.76 (monotonicity). *A determiner* det *is*

- left upward monotone *(↑MON) iff*

$$\forall a, a' \in D_n \forall b \in D_{t_p}[(a \sqsubseteq a') \rightarrow ([\![\text{det}]\!](a)(b) \Vdash [\![\text{det}]\!](a')(b))]$$

- left downward monotone *(↓MON) iff*

$$\forall a, a' \in D_n \forall b \in D_{t_p}[(a' \sqsubseteq a) \rightarrow ([\![\text{det}]\!](a)(b) \vdash [\![\text{det}]\!](a')(b))]$$

- right upward monotone *(MON↑) iff*

$$\forall a \in D_n \forall b, b' \in D_{t_p}[(b \sqsubseteq b') \rightarrow ([\![\text{det}]\!](a)(b) \Vdash [\![\text{det}]\!](a)(b'))]$$

- right downward monotone *(MON↓) iff*

$$\forall a \in D_n \forall b, b' \in D_{t_p}[(b' \sqsubseteq b) \rightarrow ([\![\text{det}]\!](a)(b) \Vdash [\![\text{det}]\!](a)(b'))]$$

Recall that for each specific determiner, the ordering and proof-theoretic consequence employed are according to whether it is positive or negative.

Lemma 10.7.3 (order). *Suppose $a \sqsubseteq b$. Then, one of the following cases holds.*

1. $a = \lambda \mathbf{j}.[\![\mathbf{j} \text{ isa } X]\!]$, $b = \lambda \mathbf{j}.[\![\mathbf{j} \text{ isa } Y]\!]$, *and for every* Γ, *if* $\Gamma \vdash \mathbf{j}$ isa X, *then* $\Gamma \vdash \mathbf{j}$ isa Y.

2. $a = \lambda \mathbf{j}.[\![S[\mathbf{j}]]\!]$, $b = \lambda \mathbf{j}.[\![S'[\mathbf{j}]]\!]$, *and for every* Γ, *if* $\Gamma \vdash S[\mathbf{j}]$, *then* $\Gamma \vdash S'[\mathbf{j}]$.

Proof: Suppose $a \sqsubseteq b$. The proof is by induction on '$\sqsubseteq$'. I only show the proof for relativization. Adjectival modification is similar, using the adjective rules instead of the relativization rules. Consider first the unilateral case.

- For $\sqcap(a)(b) \sqsubseteq a$, $a \in D_n$ and $b \in D_{t_p}$, we have by lemma 10.3.1

    - $a(\mathbf{j}) = [\![\mathbf{j} \text{ isa } Y]\!]$, for some noun Y.
    - $b(\mathbf{j}) = [\![S[\mathbf{j}]]\!]$ for some verb-phrase $S[-]$.
    - $(\sqcap(a)(b))(\mathbf{j}) = [\![\mathbf{j} \text{ isa } Y \text{ who } S[-]]\!]$.

Therefore,

$$\frac{\Gamma : \mathbf{j} \text{ isa } Y \text{ who } S[-] \quad \Gamma, \mathbf{j} \text{ isa } Y, [S[\mathbf{j}] : \mathbf{j} \text{ isa } Y}{\Gamma : \mathbf{j} \text{ isa } Y} \ (rE)$$

- For $\sqcap(a)(b) \sqsubseteq a$, $a \in D_d$ and $b \in D_n$, we have by lemma 10.7.2

  – $a(\mathbf{j}) = [[\mathbf{j} \text{ is } A]]$, for some adjective A.
  – $b(\mathbf{j}) = [[\mathbf{j} \text{ isa } Y]]$ for some noun Y.
  – $(\sqcap(a)(b))(\mathbf{j}) = [[\mathbf{j} \text{ isa } A\ Y]]$.

Therefore,

$$\frac{\Gamma : \mathbf{j} \text{ isa } A\ Y \quad \Gamma, \mathbf{j} \text{ is } A, \mathbf{j} \text{ isa } Y : \mathbf{j} \text{ isa } Y}{\Gamma : \mathbf{j} \text{ isa } Y} \ (adjE)$$

- For $a, b \in D_{t_p}$ the claim reduces to the first case by applying **n**.

- The reflexivity and transitivity cases follow directly from the reflexivity and transitivity of '⊢'.

For the bilateral case, the proof is similar, using the bilateral relativization rules (in Figure 10.7).

10.7.1   Examples of monotonicity proofs: positive determiners

I show that every, a positive determiner, is ↓MON↑. The monotonicity of other positive determiners, agreeing with their monotonicity as GQs (in MTS), is shown in a similar way, and I omit the details.

Proposition 10.7.14. every *is* ↓MON↑.

Proof:

1. To show ↓MON, we need to show that for arbitrary $a, a' \in D_n$ and $b \in D_{t_p}$ s.t. $a' \sqsubseteq a$:

   $$[[\text{every}]](a)(b) \Vdash [[\text{every}]](a')(b) \tag{10.7.60}$$

   Similarly, to show that every is MON↑, we need to show that for arbitrary $a \in D_n$ and $b, b' \in D_{t_p}$ s.t. $b \sqsubseteq b'$:

   $$[[\text{every}]](a)(b) \Vdash [[\text{every}]](a)(b') \tag{10.7.61}$$

   By lemma 10.3.1, there are nouns X, Y and pseudo-sentences $S[\mathbf{j}], S'[\mathbf{j}]$ such that $a = \lambda \mathbf{j}.[[\mathbf{j} \text{ isa } X]]$, $a' = \lambda \mathbf{j}.[[\mathbf{j} \text{ isa } Y]]$, $b = \lambda \mathbf{j}.[[S[\mathbf{j}]]]$ and $b' = \lambda \mathbf{j}.[[S'[\mathbf{j}]]]$. (10.7.60) holds iff, for every Γ (10.7.62) holds.

   If, for a fresh **j**, $\Gamma, \mathbf{j}$ isa $X \vdash S[\mathbf{j}]]$, then, for a fresh **j**, $\Gamma, \mathbf{j}$ isa $Y \vdash S[\mathbf{j}]]$

   $$\tag{10.7.62}$$

Similarly, (10.7.61) holds iff, for every Γ (10.7.63) holds.

If, for a fresh $\mathbf{j}$, $\Gamma, \mathbf{j}$ isa $X \vdash S[\mathbf{j}]]$, then, for a fresh $\mathbf{j}$, $\Gamma, \mathbf{j}$ isa $X \vdash S'[\mathbf{j}]]$ (10.7.63)

Thus, we are to show that the two entailments in (10.7.62), (10.7.63) hold.

- To show that (10.7.62) holds, it is enough to show that if $\Gamma, \mathbf{j}$ isa $X \vdash S[\mathbf{j}]$ then $\Gamma, \mathbf{j}$ isa $Y \vdash S[\mathbf{j}]$. This is achieved by the following derivation (Weakening is implicitly used):

$$
\cfrac{
\cfrac{\Gamma, \mathbf{j}\text{ isa } X : S[\mathbf{j}]}{\Gamma : S[(\text{every } X)_{r(S[\mathbf{j}])+1}]}\,(eI)
\quad
\cfrac{\cfrac{\mathbf{j}\text{ isa } Y : \mathbf{j}\text{ isa } Y}{\mathbf{j}\text{ isa } Y : \mathbf{j}\text{ isa } X}\,{(Ax)\atop(\text{lemma }10.7.3.1)}
\quad
\cfrac{}{S[\mathbf{j}] : S[\mathbf{j}]}\,(Ax)}{\Gamma, \mathbf{j}\text{ isa } Y : S[\mathbf{j}]}
}{}\,(eE^{j})
$$

2. As for MON↑, the argument is similar, using Lemma 10.7.3.2 instead of Lemma 10.7.3.1.

10.7.2 Examples of monotonicity proofs: negative determiners

I next show that no, a negative determiner, is ↓MON↓. The monotonicity of other negative determiners, again agreeing with their monotonicity as GQs (in MTS), is shown in a similar way, and I again omit the details.

Proposition 10.7.15. no *is* ↓MON↓.

1. To show ↓MON, we need to show that for arbitrary $a, a' \in D_n$ and $b \in D_{t_p}$ s.t. $a' \sqsubseteq a$:

$$[\![no]\!](a)(b) \Vdash [\![no]\!](a')(b) \tag{10.7.64}$$

Similarly, to show that no is MON↓, we need to show that for arbitrary $a \in D_n$ and $b, b' \in D_{t_p}$ s.t. $b' \sqsubseteq b$:

$$[\![no]\!](a)(b) \Vdash [\![no]\!](a)(b') \tag{10.7.65}$$

By lemma 10.3.1, there are nouns X, Y and pseudo-sentences $S[\mathbf{j}], S'[\mathbf{j}]$ such that $a = \lambda\mathbf{j}.[\![\mathbf{j}\text{ isa } X]\!]$, $a' = \lambda\mathbf{j}.[\![\mathbf{j}\text{ isa } Y]\!]$, $b = \lambda\mathbf{j}.[\![S[\mathbf{j}]]\!]$ and $b' = \lambda\mathbf{j}.[\![S'[\mathbf{j}]]\!]$. (10.7.64) holds iff, for every Γ (10.7.66) and (10.7.67) hold.

If, for a fresh $\mathbf{j}$, $\vdash\Gamma, +\mathbf{j}$ isa $X : -S[\mathbf{j}]]$, then, for a fresh $\mathbf{j}$, $\vdash\Gamma, +\mathbf{j}$ isa $Y : -S[\mathbf{j}]]$ (10.7.66)

and

If $\vdash\Gamma : +\mathbf{j}$ isa Y and $\vdash\Gamma : +S[\mathbf{j}]]$, then, $\vdash\Gamma : +\mathbf{j}$ isa X and $\vdash\Gamma : +S[\mathbf{j}]]$ (10.7.67)

Similarly, (10.7.65) holds iff, for every Γ (10.7.68) and (10.7.69) hold.

If, for a fresh $\vdash\mathbf{j}$, $\Gamma, +\mathbf{j}$ isa $X : -S'[\mathbf{j}]]$, then, for a fresh $\mathbf{j}$, $\vdash\Gamma, +\mathbf{j}$ isa $X : -S[\mathbf{j}]$ (10.7.68) and

If $\vdash\Gamma : +\mathbf{j}$ isa X and $\vdash\Gamma : +S'[\mathbf{j}]]$, then, $\vdash\Gamma : +\mathbf{j}$ isa X and $\vdash\Gamma : +S[\mathbf{j}]]$ (10.7.69)

Thus, we are to show that the entailments in (10.7.66), (10.7.67), (10.7.68) and (10.7.69) hold.

I only show the derivation establishing (10.7.66) is a natural adaptation of the one above to the bilateral rules, where I ignore scope marking for better readability.

$$\cfrac{\cfrac{\Gamma, [+\mathbf{j} \text{ isa } X]_i \vdash +S[(\mathbf{j})]}{\Gamma \vdash +S[(\text{no } X)]}\ (nI^+) \quad \cfrac{+\mathbf{j} \text{ isa } Y \vdash +\mathbf{j} \text{ isa } Y}{+\mathbf{j} \text{ isa } Y \vdash +\mathbf{j} \text{ isa } X}\ (Lemma\ 10.7.3.1) \quad [+S[\mathbf{j})]]_j \vdash +S[\mathbf{j})]}{\Gamma, +\mathbf{j} \text{ isa } Y \vdash -S[\mathbf{j}]}\ (nE^-)$$
(10.7.70)

10.8   Summary

This chapter studied properties of determiners based on their proof-theoretic semantic meaning. By providing a PTS analogy of the traditionally MTS-based property of conservativity (by transferring to PTS the MTS concepts used in its traditional definition), I was able to *prove* that *every* determiner is conservative in at least one of its arguments. The proof is based solely on the proof-theoretically interpreted type of a determiner, both positive and negative. This type restricts determiners meanings much stronger than the GQ-based denotation employed by MTS, and hence the provability of the result only stipulated as a semantic universal in MTS. Note that this theorem makes conservativity devoid of any characterization power as a tool for determination what can be realized as an actual NL determiner's meaning. PTS properties such as harmony and conservativity serve better for this purpose, as exemplified by '*donk*'.

Chapter 11

Opacity: Intensional transitive verbs

11.1 Introduction

In this chapter I present a proof-theoretic semantics for sentences headed by *intensional transitive verbs (ITVs)*, notorious for its difficulty in specification and formulation of model-theoretic truth-conditions for sentences exhibiting it (see Section 11.1.4). Those verbs are also known as *opaque verbs*, and are traditionally characterized by the following three properties:

Admittance of non-specific (notional) objects

Resistance to substitutability of coextensives

Suspension of existential commitment

Let me state on the onset that I am not considering any specific *lexical* properties or world-knowledge of members of this class of verbs transcending the above characteristics. I take those three properties as the *full* characterization of the constructions studied. Whenever seek is used in examples, no attempt is made to "really" understand what seeking is about, and any other ITV could serve the purpose of the example. So, approaches reducing seek to try to find are not appropriate on the onset. As not all transitive verbs admit notional arguments, I let $\hat{R}$ range over those which do, which are assumed to be a *given* class of words. This family includes[1] verbs like seek, need, look for and want, differing from the extensional transitive verbs like kiss, or find, the latter lacking the above characteristics.

---

[1]In Moltmann [121] there is a detailed presentation of the landscape of families of such verbs.

One more thing that I would like to mention right here is that I will confine my discussion to meaning of single sentences, not in the role ITVs play in discourse. So, phenomena like the absence of anaphora support, that are also mentioned as characterizing ITVs (e.g., Moltmann [119]), are outside the scope of this discussion. Still, I do believe that extending proof-theoretic semantics to discourse in general (and to dialogs) is a worthy task.

Before formalizing the proof-theoretic characterization of ITVs, the three characteristics are discussed in more detail.

11.1.1   Non-specific objects

In the fragment studied here, this is taken as admitting as objects of ITVs indefinite[2]. DPs with a non-specific reading.

So, those verbs are ambiguous, having also a meaning allowing specific readings parallel to more ordinary extensional verbs. Other quantifiers can also stand in the object position of such verbs: one may look for all/ most sheep, still without looking for any specific one.   I focus here on the existential quantifier embodied as an indefinite DP (but see also the treatment of other existential determiners on page 333).

Consider, for example,

$$\text{Jacob seeks a secretary.} \qquad (11.1.1)$$

in contrast to its extensional counterpart

$$\text{Jacob finds a secretary.} \qquad (11.1.2)$$

While in (11.1.2) Jacob finds[3] *some specific secretary*, there is no specific secretary Jacob necessarily seeks in one reading of (11.1.1), though there is also a reading of (11.1.1) where he does seek a specific secretary, analogously to finding one. Note that this ambiguity in the meaning of sentences like (11.1.1) is not a problem of quantifier-scope ambiguity: the sentence contains just one quantifier. This issue has been noticed already in mediaeval times, where Buridan considered

$$\text{Debeo tibi equum, namely, I owe you a horse} \qquad (11.1.3)$$

The distinction between specific and non-specific meaning of indefinite DPs is reflected in the proof-theory by the difference in the way they are introduced into the object position by means of applying an *I*-rule of the meaning-conferring natural-deduction proof-system. This contrasts with the model-theoretic semantics (MTS) rendering of this distinction.

---

[2] I use a and some interchangeably

[3] Thus, the sentence can be paraphrased as *there is some secretary s.t. Jacobs finds her/him*, like in *de re* readings.

11.1.2 Resistance to substitutability of coextensives

Here, the problem can be explained with reference to validity of the following argument.

> John seeks Mary
> Mary is the daughter of the dean
> $- - - - - - - - - - - - - - - -$
> John seeks the daughter of the dean

$(11.1.4)$

In MTS, the issue is referred to as resistance to substitutivity of coextensive expressions. The copula is is interpreted as Mary and the daughter of the dean having the same extension (i.e. denotation) by referring to the same element in every model. In PTS, is is interpreted via the I/E-rules for identity (repeated in Figure 11.1 below). The argument (11.1.4) is considered invalid in general, by an appeal to John's possible unawareness of Mary being identical to the daughter of the dean. A major task of MTS is to reflect this awareness in models. Here, the proof-theoretic task is to enforce the non-derivability of the conclusion of the argument (11.1.4) from its premises by the rules.

Let me note in passing, that there is no general consensus that resistance to substitutivity is a correct characteristic of ITVs (and other opaque sentences, most notably propositional attitudes). For example, Salmon [181] claims that resistance to substitutivity is only apparent, originating not from meaning, but rather from the pragmatics of use. Soames [204] also rejects failure of substitutivity, as well as others. I find myself in this camp, though for totally different reasons, not connected to "direct reference", an MTS concept. Rather, I believe that when judging consequence, or validity of arguments, only literal meaning is to be taken into account, not incorporating details of the situation a premise is "asserted", such as agent intentions/plans, or even the intentions/plans of the evaluator of the argument. Consequence is an objective relation once meaning has been defined.

However, since resistance to substitutivity has been taken as a characteristic of ITVs by many, I show here how it is enforceable proof-theoretically. Fortunately, as I show in Section 11.2.3, there is a simple "on/off switch" that can chose between enforcing resistance to substitutivity, or not.

11.1.3 Suspension of existential commitment

A sharpening of the intensionality problem occurs in sentences like

> Jacob seeks a unicorn

$(11.1.5)$

where unicorns need not exist at all in order to be sought (evading the debatable issue of non-existent objects). In MTS, existence is taken as reference to an entity in the domain of the model. Some means is needed to be introduced in order to interpret proof-theoretically existence (other than resorting to a free logic); in contrast to model-theoretic semantics, there is no access here to elements in models. This is explained in Section 11.2.4.

11.1.4   Model-theoretic semantics of ITVs

Defining the meaning of ITVs is considered a semantic problem notorious for its difficult specification of model-theoretic truth-conditions for sentences exhibiting it (e.g., Zimmermann [238], [119], [120] and [121] - draft on her home-page). There is no consensus regarding the truth-conditions of opaque sentences; not even regarding the semantic type of the object DP of the intensional verb, e.g., a quantifier (Zimmermann [238]), a property (Moltmann [121]), or a minimal situation (Moltmann [119]). There is also an indirect interpretation via *decomposition* of the intensional verb (Larson [102]). Another approach (Forbes [40]) appeals to Davidsonian event semantics and its thematic roles. Recently, a *dynamic* semantics was proposed in Sznadjder [211], where an update via ITVs is proposed.

11.2   A Proof-Theoretic Meaning for Opacity

11.2.1   The basic rules for non-specifity

The starting point is to augment the proof-language with another family of parameters, referred to as *notional parameters*, in addition to the individual parameters used for the extensional fragment. Also, the proof language is extended with a family of *notions*, a proof-theoretic counterpart of properties. Notions are formed from nouns (either lexical or compound by modification), from adjectives and from verb-phrases (the latter explained in detail in Section 11.2.2). Recall that such an extension, being another proof-language artefact, does not carry any ontological commitments, in contrast to the various entities assumed by MTS to populate models. The purpose of those additional notional parameters is to allow associating notions with them and the DP Some X (for a noun X) *to be introduced in a different way*, not assuming any individual parameter being an X, thus exhibiting also a reading escaping specificity. Note that I-rules were shown above as explaining quantifier-scope ambiguity as a difference in the order of introduction of the subjects DP and the object DP. I-rules constitute a major tool available in PTS, with a strong explanatory power.

Let **B** be a countable set, disjoint from **P**, of *notional parameters*, ranged over by (possibly indexed) meta-variables **n, m**. I need an analog of the predication **j** isa X, X a (possibly compound) noun, associating a notional parameter with the notion of being an X. I thus extend the proof-language with ground pseudo-sentences of the following forms:

nominal: n is being a X, X a (possibly compound) noun. For example, **n is being a secretary**.

adjectival: n is being A, A an adjective. For example, **n is being clever**.

verbal: n is Ving, V the gerundive form of a VP. For example, **n is singing, n is loving k** and **n is j loving** (the latter more colloquially expressed by passivisation, namely **being loved by j** (see Section 8.4)). I refer to such notions as *verbal*.

The rules for forming compound notions via modification by (intersective) adjectives and relativization, which use adjectival and verbal notions, are presented in Section 11.2.2.

Note the difference between the two ways of connecting a parameter and a noun:

j isa X: expresses usual predication.

n is being a X: associates a nominal parameter with a nominal notion.

Note that X can be a compound noun, so we may have notions like **being a secretary, being a talented secretary, being a secretary who seeks a unicorn, being a girl who loves every boy**, etc.

For an opaque verb $\hat{R}$, we have now an additional ground pseudo-sentence of the form j $\hat{R}$ n. So, there are now two ways of $\hat{R}$-relating an individual parameter **j** to an object of a transitive verb:

- to another individual parameter, say **k**: j $\hat{R}$ k. For example, **j seeks k**.

- to a nominal notion or a notional parameter, say **n**: j $\hat{R}$ n. For example, **j seeks n, j seeks a secretary**.

It is important to observe that the following are *ill-formed*:

$$\text{j is being a } X, \quad \text{n isa } X$$

The first argument of **is being** has to be a notional parameter, while the first argument of **isa** must be an individual parameter.

I now can formulate the following I-rule, introducing some X *notionally* into an object position of an ITV $\hat{R}$ with a non-specific reading.

$$\frac{\text{n is being a } X \quad \text{j } \hat{R} \text{ n}}{\text{j } \hat{R} \text{ some } X} \ (s_n I)$$

$$(11.2.6)$$

The non-specificity of some X is reflected by the absence of a predicative premise of the form j isa X. This is in contrast to the introduction of some X in its specific

reading via the I-rules (sI), repeated below in a sufficiently simplified form, that has
k isa X as a premise, expressing specificity.

$$\frac{\textbf{k isa } X \quad \textbf{j } R \textbf{ k}}{\textbf{j } R \text{ (some } X)_1} \ (sI_1) \qquad \frac{\textbf{k isa } X \quad \text{(some } Y)_1 \, R \, \textbf{k}}{\text{(some } Y)_1 \, R \text{ (some } X)_2} \ (sI_2) \qquad\qquad (11.2.7)$$

$$\frac{\textbf{k isa } X \quad \text{(every } Y)_1 \, R \, \textbf{k}}{\text{(every } Y)_1 \, R \text{ (some } X)_2} \ (sI_3)$$

Note that the $(s_n I)$-rule enforces the introduction of the notional object some X only
at the lowest scope level. This again contrasts the introduction in the specific reading
(sI), that can introduce some X at any scope level. has one reading (therefore, one
derivation) only, with an object narrow scope.

$$\frac{\begin{array}{c}[\textbf{r isa } X]_i \\ \dfrac{\mathcal{D}_1}{\textbf{r } \hat{R} \textbf{ n}} \quad \dfrac{\mathcal{D}'_2}{\textbf{n is being a } Y} \\ \hline \textbf{r } \hat{R} \text{ (some } Y)_1 \end{array}}{\text{(every } X)_2 \, \hat{R} \text{ (some } Y)_1} \begin{array}{l} (sI) \\[4pt] (eI^i) \end{array} \qquad\qquad (11.2.8)$$

Let me summarise the difference between the extensional (sI)-rules and the intensional $(s_n I)$-rule, a difference constituting a major issue in the whole approach of
PTS to ITVs. I will use the more familiar model-theoretic analogues to explain the
differences.

Scope: There are three extensional (simplified) rules, allowing the introduction of
some X into an object position of R in three cases:

1. The subject is an individual parameter: in this case, some X obtains the
   lowest scope.
2. The subject has already an introduced DP (either some Y or every Y), a
   DP which has already the lowest scope level, and so the object some X
   obtains the higher scope level.

On the other hand, there is *only one* intensional object introduction rule, "insist-
ing", so to speak, that the subject of $\hat{R}$ be an individual parameter. As a result,
the object some X of $\hat{R}$ can only obtain the lowest scope level.

Specificity: Here the difference of interpretation resides in the way parameters are
associated with nouns.

1. In **j** isa X, the notation is just a sugaring of the standard first-order logic
   notation $P(a)$, which model-theoretically would imply that a is referring
   and P has an extension containing the object referred to by a.
2. On the other hand, **n** is being a X is not first order expressible. Rather, it
   would correspond to a second-order variable bound to a predicate name.
   Model theoretically, there is no reference involved to an element of the
   domain.

The harmoniously induced GE-rule is

$$\frac{j \, \hat{R} \text{ some } X \qquad \begin{array}{c} [\mathbf{n} \text{ is being a } X]_i, [\mathbf{j} \, \hat{R} \, \mathbf{n}]_j \\ \vdots \\ S \end{array}}{S} \, (s_n E^{i,j}) \qquad \mathbf{n} \text{ fresh} \qquad (11.2.9)$$

11.2.2 Monotonicity of Opacity

I introduce a partial ordering among notions, reflecting their *generality*. I will express the fact that being a Y is more general than being a X by

$$\text{being a } X \leq_n \text{ being a } Y \qquad (11.2.10)$$

The source of the partial order $\leq_n$ is from modification of notions by means of (intersective) adjectives and relative clauses[4], as described below. Again, it is important to realize that this partial order is internal to the proof-system, and does not reflect any ordering of something populating models[5].

The meaning of sentences with opaque verbs admits *upward monotonicity* in the object argument (Zimmermann [239]), as exhibited by the following instance of the general inference involved.

$$\frac{\text{Jacob seeks a clever secretary} \quad \text{every clever secretary isa secretary}}{\text{Jacob seeks a secretary}} \qquad (11.2.11)$$

The monotonicity is w.r.t. the partial order $\leq_n$. In the example above, the result will follow by introducing an appropriate rules and by showing being a clever secretary $\leq_n$ being a secretary is provable.

First, I introduce the rules connecting a notion its with modification via (intersective) adjectives like $\mathbf{n}$ is being white. I assume here that the intersectivity of intersective adjectives holds also under the scope of opacity[6].

The I/E-rules are the following.

Adjectival modification: The I-rule is

$$\frac{\mathbf{n} \text{ is being a } X \quad \mathbf{n} \text{ is being } A}{\mathbf{n} \text{ is being a } A \, X} \, (a_n I) \qquad (11.2.12)$$

---

[4]In a more extensive fragment, there may be more sources for this partial order.

[5]Contrast this partial order with the one in Zimmermann [239], .

[6]As far as I am aware of, this issue is not discussed in the literature. There are also opinions denying the monotonicity of opacity. A way to block it might be prohibiting intersectivity under the scope of opacity. I deal here only with the case where monotonicity *is* assumed to hold.

A notional parameter associated with two (or more) notions is an analog to predicating two (or more) predicates on an individual parameter. For example, an instance of $(a_n I)$ is

$$\frac{\text{n is being a secretary} \quad \text{n is being clever}}{\text{n is being a clever secretary}} \ (a_n I)$$

The E-rule is

$$\frac{\text{n is being a } X}{S} \quad \begin{array}{c} [\text{n is being a } X]_i, [\text{n is being } A]_j \\ \vdots \\ S \end{array} \ (a_n E^{i,j}) \qquad (\textbf{n} \text{ fresh}) \quad (11.2.13)$$

Relativization: Recall that the fragment allows three kinds of relativization. I will collapse all relative pronouns like who, whom, which etc. to that. I assume that all[7] relativisation is carried out by means of verbal notions.

Intransitive subject: For example,

$$\text{John seeks a secretary that smiles} \qquad (11.2.14)$$

Transitive subject: For example,

$$\text{John seeks a secretary that knows Mary} \qquad (11.2.15)$$

Transitive object: For example,

$$\text{John seeks a secretary that Mary knows} \qquad (11.2.16)$$

The rule for the compound noun has as premises being a notion corresponding to the unmodified noun (secretary in the examples), and a notion corresponding to a verbal notion. For the third case (transitive object), such a notion is usually expressed by using the passive gerundive form (see Section 8.4). In the examples, the three notions are smiling, knowing Mary, being known by Mary.

The notional relatvization rules are the following.

$$\frac{\text{n is being a } X \quad \text{n is } V\text{ing}}{\text{n is being a } X \text{ that } V} \ (r_n I) \qquad (11.2.17)$$

Here V is the verbal form from which the verbal notion is Ving is formed. For example,

$$\frac{\text{n is being a secretary} \quad \text{n is knowing k}}{\text{n is being a secretary that knows k}} \ (r_n I)$$

and

$$\frac{\text{n is being a secretary} \quad \text{n is k knowing}}{\text{n is being a secretary that k knows}} \ (r_n I)$$

---

[7]This is just a simplification of presentation. Nothing prevents consideration, for example, of j seeks a secretary that is clever, which is more colloquial expressed as j seeks a clever secretary.

The E-rule is

$$\frac{\mathbf{n} \text{ is being a } X \text{ that } V \qquad \overset{[\mathbf{n} \text{ is being a } X]_i, [\mathbf{n} \text{ is } V\text{ing}]_j}{\overset{\vdots}{S}}}{S} \; (r_n E^{i,j}) \quad , \mathbf{n} \text{ fresh} \quad (11.2.18)$$

There is a special case of relativization worth noting. The verbal notion Ving may itself be based on an ITV, forming *nesting* of ITVs. Consider, for example, the following instance of $(r_n I)$:

$$\frac{\mathbf{n} \text{ is being a secretary} \quad \mathbf{n} \text{ is seeking a laptop}}{\mathbf{n} \text{ is being a secretary that seeks a laptop}} \; (r_n I)$$

Combined with $(s_n I)$, the following interpretation of a nested ITV results.

$$\frac{\mathbf{j} \text{ seeks } \mathbf{n} \quad \dfrac{\mathbf{n} \text{ is being a secretary} \quad \mathbf{n} \text{ is seeking a laptop}}{\mathbf{n} \text{ is being a secretary that seeks a laptop}} \; (r_n I)}{\mathbf{j} \text{ seeks a secretary that seeks a laptop}} \; (s_n I) \quad (11.2.19)$$

We see that there is nothing special about this nesting. The above sentence is interpreted the same way as its following counterpart, that has an extensional TV instead of the ITV:

$$\mathbf{j} \text{ seeks a secretary that has a laptop}$$

In particular, the nested **seek** above is not applied in (11.2.19) to any individual parameter as a subject – no "unspecified seeker" arises ...

I now turn to the rules pertaining '$\leq_n$', the partial ordering of notions. The first source of $(\leq_n)$ is from an explicit sentence in the extensional fragment, relating two nouns, X and Y: every X isa Y. This induces the following I/E-rules for notional comparison (omitting Γ for brevity).

$$\frac{\text{every } X \text{ isa } Y}{\text{being a } X \leq_n \text{being a } Y} \; (\leq_n I_1) \qquad \frac{\text{being a } X \leq_n \text{being a } Y}{\text{every } X \text{ isa } Y} \; (\leq_n E_1) \quad (11.2.20)$$

And similarly

$$\frac{\text{every } X \; V}{\text{being a } X \leq_n V\text{ing}} \; (\leq_n I_2) \qquad \frac{\text{being a } X \leq_n V\text{ing}}{\text{every } X \; V} \; (\leq_n E_2) \quad (11.2.21)$$

Based on the $(a_n I)$ and $(r_n I)$ rules, the following additional I-rules for the notional partial order (with no premises) are introduced.

$$\frac{}{\text{being a } A \; X \leq_n \text{being a } X} \; (\leq_n I_3) \qquad \frac{}{\text{being a } A \; X \leq_n \text{being } A} \; (\leq_n I_4) \quad (11.2.22)$$

$$\frac{}{\text{being a } X \text{ that } V \leq_n \text{being a } X} \; (\leq_n I_5) \qquad \frac{}{\text{being a } X \text{ that } V \leq_n V\text{ing}} \; (\leq_n I_6)$$
$$(11.2.23)$$

Let $\leq_n$ be the least partial order (over notions) extending the base cases above. The following lemma is a characterisation of $(\leq_n)$ (for the fragment under consideration).

Lemma 11.2.4 (notional ordering). *For any two notions* being a X *and* being a Y, $\vdash\Gamma$: being a $X \leq_n$ being a Y *iff one of the following cases applies.*

1. *X, Y are (possibly compound) nouns and $\vdash\Gamma$: every X isa Y.*

2. *For some adjective A and (possibly compound) noun Z, X is A Z and Y is Z.*

3. *For some noun Z and some V, X is a Z that V and Y is Z.*

Similar clauses apply for the larger notion being adjectival or verbal, which I omit since they are not used in the sequel.

I now attend to some general issues related to monotonicity of opacity. The upward monotonicity[8] exemplified by the following additional derived I-rule for non-specific objects is mostly accepted in the literature (in its model-theoretic guise); see, for example, Moltmann [119] (p.7, fn. 4) and Zimmermann [239].

$$\frac{\mathbf{j} \ \hat{R} \text{ some } X \quad \text{being a } X \leq_n \text{ being a } Y}{\mathbf{j} \ \hat{R} \text{ some } Y} \ (s_n I_2)$$

$$(11.2.24)$$

The derivation of $(s_n I_2)$ follows the cases of the notional ordering lemma. I show one of the cases, the one based on $(\leq_n I_2)$ (omitting the context Γ). I have to show

$$\vdash\mathbf{j} \ \hat{R} \text{ some } A \ X : \mathbf{j} \ \hat{R} \text{ some } X \qquad (11.2.25)$$

of which (11.2.11) is an instance. The derivation is

$$\frac{\mathbf{j} \ \hat{R} \text{ some } A \ X \quad \dfrac{[\mathbf{j} \ \hat{R} \ \mathbf{n}]_1 \quad \dfrac{[\mathbf{n} \text{ is being a } A \ X]_2 \quad \dfrac{\dfrac{[\mathbf{n} \text{ is being } A]_3 \quad [\mathbf{n} \text{ is being a } X]_4}{\mathbf{n} \text{ is being a } X} \ (a_n E^{3,4})}{\mathbf{n} \text{ is being a } X} \ (s_n I)}{\mathbf{j} \ \hat{R} \text{ some } X} \ (s_n E^{1,2})}{\mathbf{j} \ \hat{R} \text{ some } X}$$

$$(11.2.26)$$

Similarly, one can derive

$$\vdash\mathbf{j} \ \hat{R} \text{ some } X \text{ that } V : \mathbf{j} \ \hat{R} \text{ some } X \qquad (11.2.27)$$

Note that accepting the $(s_n I)$-rule detaches **seek**, for example, from anything like "the intentional state of the seeker", occasionally appealed to in some model-theoretic treatments, e.g.,Richard [178] (not in connection with monotonicity). The second premise is an objective statement about the relationship between being an X and being a Y, independent of the seeker. I consider this right, as long as no more specific lexical semantics for **seek** is given. With the above rules, one can derive, for example, (11.2.11).

---

[8]The downwards monotonicity is generally agreed not to hold.

I assume the natural assumption that the partial ordering of notions admits upper bounds. That is, for every X, Y there is some Z such that being a $X \leq_n$ being a Z and being a $Y \leq_n$ being a Z. Thus, a common upper bound of being a sweater and being a pen might be being a material object, or just being a sweater or a pen if coordination is added to the fragment. The issue of the existence *least* upper bound on notions is immaterial here. Note that I take the existence of upper bounds as an assumption, since the theory here does not include a lexical semantics theory, from which such an assumption might be derived as a fact.

Another issue related to the monotonicity of notions results from coordination. In the extensional fragment, there are two families of coordination rules related to determiners (besides the familiar logical rule for sentential coordination). For simplicity, I formulate only a simplified version pertaining to subjects of transitive verbs and an indefinite object. For two determiners δ_1, δ_2 and two (possibly compound) nouns X, Y, Z,

Determiner phrase coordination:

$$\frac{\delta_1 X \ R \ \text{some} \ Z \quad \delta_2 Y \ R \ \text{some} \ Z}{\delta_1 X \wedge \delta_2 Y \ R \ \text{some} \ Z} \ (DP \wedge I) \tag{11.2.28}$$

Determiner coordination:

$$\frac{\delta_1 X \ R \ \text{some} \ Z \quad \delta_2 X \ R \ \text{some} \ Z}{\delta_1 \wedge \delta_2 X \ R \ \text{some} \ Z} \ (D \wedge I) \tag{11.2.29}$$

There is an important issue resulting from accepting upward monotonicity, and in particular the existence of upper bounds, known as *the common objective* issue. It is exemplified in the following derivability claim. Let Z be an upper bound on X and Y, and suppose **j** and **k** are John and Mary, respectively.

$$\vdash \mathbf{j} \ \text{seeks some} \ X, \ \mathbf{k} \ \text{seeks some} \ Y, \text{being an} \ X \leq_n \text{being a} \ Z, \tag{11.2.30}$$
$$\text{being a} \ Y \leq_n \text{being a} \ Z : \mathbf{j} \ \text{and} \ \mathbf{k} \ \text{seek some} \ Z$$

The derivation of (11.2.30) is the following.

$$\frac{\dfrac{\mathbf{j} \ \text{seeks some} \ X \quad \text{being a} \ X \leq_n \text{being a} \ Z}{\mathbf{j} \ \text{seeks some} \ Z} \ (s_n I_2) \quad \dfrac{\mathbf{k} \ \text{seeks some} \ Y \quad \text{being a} \ Y \leq_n \text{being a} \ Z}{\mathbf{k} \ \text{seeks some} \ Z} \ (s_n I_2)}{\mathbf{j} \ \text{and} \ \mathbf{k} \ \text{seek some} \ Z} \ (\wedge_{DP} I) \tag{11.2.31}$$

An instance of (11.2.30) is, for example,

$$\vdash \mathbf{j} \ \text{seeks a green sweater}, \mathbf{k} \ \text{seeks a red sweater},$$
$$\text{being a green sweater} \leq_n \text{being a sweater}, \text{being a red sweater} \leq_n \text{being a sweater}$$
$$: \mathbf{j} \ \text{and} \ \mathbf{k} \ \text{seek a sweater}$$

A similar inference obtains for more general subjects, e.g.,

$$\text{every girl seeks a green sweater}, \text{some boy seeks a red sweater},$$
$$\vdash \ \text{being a green sweater} \leq_n \text{being a sweater}, \text{being a red sweater} \leq_n \text{being a sweater}$$
$$: \text{every girl and some boy seek a sweater}$$

Also, using $(D \wedge I)$ instead of $(DP \wedge I)$,

$$\vdash \begin{array}{c} \text{at least two girls seeks a green sweater, at most five girls seeks a green sweater,} \\ \text{being a green sweater} \leq_n \text{being a sweater,} \\ : \text{at least two and at most five girls seek a sweater} \end{array}$$

Note that

$$\nvdash \mathbf{j} \, \hat{R} \text{ some } X, \mathbf{k} \, \hat{R} \text{ some } X : \mathbf{j} \, \hat{R} \text{ some } X \text{ that } \mathbf{k} \, \hat{R} \qquad (11.2.32)$$

For example,

$$\nvdash \mathbf{j} \text{ seeks a secretary}, \mathbf{k} \text{ seeks a secretary} : \mathbf{j} \text{ seeks a secretary that } \mathbf{k} \text{ seeks} \quad (11.2.33)$$

Another inference associated with monotonicity can be exemplified by

$$\frac{\mathbf{j} \text{ seeks a secretary}}{\mathbf{j} \text{ seeks something}} \qquad (11.2.34)$$

The extensional fragment had only binary quantification, and **something** was not included in it. An easy extension including it is obtained by adding the following (simplified) rules for the transitive case.

$$\frac{\mathbf{j} \, R \, \mathbf{k}}{\mathbf{j} \, R \text{ something}} \; (stI) \qquad \frac{\mathbf{j} \, R \text{ something}}{\mathbf{j} \, R \, \mathbf{k}} \; (stE) \text{, } \mathbf{k} \text{ fresh} \qquad (11.2.35)$$

The difference between (11.2.35) and (11.2.7) is the absence in the former of any nominal qualification X of the object $\mathbf{k}$ of R. This extension gives rise to a notion **being something**. By considering this notion to be the greatest notion under the $\leq_n$ partial order, namely, for every X,

$$\frac{}{\text{being a } X \leq_n \text{being something}} \; (\leq_n I_{max}) \qquad (11.2.36)$$

by which the inference in (11.2.34) results using monotonicity.

Consider next the notional interpretation of

$$\mathbf{j} \text{ and } \mathbf{k} \text{ seek something} \qquad (11.2.37)$$

The sentence (11.2.37) is ambiguous between *two notional readings* (in addition to the obvious specific reading):

1. both $\mathbf{j}$ and $\mathbf{k}$ seek *the same* unspecific thing.

2. each of $\mathbf{j}$ and $\mathbf{k}$ seek a possibly different unspecific thing.

The difference in meaning between the two readings is captured by means of the two ways the readings are derived.

1. The first reading results by

$$\frac{\textbf{j}\ \text{seeks something}\quad \textbf{k}\ \text{seeks something}}{\textbf{j}\wedge\textbf{k}\ \text{seek something}}\ (DP\wedge I) \tag{11.2.38}$$

where both premises result from seeking the *same* notion **n**.

2. On the other hand, the second reading results by

$$\frac{\dfrac{\textbf{j}\ \text{seeks some}\ X\quad \text{being a}\ X\leq_n\ \text{being something}}{\textbf{j}\ \text{seeks something}}\ (s_n I_2)\qquad \dfrac{\textbf{k}\ \text{seeks some}\ Y\quad \text{being a}\ Y\leq_n\ \text{being something}}{\textbf{k}\ \text{seeks something}}\ (s_n I_2)}{\textbf{j}\ \text{and}\ \textbf{k}\ \text{seek something}}\ (DP\wedge I) \tag{11.2.39}$$

Here two generalisations to a common objective take place.

We saw already the effect of derivation method on proof-theoretic meaning in the discussion of quantifier-scope ambiguity above.

Let me mention that in her model-theoretic semantics of ITVs [119], Moltmann accepts the argument to the common objective as valid. On the other head, Zimmermann [239] rejects it (with the conclusion John seeks something which Mary seeks), based on considerations involving a principle he calls 'exact matching', a principle arising from the finer lexical semantics of seek. Since, as I mentioned above, I exclude more specific lexical meanings of ITVs from consideration here, I hold (11.2.30) valid.

I end the discussion of monotonicity of ITVs by a brief consideration of another source for monotonicity, arising when other determiners, besides some, are allowed as objects of $\hat{R}$. A fuller discussion involves a PTS for plurality, which I do not have at this stage. Consider the family of positive[9] *existential* determiners of the form at least n, $n\geq 1$. The (simplified) I/E-rules for this family in the extensional fragment are the following. For simplicity, I formulate the rules only for the object position of an extensional transitive verb.

$$\frac{\{\textbf{k}_i\ \text{isa}\ X\}_{1\leq i\leq m}\quad \{\textbf{j}\ R\ \textbf{k}_i\}_{1\leq i\leq m}}{\textbf{j}\ R\ \text{at least}\ m\ X}\ (\geq mI)\qquad \textbf{k}_i\ \text{pairwise distinct} \tag{11.2.40}$$

$$\frac{\textbf{j}\ R\ \text{at least}\ m\ X\qquad \begin{array}{c}[\{\textbf{k}_i\ \text{isa}\ X\}_{1\leq i\leq m}]_{l_1},[\{\textbf{j}\ R\ \textbf{k}_i\}_{1\leq i\leq m}]_{l_2}\\ \vdots\\ S\end{array}}{S}\ (\geq mE^{l_1,l_2})\qquad \textbf{k}_i\ \text{pairwise distinct and fresh} \tag{11.2.41}$$

For δ a determiner in this family, let $m(\delta)$ be the corresponding number of premises $\textbf{k}_i$ isa X in its δI-rule. Each such δ gives rise to a notion being $\delta\ X$. For example, being at least three secretaries. Those existential notions are ordered by

$$\frac{}{\text{being}\ \delta_1\ X\leq_n\ \text{being}\ \delta_2}\ (\leq I_\delta)\quad \text{iff}\ m(\delta_1)\geq m(\delta_2) \tag{11.2.42}$$

---

[9]A discussion of the classification of determiners into positive and negative and a proof-theoretic treatment of the negative ones is presented in Chapter 10.

$$[S[\mathbf{j}]]_1$$
$$\vdots$$
$$\frac{S[\mathbf{k}]}{\mathbf{j} \text{ is } \mathbf{k}} \; (isI^1) \quad \text{where } S \text{ is fresh}$$

$$[S[\mathbf{k}]]_1$$
$$\vdots$$
$$\frac{\mathbf{j} \text{ is } \mathbf{k} \quad S[\mathbf{j}] \quad S'}{S'} \; (isE^1)$$

Figure 11.1: Rules for the identity

(note the reversal of the ordering). Thus, the following is a correct inference:

$$\frac{\mathbf{j} \, \hat{\mathbf{R}} \, \delta_1 \, \mathbf{X} \quad \text{being } \delta_1 \, X \leq_n \text{being } \delta_2}{\mathbf{j} \, \hat{R} \, \delta_2 \, X}$$

For example,

being at least seven secretaries $\leq_n$ being at least two secretaries

Thus, the following is a correct inference[10].

$$\frac{\mathbf{j} \text{ seeks at least two secretaries}}{\mathbf{j} \text{ seeks a secretary}}$$

where both objects of **seek** have notional reading.

11.2.3   Rules for Resistance to substitutability

The argument in (11.1.4) is an instance of the following schematic basic argument **BA**, embodying substitutivity of the analogue of coextensives:

$$\begin{array}{c} \mathbf{j} \, \hat{R} \, \mathbf{k} \\ \mathbf{k} \text{ is } \mathbf{r} \\ \text{-- -- --} \\ \mathbf{j} \, \hat{R} \, \mathbf{r} \end{array} \qquad (11.2.43)$$

Recall that is is the expression of identity in the fragment. The I/E-rules for identity are presented in Figure 11.1. Here $S[-]$ is an arbitrary sentence in the fragment, where '-' is some position within S that can be filled by a parameter. An immediately derived E-rule, obtained by choosing S' as $S[\mathbf{k}]$ itself, is

$$\frac{\mathbf{j} \text{ is } \mathbf{k} \quad S[\mathbf{j}]}{S[\mathbf{k}]} \; (is\hat{E})$$

which express the substitutivity of identicals.

---

[10]I ignore here the plural morphology.

A derivation corresponding to the argument in (11.2.43) would be a direct application of $(is\hat{E})$, the derived identity elimination rule, by taking S to be $\mathbf{j}\,\hat{R}\,\mathbf{k}$. We get

$$\frac{\mathbf{k}\text{ is }\mathbf{r} \quad \mathbf{j}\,\hat{R}\,\mathbf{k}}{\mathbf{j}\,\hat{R}\,\mathbf{r}}\ (is\hat{E})$$

All that is needed to block **BA** is to restrict (isE) (and hence $(is\hat{E})$ too) to apply only to sentences with extensional verbs! Thus, the rule (isE) can serve as a switch between accepting or rejecting the (failure of) substitutivity.

If we let notional equality to be defined in terms of the partial ordering

being a X is being a Y iff being a X $\leq_n$ being a Y $\wedge$ being a Y $\leq_n$ being a X
$$(11.2.44)$$

subsitutivity can be extended or blocked, as desired, also directly to notions.

$$\frac{\mathbf{j}\,\hat{R}\text{ some }X \quad \text{being a }X\text{ is being a }Y}{\mathbf{j}\,\hat{R}\text{ some }Y} \qquad (11.2.45)$$

This can arise, for example, under a context Γ including both every X isa Y and every Y isa X.

Note, in passing, that while resistance of ITVs to substitutivity of identicals in their object, notionally read, argument may be disputed, there is no dispute about such a substitutivity in the subject argument. Thus, the following is certainly a valid argument.

$$\frac{\mathbf{j}\text{ is }\mathbf{k} \quad \mathbf{j}\,\hat{R}\,\mathbf{r}}{\mathbf{k}\,\hat{R}\,\mathbf{r}}\ (is\hat{E})$$

For example,

$$\frac{\text{John is the chairman} \quad \text{the chairman seeks a secretary}}{\text{John seeks a secretary}}\ (is\hat{E})$$

Thus, any theory relying on "seekers intentions" (or the like) has to explain how identicals obtain "identical intentions".

11.2.4   Existential Commitment

Existential commitment here is expressed via a use of an individual parameter. So, suspension of existence is manifested directly by the form of the $(s_n I)$-rule introducing a non-specific indefinite DP into the object position, that does not have a premise using an individual parameter.

Suppose secretaries exist and unicorns do not exist. This would be embodied at the level of ground (i.e., atomic) sentences, so that $\mathbf{k}$ is a secretary could surface as a possible premise, while **k** is a unicorn will not. This difference between secretaries and unicorns has no bearing on the notions **being a secretary** and **being a unicorn**. At the atomic level, there is no difference in the status of **n** is being a secretary and **n** is being a unicorn as premises. Associating a name to a notion has no existential commitment. As a result, **j** seeking a (non-specific) secretary and **j** seeking a (non-specific) unicorn are derived in exactly the same way:

$$\frac{\text{n is being a secretary} \quad \text{j seeks n}}{\text{j seek some secretary}} \ (s_nI) \qquad \frac{\text{n is being a unicorn} \quad \text{j seeks n}}{\text{j seek some unicorn}} \ (s_nI)$$

This can be contrasted with the derivation of **j** finds a unicorn, which needs a premise appealing to an individual parameter.

$$\frac{\text{k isa unicorn} \quad \text{j finds k}}{\text{j finds some unicorn}} \ (sI)$$

Let me turn now to the proof-theoretical reified meanings of ITVs. The main observation is about the difference between the notional reading and the predicative reading of the following two generic sentences.

$$(*) \ [\![\mathbf{j} \ \hat{R} \text{ some } X]\!] \quad (**) \ [\![\mathbf{j} \ R \text{ some } X]\!] \tag{11.2.46}$$

Any canonical derivation of $(*)$ essentially ends with one of the rules (s_nI) and (s_nI_2) (the latter if monotonicity is appealed to). In both cases, there is a notional premise being a X involved. On the other hand, any canonical derivation of $(**)$ essentially ends with an application of (sI), with a predicative premise isa X.

Thus, the reified proof-theoretic meanings exhibit all the properties discussed above.

11.3 Passivisation

For ITVs, there is an interesting point to note about passivization, so I present rules for determining its meaning for ITVs. The emerging prediction from the rules is that the non-specific reading of indefinite DPs remains invariant under passivisation *but shifts to the DP in the subject position.*

I will use $\hat{R}^{-1}$ as the passive form of the ITV $\hat{R}$. The I/E-rules for passivisation are the following.

$$\frac{\mathbf{j} \ \hat{R} \ \mathbf{n}}{\mathbf{n} \ \hat{R}^{-1} \mathbf{j}} \ (passI) \qquad \frac{\mathbf{n} \ \hat{R}^{-1} \mathbf{j}}{\mathbf{j} \ \hat{R} \ \mathbf{n}} \ (passE) \tag{11.3.47}$$

For example,

$$\frac{\text{j seeks n}}{\text{n is sought by j}} \ (passI)$$

The passive counterpart of (s_nI) is

$$\frac{\mathbf{n} \ \hat{R}^{-1} \ \mathbf{j} \quad \text{n is being a } X}{\text{some } X \ \hat{R}^{-1} \ \mathbf{j}} \ (s_npassI) \tag{11.3.48}$$

From $(passI)$, we easily derive the following derived rule $(pass\hat{I})$, lifting passivisation to the DP-level.

$$\frac{\textbf{j}\ \hat{R}\ \text{some}\ X}{\text{some}\ X\ \hat{R}^{-1}\ \textbf{j}}\ (pass\hat{I})$$

                                                                        (11.3.49)

For example,

$$\frac{\textbf{j}\ \text{seeks some secretary}}{\text{some secretary is sought by}\ \textbf{j}}\ (passI)$$

(with both premise and conclusion retaining the non-specific reading).

The derivation of $(s_n\hat{I})$ is (with Γ omitted)

$$\cfrac{\textbf{j}\ \hat{R}\ \text{some}\ X \qquad \cfrac{\cfrac{[\textbf{j}\ \hat{R}\ \textbf{n}]_1}{\textbf{n}\ \hat{R}^{-1}\ \textbf{j}}\ (passi) \qquad [\textbf{n is being a}\ X]_2}{\text{some}\ X\ \hat{R}^{-1}\ \textbf{j}}\ (s_npassI)}{\text{some}\ X\ \hat{R}^{-1}\ \textbf{j}}\ (s_nE^{1,2})$$

Here one can see that that passivisation inherits the non-specific reading of the indefinite DP, this time in the subject position, from the non-specific reading of the DP in the object position in the original active sentence.

11.4   Summary

This chapter expands PTS to cover also an intensional NL fragment, based on intensional transitive verbs (ITVs). The main lesson is the used of notional parameters for expressing the non-specific reading of indefinite objects of the verb, a syntactic tool without any ontological burdens arising from decisions as to how to populate models in order to appeal to MTS. All the ITV's characteristics are enforced by I/E-rules of a meaning-conferring ND-system. Once again, the power of an appeal to the way of introduction a DP into its position manifests itself.

Chapter 12

Contextual Meaning Variation

12.1 Introduction

A peculiar property of NL is that meaning can vary with context. From [205]:

> The problem of context dependence is the problem of explaining how context contributes to interpretation ...

See Stanley and Gendler Szabó [205] for a discussion of a variety of special cases of the general problem of meaning variation with context. The purpose of this chapter is to provide a proof-theoretic semantics (PTS) for two special cases of the general dependence problem, the contextually varying meaning. It may well be the case that the proof-theoretic interpretation of other kinds of expressions with contextually varying meanings will require different proof-theoretic techniques than the one used here. I focus on the problems below as they fit naturally into the fragment of NL for which a PTS has already been proposed.

Contextual Domain Restriction (QDR): This problem has a rich history (see [205] for references to earlier work), all carried out under the *model-theoretic semantics (MTS)* theory of meaning.

To explain the QDR-problem, consider the following example sentence (from [205]).

$$\text{every bottle is empty} \tag{12.1.1}$$

The literal model-theoretic meaning of (12.1.1), involving quantification and predication, attributes the property of emptiness to every entity in a model falling under the extension[1] of bottle. This truth-condition is usually expressed, alluding to the generalized quantifiers theory, as the inclusion of the extension of bottle in the extension of empty. The general consent is, however, that in different circumstances, to be captured by contexts, the domain of quantification is not over the whole extension of bottle (all bottles in the universe); rather, it is over a restriction of this extension to one determined by a context, e.g., every bottle in a room where some party takes place in one context, or bottles in some chemistry laboratory in another context. Similarly,

$$\text{some bottle is empty} \qquad (12.1.2)$$

is taken also to have contextually varying meaning, asserting that some bottle, determined by a given context, is empty, not that some bottle in the universe is empty.

A more radical approach, called *contextualism*, claims that *there is no quantification which is not contextually restricted*! Even apparently unrestricted quantification as expressed by everything or something are contextually restricted. See, for example, Glanzberg [65].

Contextual Definiteness: Note that there are in the literature several interpretations of definiteness, fitting different uses of the definite article. Here, my main concern is not a comprehensive proof-theoretic definition of all those uses. Rather, the focus is on the proof-theoretic *strategy* of moving from absolute definiteness to contextual definiteness. Therefore, one interpretation is chosen, following Russell's contextual[2] definition, that lends itself conveniently to such a meaning shift. It also fits nicely with the general contextual definition of other determiners, as explained in Chapter 7. Other approaches to definiteness, for example involving *familiarity* instead of (relative) uniqueness, can be treated by a somewhat different structure of a context and are not considered here further.

Consider the following example sentence.

$$\text{The bottle is empty} \qquad (12.1.3)$$

Following Russell's famous formalization of

$$\text{The (current) king of France is bald} \qquad (12.1.4)$$

involving a conjunction of *existence* and *uniqueness*, in an *absolute* sense, and then predication. The literal model-theoretic meaning of (12.1.1), given a model, first attributes uniqueness to the extension of bottle in the given model, and then attributes the property of emptiness to that unique entity in a model falling under that extension of bottle by requiring that unique element to be a member of the extension of empty in that model. Thus, this literal meaning fits the general scheme of generalized quantifiers theory. Here too, the general consent is, however, that in different circumstances, to be captured by contexts, again giving rise to contextual meaning variation, the uniqueness is not over the whole

---

[1] As noted by Glanzberg (in [65]), it suffices to conduct this study in an extensional fragment of NL, as intentionality seems orthogonal to QDR-problem.

[2] Note that 'context' is used here in two different ways; no confusion should arise.

extension of bottle (a single bottle in the universe); rather, it is over a restriction of this extension to one determined by a context, e.g., a unique bottle in a room where some party takes place in one context, or a unique bottle in some chemistry laboratory in another context.

Note that MTS in general adheres to a compositional sentential meaning assembly. The primary carriers of meaning are words, interpreted as having denotations in models (that can be rather complex), and semantic composition generates meanings for phrases, until the meaning of a whole (affirmative) sentence is determined. According to this methodology, (some of) the word denotations are context-dependent, a dependence propagated to larger phrases as the interpretation process advances. I'll return to this issue in the sequel.

The *general* semantic problem faced in an attempt to model the variance of literal meaning with context has, according to Stanley and Gendler Szabó [205], two facets.

Descriptive: Deriving the interpretation of some phrase relative to a context, given prior characterization of what features of a context have a bearing on the meaning of that phrase.

Fundamental: Specifying the above mentioned characterization, namely what it is about a context in virtue of which the derivation of the interpretation yields the correct meaning in that context. This specification involves some *explicit definition* of a context.

Thus, for (12.1.1), the descriptive meaning is the proper derivation of the restricted domain of quantification given a context, while the fundamental issue is what in the structure of a context determines the appropriate domain restriction.

In general, MTS has many difficulties in adequately solving the foundational aspect of contextual variance of truth-conditions. It is far from clear where to "locate" contexts w.r.t. a model. In Stanley and Gendler Szabó [205], for example, the authors admittedly (p. 222) avoid giving a formal characterization of the notion of a context. They just stipulate (for the QDR problem) a certain marking *in the syntactic tree* (the "logical form") that interfaces in a certain way with a context, and provide a description of the way this marking participates in meaning derivations (by intersecting the extension of the head noun with a set "pointed to" by the above mentioned marking). More specifically, Stanley and Gendler Szabó [205] posits as the lexical entry of a noun, say man (in the appropriate leaf of a syntactic tree) the following compound expression.

$$\langle man, f(i) \rangle \tag{12.1.5}$$

where man is the usual extension of man (in a model), i is an anchor for an object to be provided by a context, and f is also an anchor, to a function from objects to objects, also to be provided by context. The rule for computing the extension of man in a given context c is the following (in a slightly modified notation).

$$[[\langle man, f(i) \rangle]]_c =^{df.} [[man]] \cap \{x \mid x \in c[[f]](c[[i]])\} \tag{12.1.6}$$

For an argument for a different location (in the syntactic tree) of that marking (and for a rebuttal of the rejection of this location by Stanley and Gendler Szabó [205]) see Pelletier [139]. There are also views locating this marker on the determiner node, e.g., Westerståhl [234]. Note that in all those approaches, there is no constraint at all imposed on the set $\{x \mid x \in c[\![f]\!](c[\![i]\!])\}$. In particular, as is traditional in generalized quantifier theory, this set need not be the extension (in the model at hand) of any NL phrase.

I would like to claim that this degree of freedom regarding the contextual restriction set is a drawback of all the above approaches to QDR. In general, a context (for meaning variation) can be seen either as external to the interpreted sentence (e,g., a context of utterance), or explicitly contributed by some phrase in the sentence itself. For example, (12.1.1) can be seen as uttered in context of bottles on some table; however, the location of the bottles can be explicitly given in the sentence itself, say by means of a preposition phrase, as in

$$\text{every bottle on the table is empty} \tag{12.1.7}$$

I would like to posit the following *context incorporation principle* as a characterization of contextually varying meaning (as far as QDR is concerned). I see this principle as originating from the semantic concept of ÔmeaningÕ, as far as it relates to contextual meaning variation, and not from any empirical fact about this variation. One certainly can conceive of ÒcontextsÓ not having any linguistic expression. As I see it, while such contexts can contribute to other dimensions of language use, alluding to them is not part of meaning.

The context-incorporation principle (CIP): For every quantified sentence S with meaning depending only on a context c, there exists a (not necessarily unique) sentence S', s.t.

$$[\![S]\!]_c = [\![S']\!] \tag{12.1.8}$$

In other words, the effect of an external context c in terms of QDR in S is always expressible by S' the meaning of which is independent of c (all in the same language, or fragment thereof). Clearly, while (**CIP**) is compatible with (12.1.6), is not enforced by (12.1.6).

It is important to realize what is not the semantic problem discussed here, namely the determination of which context is the ÒrightÓ context for any given token of a contextually-dependent meaning of a sentence. The latter issue is always determined by extra-linguistic means, independently of whether MTS or PTS are employed as the theory of meaning. Rather, the issue is how to handle contextual meaning variation once a context has been determined. For a convincing criticism of the ability to theorize about the actual determination of the intended restriction, see Pupa [157].

Finally, the *consequences* that can be drawn from the contextually-varying meaning of an (affirmative) sentence (i.e., (affirmative) sentences *entailed* by a sentence with contextually-varying meaning, which themselves have meanings varying with context, are hardly ever considered in MTS-based discussions. I will relate to them in the proposed PTS via E-rules.

Note that I adopt here the view expressed by Stanley and Gendler Szabò in [205] that contextual meaning variation is a *semantic* phenomenon, and not a syntactic (ellipsis) or pragmatic (agent related) one. I would like to stress that I am investigating *what (affirmative) sentences mean, and how this meaning varies with context*, and *not* with what an agent means by asserting a sentence in a given context (cf. p. 240); the latter, involving intentions, plans etc., I do see as pragmatic.

Why adhere to **CIP**?

- One can see the semantic view of the QDR phenomenon alluded above as a (partial) justification of **CIP**, that relates to linguistically expressible contexts. In a performative, agent related, use of a sentence with a contextually varying meaning, it is conceivable that other kinds of contextual information, not language oriented, may have an effect. For example, complex visual information in a common ground of speaker and hearer. This is certainly true for contextual resolution of deictic elements in a sentence. This would pertain to context dependence of meaning that fits a more traditional view of it, as pragmatic, not semantic.

- While I am concerned here with meanings of single (affirmative) sentences, there is clearly much semantic interest in *dialogues*, or *discourses*, which are multi-sentential linguistic entities. Adhering to CIP is strongly compatible with identifying the context for meaning variation with the contents of sentences by other participants in a dialogue, or preceding sentences in a discourse. From my proof-theoretic point of view, the PTS for such multi-sentential linguistic constructs is, at best, in its infancy. Principles like **CIP** may encourage further proof-theoretic investigations of such constructs.

There is also a certain proof-theoretic innovation in the concept of parametric family of I-rules (in a natural-deduction system), which is not directly connected to the NL set-up aimed to in the chapter.

For each of the variants of contextual meaning variation considered, first an extension of FOL incorporating that variant, and only then is the variant studied in the NL-fragment considered in this book. The reason of structuring the presentation that way is that most readers will be familiar with FOL and grasp easier the ideas behind the proposed PTS.

12.2   Logic with contextual domain restriction

In this section, I present a version of 1st-order logic (FOL) in which quantifiers are interpreted in a contextually-dependent way. While there is not much interest in such a logic per se, it serves as a vehicle for a clear presentation of the ideas underlying the application of the approach to natural language. It also provides a natural host for the novel proof-theoretic concept of a *parameterized family of I-rules* in the intended natural-deduction meaning-conferring proof system.

12.2.1   First-order logic with contextual domain restriction

I assume the usual object language for FOL, with the usual definition of free/bound variables. For simplicity, a language without constant or function symbols is considered.

Definition 12.2.77 (**R**-context). *An R-context (restricting context) c is a finite collection Γ_c of open formulas with one free variable only; $\Gamma_{c,x}$ is the sub-collection of Γ_c the free variable of which is x. Let C be the collection of all R-contexts.*

This definition of an R-context is certainly not the most general one for a context affecting sentential meanings, but it is intended to capture contexts as providing restriction on quantifiers (also, in Section 12.2, on definiteness), for which purpose this definition suffices. Let $\wedge_{\Gamma_{c,x}}$ be the conjunction of all open formulas in $\Gamma_{c,x}$ (that have x free). I use $\Gamma_{c,x}$ and $\wedge_{\Gamma_{c,x}}$ interchangeably. I use $\Gamma_{c,x}(y)$, $\wedge_{\Gamma_{c,x}}(y)$ to indicate the application of the condition on x to another variable, y.

The main idea, to be captured by the rules below, is that an R-context provides an assumption, (possibly) *dischargeable* in the universal case, in the premise of the I-rule of a quantifier. First, recall the standard I/E-rules for the universal and existential quantifiers in an ND-system for FOL.

$$\frac{\Gamma : \varphi(x)}{\Gamma : \forall x.\varphi(x)} \ (\forall I) \ , \ x \notin free(\Gamma) \qquad \frac{\Gamma : \forall x.\varphi(x)}{\Gamma : \varphi(y)} \ (\forall E) \qquad (12.2.9)$$

$$\frac{\Gamma : \varphi(y)}{\Gamma : \exists x.\varphi(x)} \ (\exists I) \qquad \frac{\Gamma : \exists x.\varphi(x) \quad \Gamma, \varphi(y) : \chi}{\Gamma : \chi} \ (\exists E) \ , \ y \notin free(\Gamma, \chi) \quad (12.2.10)$$

where $\varphi(y)$ is the result of substituting y for all free occurrences of x in $\varphi(x)$. I now introduce a revised ND-system, derivability in which is indicated as '$\vdash_c$' (in contrast to '$\vdash$' indicating the derivability in the standard system). The context c is recorded in a sequent on its separator, in the form $\Gamma :_c \varphi$.

Restricting the universal quantifier: Recall that the intuition behind the usual $(\forall I)$-rule is that since x does not occur free in Γ, it can be seen as standing for an *arbitrary* value, unrestricted in any way by Γ, hence supporting the universal generalization embodied in the $(\forall I)$-rule. The idea behind the I-rule below is to restrict the generalization to those values of x satisfying the contextual restriction imposed by $\Gamma_{c,x}(x)$ for a given R-context c. Thereby, the *same* formula $\forall x.\varphi(x)$ is read differently in different R-contexts. This is achieved by using $\Gamma_{c,x}(x)$ as a discharged assumption in the premise of the rule.

$$\frac{\Gamma, \Gamma_{c,x}(x) :_c \varphi(x)}{\Gamma :_c \forall x.\varphi(x)} \ (\forall I_C), \ x \notin free(\Gamma) \qquad \frac{\Gamma :_c \forall x.\varphi(x) \quad \Gamma :_c \wedge_{\Gamma_{c,x}}(y)}{\Gamma :_c \varphi(y)} \ (\forall E_C)$$

(12.2.11)

Here $\forall I_C$ is a parameterized family of I-rules indexed by R-contexts. Every application of this rule is always by appealing to some given R-context $c \in C$. In the interesting cases, $\Gamma_{c,x} \neq \varnothing$ will hold, though there might be "vacuous" R-contexts not affecting the meaning of a universal sentence. Similarly, $(\forall E_C)$ is a family of parametric E-rules indexed by R-contexts. The conclusion drawn from $\forall x.\varphi(x)$ deduced relative to an R-context c is read as $\wedge_{\Gamma_{c,x}}(y) \to \varphi(y)$, namely that y satisfies both $\varphi(x)$ and the contextual restriction $\wedge_{\Gamma_{c,x}}(x)$.

Restricting the existential quantifier: As for existential quantification, the contextual rules are presented below.

$$\frac{\Gamma :_c \varphi(y) \quad \Gamma :_c \wedge_{\Gamma_{c,x}}(y)}{\Gamma :_c \exists x.\varphi(x)} \ (\exists I_C) \qquad \frac{\Gamma :_c \exists x.\varphi(x) \quad \Gamma, \varphi(y), \wedge_{\Gamma_{c,x}}(y) :_c \chi}{\Gamma :_c \chi} \ (\exists E), \ y \notin free(\Gamma, \chi)$$

(12.2.12)

The I-rule requires that for some y that satisfies the contextual restrictions imposed by $\Gamma_{c,x}$, $\varphi(y)$ is derived, in order to deduce that the contextually-restricted (by c) existential conclusion be derived. Recall that, like in the standard $(\exists I)$-rule, y may, (and in general, will) appear free in Γ. So, the rule forces y to also fall under the contextual restriction imposed by c. The E-rule, like the standard $(\exists E)$-rule, allows the derivation of an arbitrary conclusion χ, under the assumption that φ and the contextual restriction themselves derive χ (for a fresh y).

Remark: From the above rules, it is evident that **(CIP)** holds for FOL with QDR. This is true since $\Gamma_{c,x}$, and consequently, $\wedge_{\Gamma_{c,x}}$, consist only of formulas in the language.

Theorem 12.2.14 (context-incorporation).

1. $\vdash_c \Gamma :_c \forall x.\varphi(x) \ iff \vdash \Gamma : \forall x.\wedge_{\Gamma_{c,x}}(x) \to \varphi(x)$.

2. $\vdash_c \Gamma :_c \exists x.\varphi(x) \ iff \vdash \Gamma : \exists x.\wedge_{\Gamma_{c,x}}(x) \wedge \varphi(x)$.

Proof.

1. (a) Assume $\vdash_c \Gamma :_c \forall x.\varphi(x)$ is derived by means of $(\forall I_C)$. By an inductive argument, the premise of $(\forall I_C)$, namely $\vdash_c \Gamma, \Gamma_{c,x}(x) :_c \varphi(x)$ (with $x \notin free(\Gamma)$), implies that $\vdash \Gamma, \Gamma_{c,x}(x) : \varphi(x)$. Therefore, by using $(\to I)$, $\vdash \Gamma : \wedge_{\Gamma_{c,x}(x)} \to \varphi(x)$, and by applying $(\forall I)$ (since $x \notin free(\Gamma)$), we get $\vdash \Gamma : \forall x.\wedge_{\Gamma_{c,x}(x)} \to \varphi(x)$.

   (b) Conversely, suppose $\vdash \Gamma : \forall x.\wedge_{\Gamma_{c,x}(x)} \to \varphi(x)$ is derived via $(\forall I)$ with the premise $\vdash \Gamma : \wedge_{\Gamma_{c,x}(x)} \to \varphi(x)$, where $x \notin free(\Gamma)$. Thus, also $\vdash \Gamma, \Gamma_{c,x}(x) : \varphi(x)$ (due to $(\to I)$). By an application of $(\forall I_C)$ the result follows.

2. The argument for existential quantification is similar and omitted.

□

Remarks:

1. In the R-context c, $\forall x.\varphi(x)$ is read as $\forall x.\wedge_{\Gamma_{c,x}}(x)\to\varphi(x)$. When $\varphi(x)$ is itself an implication, say $\alpha(x)\to\beta(x)$, then the result is equivalent to conjoining the antecedent with the contextual restriction, $\forall x.\alpha(x)\wedge\wedge_{\Gamma_{c,x}}(x)\to\beta(x)$.

2. Similarly, in the R-context c, $\exists x.\varphi(x)$ is read as $\exists x.\wedge_{\Gamma_{c,x}}(x)\wedge\varphi(x)$.

Here are some examples for the more interesting direction.

Example 12.2.63. *Suppose (12.1.1) is regimented by the FOL-formula* $\forall x.B(x)\to E(x)$ *(with $B(x)$ interpreted as x* is a bottle *and $B(x)$ as x* is empty*). Let c_{room} be an R-context of some room, with $\Gamma_{c_{room},x}=\{R(x)\}$ (with $R(x)$ interpreted as x* is in the room*). Then,*

$$\frac{\Gamma,R(x):_{c_{room}}B(x)\to E(x)}{\Gamma:_{c_{room}}\forall x.B(x)\to E(x)}\ (\forall I_C),\ x\notin free(\Gamma)$$

Allows the derivation of a reading of (12.1.1) as $\forall x.R(x)\to(B(x)\to E(x))$*, equivalent to* $\forall x.B(x)\wedge R(x)\to E(x)$*; that is,* every bottle in the room is empty*. This can be seen as incorporating the R-context into the sentence. Note that the contextually-derived universally quantified sentence does not carry its contextually meaning "on its nose". To obtain the required reading, one has to know the R-context in which the sentence was derived (c_{room} in this example), and consult $\Gamma_{c_{room},x}$ to obtain this reading. Similarly,*

$$\frac{\Gamma:_{c_{room}}\forall x.B(x)\to E(x)\quad \Gamma:R(y)}{\Gamma:_{c_{room}}B(y)\to E(y)}\ (\forall E_C)$$

allows drawing from (12.1.1) derived in the R-context c_{room} the conclusion $R(y)\to(B(y)\to E(y))$*, equivalent to* $B(y)\wedge R(y)\to E(y)$*; namely a reading corresponding to* if y is a bottle in the room then y is empty*, a correct reading of the conclusion in the context c_{room}.*

Example 12.2.64. *Following in (12.2.14) is another example, establishing*

$$\vdash_c\forall x.W(x)\wedge I(x)\to S(x),\ \forall y.W(y)\wedge S(y)\to B(y):_c\forall z.W(z)\to B(z)\quad (12.2.13)$$

in an R-context c with $\Gamma_{c,z}=I(z)$. I'll return to this example.

Example 12.2.65. *The next example 12.2.15 is of two independent QDRs by a context. It shows why the premises of the I_C-rules have themselves to use '$\vdash_c$', and not just '$\vdash$'. where $\Gamma_{c,x}=Y(x)$, $\Gamma_{c,y}=I(y)$. Let* **I, II** *abbreviate, respectively, the two premises. I treat '$\wedge$' as having arbitrary arity, see 12.2.16.*

$$
\cfrac{
  \cfrac{
    [W(z)]_1 \quad [I(z)]_2
  }{
    \cfrac{
      \cfrac{W(z) \land I(z)}{}\,(\land I) \quad \cfrac{W(z) \land I(z) \rightarrow S(z)}{}\;\cfrac{\forall x.W(x) \land I(x) \rightarrow S(x)}{}(\forall E)
    }{S(z)}\,(\rightarrow E)
  }{
    \cfrac{W(z) \lor S(z)}{}\,(\lor I)
    \qquad
    \cfrac{[W(z)]_1}{W(z) \lor S(z)}\,(\lor I)
  }
}{
  \cfrac{
    \cfrac{
      B(z) \quad \cfrac{W(z) \lor S(z) \rightarrow B(z)}{}\;\cfrac{\forall y.W(y) \lor S(y) \rightarrow B(y)}{}(\forall E)
    }{(\rightarrow E)}
  }{
    \cfrac{W(z) \rightarrow B(z)}{}\,(\rightarrow I^1)
  }
  \quad
  \cfrac{\forall z.W(z) \rightarrow B(z)}{}\,(\forall I_C^2)
}
\tag{12.2.14}
$$

$$\vdash_c \forall x \forall y. M(x) \wedge Y(x) \wedge W(y) \wedge S(y) \to L(x,y), \ \forall z. W(z) \wedge I(z) \to S(z) :_c \forall x \forall y. M(x) \wedge V(x) \wedge W(y) \to L(x,y)$$

$$\tag{12.2.15}$$

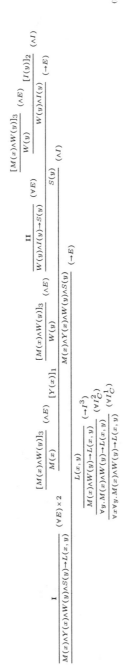

$$(12.2.16)$$

The following proposition expresses a property of the QDR-rules that will be useful below. It says that it does not matter which variable is used to express the contextual restriction, as long as it is amenable to universal generalization.

Proposition 12.2.16. *If* $\vdash_c \Gamma, \Gamma_{c,x}(x) :_c \varphi(x)$ *and* $y \notin free(\Gamma)$, *then also* $\vdash_c \Gamma, \Gamma_{c,x}(y) :_c$ $\varphi(y)$.

I now turn to the definition of the (reified) contextually-varying sentential meanings, following the ideas in Chapter 5. The notion of a derivation $\mathcal{D}$ of φ from Γ is the usual recursively defined one. Similarly, the definition of a cancel derivationn is inherited from the previous discussion, relativized to contexts. Let '$\vdash_c^c$' denote c-canonical derivability in a context c.

Let $[[\varphi]]_\Gamma^c$ denote[3] the (possibly empty) collection of all c-canonical derivations of φ from Γ.

Definition 12.2.78. *(**reified meanings**) The* contextually-varying *(reified) meaning of* φ *is given by*

$$[[\varphi]] =^{df.} \lambda\Gamma.[[\varphi]]_\Gamma^c \qquad (12.2.17)$$

We can now see the difference between "ordinary" meanings and their contextually-varying counterpart. For the context independent meaning of $\forall x.\varphi(x)$, all the canonical derivations end essentially with an application of the *same* $(\forall I)$-rule, while for the meaning of $\forall x.\varphi(x)$ in an R-context c, all c-canonical derivations end essentially with an application of $(\forall I_C)$, varying with c.

As was already observed in Section 5.2.4, this reified meaning is very fine[4], and a certain relaxation of it is found useful. Note that the **CIP** requires (strict) sameness of meaning between a contextually restricted quantified sentence and its context incorporated counterpart. However, while the relationship of canonical derivations of both are very similar – the former essentially ending with application of $(\forall I_C)$ (in the universal case) whenever the latter ends with $(\rightarrow I)$ immediately followed by $(\forall I)$ – they are strictly not the same! We can obtain a natural coarsening fitting also the current needs (for the **CIP**), still fine enough as for not identifying the meanings of logically equivalent sentences as done in MTS, by introducing *grounds (for assertion)* of sentences (see Francez and Dyckhoff [53] for a discussion of the role of those grounds in the PTS programme).

Definition 12.2.79 (grounds for contextual assertion). *The* grounds for contextual assertion *of* φ, *denoted by* $G^c[[\varphi]]$, *are given by*

$$G^c[[\varphi]] =^{df.} \{\Gamma \mid \vdash_c^c \Gamma :_c \varphi\} \qquad (12.2.18)$$

I now introduce an equivalence relation on meaning based on *sameness of contextual grounds (for assertion)*, that captures the CIP requirement in a natural way.

---

[3]The superscript 'c' here relates to canonicity, and should not be confused with an R-context.
[4]For example, it is shown in Section 5.2.4 that $[[\varphi \wedge (\psi \wedge \chi)]] \neq [[(\varphi \wedge \psi) \wedge \chi]]$.

Definition 12.2.80 (contextual grounds equivalence).

$$\varphi \equiv_{G^c} \psi \text{ iff } G^c[\![\varphi]\!] = G^c[\![\psi]\!] \tag{12.2.19}$$

Obviously, '$\equiv_{G^c}$' is an equivalence relation on contextually varying meanings. An easy inspection of the proof of the context incorporation theorem shows that the meanings of the contextual-incorporated counterparts of contextually-restricted quantified sentences are contextual grounds equivalent.

12.2.2 Harmony of the contextual domain restriction rules

Below I show the reductions needed for the intrinsic harmony of the rules for '$\vdash_c$'.

Universal contextually-restricted quantification:

$$\frac{\Gamma, \Gamma_{c,x}(x) :_c \varphi(x)}{\Gamma :_c \forall x.\varphi(x)} (\forall I_C) \quad \frac{\Gamma :_c \wedge \Gamma_{c,x}(y)}{\Gamma :_c \varphi(y)} (\forall E_C) \quad \leadsto_r \quad \Gamma[x := y], \Gamma_{c,x}(y) :_c \varphi(y)$$
$$\tag{12.2.20}$$

Note that since $x \notin free(\Gamma)$, $\Gamma[x := y] = \Gamma$. The result follows by Proposition 12.2.16. A clearer depiction of the reduction uses the simply presented $\mathcal{D}$s.

$$\frac{\begin{array}{c} [\wedge \Gamma_{c,x}(x)]_i \\ \mathcal{D} \\ \hline \forall x.\varphi(x) \end{array} (\forall I_C^i) \quad \begin{array}{c} \mathcal{D}' \\ \wedge \Gamma_{c,x}(y) \end{array}}{\varphi(y)} (\forall E_C) \quad \leadsto_r \quad \frac{\mathcal{D}[\wedge \Gamma_{c,x}(y) := \wedge \Gamma_{c,x}(y)]}{\varphi(y)}$$
$$\tag{12.2.21}$$

where the substitution $\wedge \Gamma_{c,x}(y) := \wedge \Gamma_{c,x}(y)$ reflects the usual composition of derivations, whereby an assumption $\wedge \Gamma_c(y)$ (in $\mathcal{D}$) is replaced by its given derivation in the second premise of $(\forall E_C)$.

The harmoniously-induced $(\forall GE_C)$ is given below.

$$\frac{\Gamma :_c \forall x.\varphi(x) \quad \Gamma :_c \wedge \Gamma_{c,x}(y) \quad \Gamma, \varphi(y) :_c \chi}{\Gamma :_c \chi} (\forall GE_C) \quad y \text{ fresh} \tag{12.2.22}$$

Existential contextually-restricted quantification: The reduction is the following.

$$\frac{\begin{array}{cc} \mathcal{D}_1 & \mathcal{D}_2 \\ \varphi(y) & \wedge \Gamma_{c,x}(y) \end{array}}{\exists x.\varphi(x)} (\exists I_C) \quad \frac{\begin{array}{c} [\varphi(z)]_i, [\wedge \Gamma_{c,x}(z)]_j \\ \mathcal{D} \\ \chi \end{array}}{\chi} (\exists E_C^{i,j}) \quad \leadsto_r$$
$$\tag{12.2.23}$$

$$\mathcal{D}[\varphi(z) := \frac{\begin{array}{cc} \mathcal{D}_1[y := z] & \mathcal{D}_2[y := z] \\ \varphi(z) & \wedge \Gamma_{c,x}(z) := \wedge \Gamma_{c,x}(z) \end{array}}{\chi}$$

$$\overline{\Gamma, \xi : \xi} \ (Ax)$$

$$\frac{\Gamma, \varphi(y) \ : \psi(y)}{\Gamma \ : \forall x.\varphi(x) \to \psi(x)} \ (\forall I) \qquad \frac{\Gamma : \varphi(y) \quad \Gamma : \psi(y)}{\Gamma : \exists x.\varphi(x) \land \psi(x)} \ (\exists I)$$

y fresh for Γ in $(\forall I)$.

$$\frac{\Gamma : \forall x.\varphi(x) \to \psi(x) \quad \Gamma : \varphi(y)}{\Gamma : \psi(y)} \ (\forall E) \qquad \frac{\Gamma \ : \exists x.\varphi(x) \land \psi(x) \quad \Gamma, \varphi(y), \psi(y)] : \xi}{\Gamma : \xi} \ (\exists E)$$

y fresh for Γ, ξ in $(\exists E)$.

Figure 12.1: A natural-deduction proof-system N_{rq} for restricted quantification

The $(\exists E_C)$-rule is in the GE-form to start with, thus harmonious in form too.

12.2.3   Quantifier domain restriction in restricted quantification

In order to make the subsequent presentation of QDR in NL more comprehensible, I exemplify the proof-theoretic approach by applying it to a fragment FOL_{rq} of FOL that comes closer to the NL-fragment to be considered. The fragment is known as having *restricted quantification* (not to be confused with contextually-restricted quantification, to be added on top). Quantified formulas in this fragment have the following form:

$$\forall x.\varphi(x) \to \psi(x), \quad \exists x.\varphi(x) \land \psi(x) \tag{12.2.24}$$

The universal quantification can be read as 'everything which is φ is ψ', and the existential quantification can be read as 'there exists something which is φ that is ψ'. That is, quantification is restricted to entities satisfying φ, to be called the *restrictor*. A more transparent syntax, closer to the natural language expression of quantification, would be

$$\forall x : \varphi(x).\psi(x), \quad \exists x : \varphi(x).\psi(x) \tag{12.2.25}$$

The expression of (12.2.25) as (12.2.24) is known as Frege's translation, that has drawn a lot of criticism as a way to capture natural language quantification. See Francez [45] for a detailed discussion. As I show in the next section, FOL_{rq}-quantification reflects more directly the latter.

The proof-system is presented in Figure 12.1. I use $\vdash \Gamma : \varphi$ in this subsection to indicate derivability of φ from Γ (in N_{rq}). A GE-rule for the universal quantifier, exhibiting harmony in form, is

$$\frac{\Gamma : \forall x.\varphi(x) \to \psi(x) \quad \Gamma : \varphi(y) \quad \Gamma, \psi(y) : \xi}{\Gamma : \xi} \ (\forall GE) \tag{12.2.26}$$

Next, I consider QDR in FOL_{rq}. The observation is that the restrictor can be interpreted differently in different R-contexts. Thus, the natural regimentation of (12.1.1) (cf. Example 12.2.63) would again be

$$\forall x.B(x) \to E(x) \tag{12.2.27}$$

where '$B(x)$' expresses x is a bottle and '$E(x)$' expresses x is empty. Here, the restrictor $B(x)$ can have a contextually-varying interpretation.

The idea for the proof-theoretic representation of contextual meaning variation is as before, where for universal quantification an R-context c provides a contextual discharged assumption. The generated contextual restriction *strengthens* the restrictor already present in the formula. The rules are shown below.

$$\frac{\Gamma, \varphi(x), \Gamma_{c,x}(x)] :_c \psi(x)}{\Gamma :_c \forall x.\varphi(x) \to \psi(x)} \ (\forall I_C) \ , \ x \notin free(\Gamma) \tag{12.2.28}$$

$$\frac{\Gamma :_c \forall x.\varphi(x) \to \psi(x) \quad \Gamma :_c \varphi(y) \wedge \wedge_{\Gamma_{c,x}}(y)}{\Gamma :_c \psi(y)} \ (\forall E_C)$$

$$\frac{\Gamma :_c \varphi(y) \quad \Gamma :_c \Gamma_{c,x}(y) \quad \Gamma : \psi(y)}{\Gamma :_c \exists x.\varphi(x) \wedge \psi(x)} \ (\exists I_C) \tag{12.2.29}$$

$$\frac{\Gamma :_c \exists x.\varphi(x) \wedge \psi(x) \quad \Gamma, (\varphi \wedge \wedge_{\Gamma_{c,x}} \wedge \psi)(y) :_c \chi}{\Gamma :_c \chi} \ (\exists E_C)$$

where $y \notin free(\Gamma, \chi)$ in $(\exists E)$.

12.3 Quantifier domain restriction in the NL-fragment

In this section, I develop the proof-theoretic semantics for QDR in its more suitable setting, that of the natural language fragment E_1^+ with context-independent meanings developed in Chapter 8.

Definition 12.3.81 (NLR-context). *An NLR-context (NL restricting context) c is a finite collection Γ_c of pseudo-sentences with one parameter only, where $\Gamma_{c,j}$ is the sub-collection[5] with the parameter **j**. Let CNL be the collection of all NLR-contexts.*

---

[5]For simplicity, I assume this sub-collection is a singleton.

The *NLR*-contexts $\Gamma_{c,j}(j)$ can be of one of the forms j isa X (X a noun), j is A (A an intersective adjective) and j P (P a verb-phrase). Note that compound contextual restrictions can also be imposed. For example, $\Gamma_{c,j} = j$ isa man whom every girl loves. Since the fragment E_1^+ has only modification by means of (intersective) adjectives and relative clauses, all the examples will be restricted to such modification. Extension, for example, to incorporate preposition phrases are not an obstacle in principle, just that it has not been done yet.

Restricting universal quantification: Again, an NLR-context c provides a *discharged* assumption for imposing its restriction.

$$\frac{\Gamma, j \text{ isa } X, \Gamma_{c,j}(j) :_c S[j]}{\Gamma :_c S[(\text{every } X)_{r(S[j])+1}]} \ (eI_{CNL}), \ j \text{ fresh for } \Gamma \qquad (12.3.30)$$

$$\frac{\Gamma :_c S[(\text{every } X)_{r(S[j])+1}] \quad \Gamma :_c k \text{ isa } X \quad \Gamma :_c \Gamma_{c,j}(k) \quad \Gamma, S[k] :_c S'}{\Gamma : S'} \ (eE_{CNL})$$

$$(12.3.31)$$

Again, a family of I/E-rules is employed, for all possible NLR-contexts.

Example 12.3.66. *Below is a derivation establishing*

$$\vdash_c \text{ every Italian woman smiles}, \qquad (12.3.32)$$

$$\text{every woman who smiles is beautiful } :_c$$

$$\text{every woman is beautiful}$$

in an NLR-context c with $\Gamma_{c,k}(k)$ = k is Italian, intended to restrict the universal quantification on women to a universal quantification on Italian women. The observant reader will notice that example 12.2.64 is a regimentation of this example in FOL. Since there is no quantifier scope ambiguity involved in this example, I omit in the derivation the scope level indicator to avoid notational clutter. Also, for typographical reasons, I abbreviate in the derivation Woman, Italian, Beautiful *and* Smiles *to* W, I, B *and* S, *respectively.*

$$\frac{[k \text{ isa } W]_1 \quad \dfrac{\dfrac{[k \text{ is } I]_2 \quad [k \text{ isa } W]_1}{k \text{ isa } I W} \ (adjI) \quad \text{every } I W S}{k S} \ (e\hat{E})}{\dfrac{\dfrac{k \text{ isa } W \text{ who } S}{\dfrac{k \text{ is } B}{\text{every } W \text{ is } B}} \ (rel I) \qquad \text{every } W \text{ who } S \text{ is } B}{}} \ (e\hat{E})$$

$$(12.3.33)$$

It is interesting to note that QDR holds in whichever scope level the restricted quantifier is.

Example 12.3.67. *Consider the following scope variants of*

$$\text{some man admires every actress} \qquad (12.3.34)$$

In the NLR-context c with $\Gamma_{c,\mathbf{k}} = \mathbf{k}$ *is Italian, intended to restrict the universal quantification on actresses to universal quantification on Italian actresses. I show the following:*

some P isa M, every P is S, every S M adm every I $A \vdash_c$ some M adm every A

$$(12.3.35)$$

under both scope variants of the conclusion. The following abbreviations are used: P for philosopher, M for man, I for Italian, A for actress. *For typographical reasons, the derivations are presented with a common sub-derivation $\mathcal{D}$ factored out.*

$$\mathcal{D} = \cfrac{\cfrac{[\mathbf{j}\ \text{isa}\ M]_1 \quad \cfrac{\cfrac{[\mathbf{j}\ \text{isa}\ P]_2 \quad \text{every}\ P\ \text{is}\ S}{\mathbf{j}\ \text{is}\ S}(e\hat{E})}{\mathbf{j}\ \text{isa}\ S\ M}(adjI)}{\mathbf{j}\ \text{adm every}\ I\ A}\quad \text{every}\ S\ M\ \text{adm every}\ I\ A}{}(e\hat{E})$$

$$(12.3.36)$$

Subject wide scope: *The derivation is (with obvious abbreviations and Γ omitted):*

$$\cfrac{\text{some}\ P\ \text{isa}\ M \quad \cfrac{[\mathbf{j}\ \text{isa}\ M]_1 \quad \cfrac{\cfrac{\cfrac{[\mathbf{k}\ \text{is}\ I]_3 \quad [\mathbf{k}\ \text{isa}\ A]_4}{\mathbf{k}\ \text{isa}\ I\ A}(adjI) \quad \cfrac{\mathcal{D}}{\mathbf{j}\ \text{adm every}\ I\ A}}{\mathbf{j}\ \text{adm}\ \mathbf{k}}(e\hat{E})}{\mathbf{j}\ \text{adm}\ (\text{every}\ A)_1}(eI^{3,4}_{CNL})}{(\text{some}\ M)_2\ \text{adm}\ (\text{every}\ A)_1}(sI)}{(\text{some}\ M)_2\ \text{adm}\ (\text{every}\ A)_1}(sE^{1,2})$$

$$(12.3.37)$$

Object wide scope: *The derivation is*

$$\cfrac{\cfrac{\text{some}\ P\ \text{isa}\ M \quad \cfrac{[\mathbf{j}\ \text{isa}\ M]_1 \quad \cfrac{\cfrac{[\mathbf{k}\ \text{is}\ I]_3 \quad [\mathbf{k}\ \text{isa}\ A]_4}{\mathbf{k}\ \text{isa}\ I\ A}(adjI) \quad \cfrac{\mathcal{D}}{\mathbf{j}\ \text{adm every}\ I\ A}}{\mathbf{j}\ \text{adm}\ \mathbf{k}}(e\hat{E})}{(\text{some}\ M)_1\ \text{adm}\ \mathbf{k}}(sI)}{(\text{some}\ M)_1\ \text{adm}\ \mathbf{k}}(sE^{1,2})}{(\text{some}\ M)_1\ \text{adm}\ (\text{every}\ A)_2}(eI^{3,4}_{CNL})$$

$$(12.3.38)$$

Example 12.3.68. *The following example from Stanley and Gendler Szabò [205] is pointed out as being difficult for MTS-handling, as it seemingly requires context-shift during meaning evaluation.*

every sailor waved to every sailor $(12.3.39)$

where the context imposes the restriction that the quantification in the subject is restricted to one "kind" of sailors, say sailors on the ship, while the object quantification is restricted, say, to sailors on the shore. Under the current approach, such examples pose no problem whatsoever. Suppose that $\mathbf{j}$ is the parameter used to introduce every sailor *in the subject, while $\mathbf{k}$ is the parameter used to introduce* every sailor *in the object (where both scope relations are equivalent). Then, all we have to do is consider a context $c_{sailors}$, with $\Gamma_{c_{sailors},j} = \mathbf{j}$ is $-$ on $-$ the $-$ ship, and $\Gamma_{c_{sailors},k} = \mathbf{k}$ is $-$ on $-$ the $-$ shore. No context shift is involved. As the full derivation is somewhat lengthy, I skip the details.*

Restricting existential quantification:

$$\frac{\Gamma : \mathbf{j} \text{ isa } X \quad \Gamma :_c \Gamma_{c,\mathbf{j}}(\mathbf{j}) \quad \Gamma :_c S[\mathbf{j}]}{\Gamma :_c S[(\text{some } X)_{r(S[\mathbf{j}])+1}]} \ (sI_{CNL})$$

$$(12.3.40)$$

$$\frac{\Gamma :_c S[(\text{some } X)_{r(S[\mathbf{j}])+1}] \quad \Gamma, \mathbf{k} \text{ isa } X, \Gamma_{c,\mathbf{j}}(\mathbf{k}), S[\mathbf{k}] :_c S'}{\Gamma : S'} \ (sE_{CNL})$$

$$(12.3.41)$$

where $\mathbf{k}$ is fresh for $\Gamma, S[\text{some } X], S'$ in (sE_{CNL}).

The reductions needed to show the intrinsic harmony of the (CNL)-rules are very similar to those for the regular rules (shown in Section 8.3.6.2) and are omitted.

Next, I show how the **CIP** is satisfied for E_1^+. Note that the fragment includes neither implication nor conjunction (on the sentential level). To express the **CIP** effect, I use the following notation. For $S[(q\ X)]$ (with q either every or some), let $S_{\Gamma_{c,\mathbf{j}}}$ be defined as follows.

$$S_{\Gamma_{c,\mathbf{j}}} = \begin{cases} S[(q\ X \text{ who isa } Y)] & \Gamma_{c,\mathbf{j}}(\mathbf{j}) = \mathbf{j} \text{ isa } Y \\ S[(q\ A\ X)] & \Gamma_{c,\mathbf{j}}(\mathbf{j}) = \mathbf{j} \text{ is } A \\ S[(q\ X \text{ who } P)] & \Gamma_{c,\mathbf{j}}(\mathbf{j}) = \mathbf{j}\ P \end{cases}$$

$$(12.3.42)$$

Theorem 12.3.15 (E_1^+ **context incorporation**).

$$\vdash_c \Gamma :_c S[(q\ X)] \text{ iff } \vdash \Gamma : S_{\Gamma_{c,\mathbf{j}}}$$

$$(12.3.43)$$

Proof. I will show only the proof of the first case, for q = every; all other cases are similar. To simplify, I also omit the scope indications.

1. Asume $\vdash_c \Gamma :_c S[(\text{every } X)]$, where $\Gamma_{c,\mathbf{j}} = \mathbf{j}$ isa Y. So, the derivation ends with (omitting scope indication)

$$\frac{\Gamma, \mathbf{j} \text{ isa } X, \mathbf{j} \text{ isa } Y :_c S[\mathbf{j}]}{\Gamma :_c S[(\text{every } X)]} \ (eI_{CNL}), \mathbf{j} \text{ fresh for } \Gamma$$

$$(12.3.44)$$

Therefore, the following derivation (with contexts omitted) can be formed, where by the induction hypothesis on the premise '$\vdash$' is used instead of '$\vdash_c$'.

$$\frac{\dfrac{[\mathbf{j} \text{ isa } X \text{ who isa } Y]_i}{\mathbf{j} \text{ isa } X} \ (relE) \quad \dfrac{[\mathbf{j} \text{ isa } X \text{ who isa } Y]_i}{\mathbf{j} \text{ isa } Y} \ (relE)}{\dfrac{S[\mathbf{j}]}{S[(\text{every } X \text{ who isa } Y)]} \ (eI^i)} \ (ass.)$$

$$(12.3.45)$$

2. Assume $\vdash\Gamma : S[(\text{every } X \text{ who isa } Y)]$. The derivation (again, omitting contexts and scope indication) ends with

$$\frac{\Gamma, \mathbf{j} \text{ isa } X \text{ who isa } Y : S[\mathbf{j}]}{\Gamma : S[(\text{every } X \text{ who isa } Y)]} \ (eI) \tag{12.3.46}$$

Let $\Gamma_{c,\mathbf{j}} = \mathbf{j}$ isa Y. Therefore, the following derivation can be formed:

$$\frac{\dfrac{[\mathbf{j} \text{ isa } X]_i \quad [\mathbf{j} \text{ isa } Y]_j}{\mathbf{j} \text{ isa } X \text{ who isa } Y} \ (relI)}{\dfrac{S[\mathbf{j}]}{S[(\text{every } X)]} \ (eI^{i,j}_{CNL})} \ (ass.) \tag{12.3.47}$$

Once again, by inspecting the rules, we obtain that

$$[[S[(q\ X)]]]_c = [[S_{\Gamma_{c,\mathbf{j}}}]] \tag{12.3.48}$$

validating the **CIP**. Note again the correspondence between derivations, where a use of (qI_{NLC}) is associated with (qI) followed by $(relI)$ (or by $(AdjI)$ in some of the cases), and similarly for the E-rules. $\qquad\qquad\square$

12.4 Logic with contextual definiteness restriction

In this section, I present another variant of 1st-order logic (FOL), this time extended with a ι operator expressing *definiteness*, first interpreted absolutely (in Section 12.4.1) and then interpreted in a contextually-dependent way (in Section 12.4.5). Once again, while there is not much interest in such a logic per se, it serves as a vehicle for a clear presentation of the ideas underlying the application of the approach to natural language (in Sections 12.5 and 12.6). It provides another natural host for the novel proof-theoretic concept of a *parameterized family of I-rules* in the intended natural-deduction meaning-conferring proof system. In Section 12.4.2 I compare our proposed ND-system for definiteness with some other proposals of such ND-systems in the literature. As an intermediate step, I first extend FOL with *absolute* definiteness (in Section 12.4.1), and then explain how to derive from it FOL with *contextual* definiteness (in Section 12.4.5).

12.4.1 First-order logic with absolute definiteness

I again assume the usual object language for FOL, with the usual definition of free/bound variables. I start from Russell's contextual definition (in his famous example (12.1.4)) in Russell [180]) of definiteness consisting of three components:

1. **existence**: in the example, there is at least one (current) king of France.

2. **uniqueness**: in the example, there is at most one (current) king of France.

3. **predication**: in the example, the only (current) king of France is bald.

This naturally generalizes to a general statement about the only φ being ψ, usually expressed as

$$\exists x(\varphi(x) \wedge \forall y(\varphi(y) \to y = x) \wedge \psi(x)) \tag{12.4.49}$$

For later use, I find it useful to consider instead the following equivalent formulation.

$$\exists x.\varphi(x) \wedge \forall u.\forall v.\varphi(u) \wedge \varphi(v) \to u = v \wedge \forall w.\varphi(w) \to \psi(w) \tag{12.4.50}$$

(avoiding a quantifier scoping over the conjunction, that makes easier the adaptation to NL shown below). The actual formulation which underlies the rules for the NL-definiteness is yet another equivalent formulation, as follows.

$$\exists x.\varphi(x) \wedge \psi(x) \wedge \forall u.\forall v.\varphi(u) \wedge \varphi(v) \to u = v \tag{12.4.51}$$

Here the existential quantifier can be presented as a binary, *restricted* quantifier, fitting the general familiar scheme of quantification in NL.

So, following Russell's idea, I extend FOL by means a term-forming operator (*subnector*) $\iota x.\varphi(x)$, where (12.4.49), (12.4.50) and (12.4.51) are expressed as the formula

$$\psi(\iota x.\varphi(x)) \tag{12.4.52}$$

where, at a first stage, ψ (the context formula) is assumed to be a unary predicate, and φ is assumed to have exactly one free variable[6], x, and all free occurrences of x in φ are bound by ιx. As usual, in the absolute definiteness setting bound variables can be renamed, and $\iota x.\varphi(x)$ is the same as $\iota y.\varphi(y)$. However, the same does not hold in case of contextual definiteness (cf.(12.4.5.1)).

Thus, the regimentation of (12.1.4) would be expressed as

$$B(\iota x.K(x)) \tag{12.4.53}$$

with the obvious interpretation of the predicate symbols used. Denote this extension of FOL by FOL_{ι_0}.

Under the model-theoretic interpretation of this subnector, when (12.4.50) holds, such a subnector term is *proper*, a referring expression (denoting an element in the domain of the model). However, there is a problem when such a term is *improper*, i.e., when (12.4.50) does not hold, in which case the term cannot refer without some further stipulation. As we shall see below, under our proof-theoretic meaning definition, the issue of being improper never arises. Whenever $\psi(\iota x.\varphi(x))$ is assertible, its I-rule guarantees that (12.4.50), being the premises of this I-rule, hold.

The semantic focus shifts from referentiality to grounds for assertion!

---

[6]One can extend the theory also to φ containing *no* free occurrences of x, like vacuous quantification in FOL; I shall not bother doing so here.

12.4.1.1 Parallel absolute definiteness

The first relaxation of the above assumptions arises from a need to regiment examples like the following.

$$\text{the king of France loves the queen of Spain} \qquad (12.4.54)$$

where the context is a binary relation love. The regimentation of (12.4.54) should look like

$$L(\iota x.K(x), \iota y.Q(y)) \qquad (12.4.55)$$

This means we have to relax the assumption that the context formula ψ is unary, and let it be n-ary, for an arbitrary $n \geq 1$. Thus, the general form of a formula with a ι-term becomes

$$\psi(\iota x.\varphi(x), \overline{a}) \qquad (12.4.56)$$

where $\overline{a}$ is a list of m additional parameters. Its Russellian contextual definition now becomes

$$\exists x.\varphi(x) \wedge \forall u.\forall v.\varphi(u) \wedge \varphi(v) \rightarrow u = v \wedge \forall w.\varphi(w) \rightarrow \psi(w, \overline{a}) \qquad (12.4.57)$$

12.4.1.2 Nested absolute definiteness

The ι-operator can be nested (embedded within another ι-term). This innocently look-ing construct raises several problems. A thorough study can be found in Kuhn [94]. For example, a regimentation of

$$\text{the girl that the boy loves smiles} \qquad (12.4.58)$$

would be

$$S(\iota x.G(x) \wedge L(\iota y.B(y), x)) \qquad (12.4.59)$$

with the obvious interpretation of the predicate symbols used. By the Russellian anal-ysis, (12.4.59) can be interpreted as the proposition that there exists exactly one girl that is loved by exactly one boy, and that girl smiles. Note that (12.4.59) does not preclude there being other girls (loved by no boy or loved by more than one boy). The general form of nested definiteness is

$$\psi(\iota x.\chi(\iota y.\varphi(y), x), \overline{a}) \qquad (12.4.60)$$

See Kuhn [94] for a criticism of the Russellian reading of (12.4.60) (and (12.4.58)) and a proposal of an alternative reading, claimed to be more plausible. Note that Kuhn [94] does not consider n-ary ψs.

I will return to this issue when introducing an ND-system for ι (see Section 12.4.1.5).

12.4.1.3   Parameterized definiteness

Next, I transcend Russell's original definition of definiteness with what I call *parameterized (absolute) definiteness*. To motivate this generalization, suppose one wants to regiment

$$\text{the king of France is bald and the king of Spain is bald} \qquad (12.4.61)$$

For that purpose, it is clear that king cannot be considered a unary predicate symbol, but a binary one, known as a *relational noun*, expressing a relation between, in this case, a person and a country. This would render the regimentation of (12.1.4) as

$$B(\iota x.K(x, France)) \qquad (12.4.62)$$

where $France$ is an individual constant standing for France. Thus, (12.4.61) is expressed as

$$B(\iota x.K(x, France)) \wedge B(\iota x.K(x, Spain)) \qquad (12.4.63)$$

Now, suppose that a situation of a plague of 'royal baldness' occurs, where one needs to express the fact that every country is such that its king is bald. For that, the second argument of K needs to be a variable, quantifiable from outside the ι-boundary. So, we relax the assumption that φ is unary, and extend the language to contain parameterized ι-terms such as

$$\iota x.\varphi(x, \overline{b}) \qquad (12.4.64)$$

where $\overline{b}$ is a list of free variables in φ, not bound by ιx. The analogue of the Russellian definition of ι-terms, becomes

$$\exists x.\varphi(x, \overline{b}) \wedge \forall u.\forall v.\varphi(u, \overline{b}) \wedge \varphi(v, \overline{b}) \rightarrow u = v \wedge \forall w.\varphi(w.\overline{b}) \rightarrow \psi(w, \overline{a}) \qquad (12.4.65)$$

We thus can express the royal baldness situation by

$$\forall y.B(\iota x.K(x, y)) \qquad (12.4.66)$$

Let this language be denoted by FOL_ι.

12.4.1.4   Definiteness and scope

As Russell was aware of, there is a scopal problem regarding the ι-terms. Consider the special case of

$$\neg\psi(\iota x.\varphi(x)) \qquad (12.4.67)$$

given by

$$\neg B(\iota x.K(x)) \qquad (12.4.68)$$

There is a scopal ambiguity in (12.4.68): it can be read as a regimentation of each of the following two sentences.

$$\text{the (current) king of France is not bald} \qquad (12.4.69)$$

it is not the case that the (current) king of France is bald (12.4.70)

Clearly, (12.4.69) and (12.4.70) are not equivalent. For example, (12.4.70) is true (by its Russellian reading) if there is no king of France, while (12.4.69) is not. To amend the situation, Whitehead and Russell (in [235]) introduced explicit scoping of the ι-term, using the following form.

$$[\iota x.\varphi(x)]\psi(\iota x.\varphi(x)) \tag{12.4.71}$$

This allows for (12.4.69) to be regimented as

$$[\iota x.K(x)]\neg B(\iota x.K(x)) \tag{12.4.72}$$

while (12.4.70) is regimented as

$$\neg[\iota x.K(x)]B(\iota x.K(x)) \tag{12.4.73}$$

However, this amended notation precludes the representation of nested definiteness. See Kuhn [94] for a criticism of the amended notation. I will adhere here to the original notation, reading it as expressing narrow scope of negation. Note that the wide scope reading for negation is marginal in an NL setting, and the issue does not arise in the NL-fragment considered below, which does not contain sentential negation.

As for the relative scope of ι-terms w.r.t. other quantifiers, the issue does arise. Here ι-terms behave like quantifiers regarding scope. The general form of which (12.4.66) is an instance is

$$\forall y.\psi(\iota x.\varphi(x, y)) \tag{12.4.74}$$

by which the universal quantifier on y has a higher scope than the definiteness binding x. Thus, the uniqueness of x is given per y. Hence, (12.4.74) differs in meaning from its "cousin"

$$\psi(\iota x.\forall y.\varphi(x, y)) \tag{12.4.75}$$

in which the scope of the definiteness binding on x is higher than the universal quantification on y. The instance $B(\iota x.\forall y.K(x, y))$ would mean that there is a unique person that is the king of every country, and that person is bald.

As for NL, where scopal relations among quantifiers are not explicit, scope ambiguity arises (cf. Section 8.2.3). The scope of definiteness in NL is handled below.

12.4.1.5   An ND-system for absolute definiteness with ι-terms

The standard ND-system for FOL is extended with I/E-rules for ι, as presented in Figure 12.2. The second premise occurs also in the I-rule suggested by Hilbert and Bernays [73], discussed below, but without explicit parameterizations. Whenever not needed, parameterization will be omitted. Note that the premises of (ιI) enforce that whenever $\iota x.\varphi(x)$ *can* be introduced, existence and uniqueness obtain. This way, the need to appeal to a free logic (see Section 12.4.2.3 below) is avoided.

$$\dfrac{\Gamma : \exists x.\varphi(x,\overline{b}) \quad \Gamma : \forall u.\forall v.\varphi(u,\overline{b}) \wedge \varphi(v,\overline{b}) \rightarrow u = v \quad \Gamma : \forall w.\varphi(w,\overline{b}) \rightarrow \psi(w,\overline{a})}{\Gamma : \psi(\iota x.\varphi(x,\overline{b}),\overline{a})} \ (\iota I)$$

$$\dfrac{\Gamma : \psi(\iota x.\varphi(x,\overline{b}),\overline{a})}{\Gamma : \exists x.\varphi(x,\overline{b})} \ (\iota E_1)$$

$$\dfrac{\Gamma : \psi(\iota x.\varphi(x,\overline{b}),\overline{a}) \quad \Gamma : \varphi(u,\overline{b}) \quad \Gamma : \varphi(v,\overline{b})}{\Gamma : u = v} \ (\iota E_2) \text{,} \quad u,v \text{ fresh for } \Gamma$$

$$\dfrac{\Gamma : \psi(\iota x.\varphi(x,\overline{b}),\overline{a})}{\Gamma : \forall z.\varphi(z,\overline{b}) \rightarrow \psi(z,\overline{a})} \ (\iota E_3)$$

Figure 12.2: I/E-Rules for absolute definiteness

For example, $\psi(\iota x.\varphi(x) \wedge \neg\varphi(x))$ cannot be introduced, since it would need as premise $\exists x.\varphi(x) \wedge \neg\varphi(x)$, not derivable in FOL from a consistent Γ.

As an example of a derivation involving the ι-rules, consider the following example.

Example 12.4.69. *We show* $\vdash \psi(\iota x.\varphi(x)), \forall x.\psi(x) \rightarrow \chi(x) : \chi(\iota x.\varphi(x))$. *To fit the page, the derivation is displayed in pieces. Let* $\mathcal{D}_2$ *be the following sub-derivation.*

$$\dfrac{\psi(\iota x.\varphi(x)) \quad \dfrac{\dfrac{[\varphi(u) \wedge \varphi(v)]_2}{\varphi(u)} \ (\wedge_1 E) \quad \dfrac{[\varphi(u) \wedge \varphi(v)]_2}{\varphi(v)} \ (\wedge_2 E)}{\dfrac{\dfrac{u = v}{\varphi(u) \wedge \varphi(v) \rightarrow u = v} \ (\rightarrow I^2)}{\forall u \forall v.\varphi(u) \wedge \varphi(v) \rightarrow u = v} \ (\forall I \times 2)} \ (\iota E_2)}{}$$

Also, let $\mathcal{D}_3$ *be the following sub-derivation.*

$$\dfrac{[\varphi(w)]_1 \quad \dfrac{\dfrac{\dfrac{\psi(\iota x.\varphi(x))}{\forall x.\varphi(x) \rightarrow \psi(x)} \ (\iota E_3)}{\varphi(w) \rightarrow \psi(w)} \ (\forall E)}{\psi(w)} \ (\rightarrow E) \quad \dfrac{\forall x.\psi(x) \rightarrow \chi(x)}{\psi(w) \rightarrow \chi(w)} \ (\forall E)}{\dfrac{\dfrac{\chi(w)}{\varphi(w) \rightarrow \chi(w)} \ (\rightarrow I^1)}{\forall w.\varphi(w) \rightarrow \chi(w)} \ (\forall I)} \ (\rightarrow E)$$

Then, the main derivation is the following.

$$\dfrac{\dfrac{\psi(\iota x.\varphi(x))}{\exists x.\varphi(x)} \ (\iota E_1) \quad \dfrac{\mathcal{D}_2}{\forall u \forall v.\varphi(u) \wedge \varphi(v) \rightarrow u = v} \quad \dfrac{\mathcal{D}_3}{\forall w.\varphi(w) \rightarrow \chi(w)}}{\chi(\iota x.\varphi(x))} \ (\iota I) \tag{12.4.76}$$

To analyze the proof-theoretic meaning of nested (absolute) definiteness, consider the following generic skeleton of a derivation of $\psi(\iota x.\chi(\iota y.\varphi(y),x))$ (with contexts omitted). To fit the page, the derivation is displayed in several parts. For readability, some renaming of bound variables took place, see 12.4.77. By inspecting the derivation of such a nested (absolute) definiteness, we see that existence and uniqueness need to hold both for $\varphi(y)$ and for $\iota x.\chi(\iota y.\varphi(y),x)$, where the latter requires a unique x w.r.t. the unique y of the former. For (12.4.58), the rules require a unique boy, and a unique girl loved by *that* boy.

12.4.2 Comparison with other ND-rules for definiteness

I now compare the proposed I/E rules for definiteness to some other proposals found in the literature.

12.4.2.1 Hilbert and Bernays

Hilbert and Bernays [73] also suggest a contextual definition of 'ι', using the following I-rule.

$$\frac{\Gamma : \exists x.\varphi(x) \quad \Gamma : \forall u.\forall v.\varphi(u)\wedge\varphi(v)\rightarrow u = v}{\Gamma : \varphi(\iota x.\varphi(x))} \ (\iota I_{HB})$$

(12.4.81)

The premises of the rule are the same as the respective two first premises of our rule (ιI), but the conclusion is weaker. It only allows a contextual inference where the main predicate (ψ in our rule) *coincides* with the predicate on which existence and uniqueness are imposed. It allows regimenting identity-expressing trivial sentences like

the king of France is a king of France   (12.4.82)

Hilbert and Bernays have no E-rule for ι. They do prove that the I-rule above is a conservative extension of the system without it.

Furthermore, those rules do not give rise to any interesting nested definiteness.

Those rules certainly are not suitable for studying definiteness in NL. While this definition can also be contextualized, I do not pursue it further due to its weakness indicated above.

12.4.2.2 Kalish and Montague

In [86], Kalish and Montague more or less adopt the rule (ιI_{HB}), but add another rule. The need of the other rule is due to the fact that they interpret $\iota x.\varphi(x)$ as a fixed element, the number zero, if existence and uniqueness do not obtain (improper

$\mathcal{D}_2$:

$$\cfrac{\exists y.\varphi(y) \quad \forall u_1.\forall v_1.\varphi(u_1)\wedge\varphi(v_1)\rightarrow u_1 = v_1 \quad \forall w_1.\varphi(w_1)\wedge\chi(w_1,x) \quad \forall u_3.\forall v_3.\varphi(u_3)\wedge\varphi(v_3)\rightarrow u_3 = v_3 \quad \forall w_3.\varphi(w_3)\rightarrow\chi(w_3,x)}{\cfrac{\chi(\iota y.\varphi(y),x)}{\exists x.\chi(\iota y.\varphi(y),x)}\,(\exists I)}\,(\iota I) \tag{12.4.77}$$

$\mathcal{D}_1$:

$$\cfrac{\exists y.\varphi(y) \quad \forall u_2.\forall v_2.\varphi(u_2)\wedge\varphi(v_2)\rightarrow u_2 = v_2 \quad \forall w_2.\varphi(w_2)\rightarrow\chi(w_2,x)}{\chi(\iota y.\varphi(y),u_2)} \qquad \cfrac{\exists y.\varphi(y) \quad \forall u_3.\forall v_3.\varphi(u_3)\wedge\varphi(v_3)\rightarrow u_3 = v_3 \quad \forall w_3.\varphi(w_3)\rightarrow\chi(w_3,x)}{\chi(\iota y.\varphi(y),v_2)}$$

$$\cfrac{\chi(\iota y.\varphi(y),u_2)\wedge\chi(\iota y.\varphi(y),v_2))}{\cfrac{u_2 = v_2}{\cfrac{\forall u_2.\forall v_2.\chi(\iota y.\varphi(y),u_2)\wedge\chi(\iota y.\varphi(y),v_2)\rightarrow u_2 = v_2}{}(\forall I)\times 2}(\rightarrow I)}(\iota I) \tag{12.4.78}$$

$$\cfrac{\exists y.\varphi(y) \quad \forall u_4.\forall v_4.\varphi(u_4)\wedge\varphi(v_4)\rightarrow u_4 = v_4 \quad \forall w_4.\varphi(w_4)\rightarrow\chi(w_4,x)}{\chi(\iota y.\varphi(y),u_4)} \tag{12.4.79}$$

$\mathcal{D}_3$:

$$\cfrac{\cfrac{\chi(\iota y.\varphi(y),u_4)}{\cfrac{\psi(u_4)}{\cfrac{\chi(\iota y.\varphi(y),u_4)\rightarrow\psi(u_4)}{\forall u_4.\chi(\iota y.\varphi(y),u_4)\rightarrow\psi(u_4)}(\forall I)}(\rightarrow I)}}{\psi(\iota x.\chi(\iota y.\varphi(y),x))}(\iota I)$$

$\mathcal{D}_2$:

$\mathcal{D}_1$:

$$\cfrac{\exists x.\chi(\iota y.\varphi(y),x) \quad \forall u_2\forall v_2.\chi(\iota y.\varphi(y),u_2)\wedge\chi(\iota y.\varphi(y),v_2)\rightarrow u_2 = v_2 \quad \forall u_4.\chi(\iota y.\varphi(y),u_4)\rightarrow\psi(u_4)}{\psi(\iota x.\chi(\iota y.\varphi(y),x))} \tag{12.4.80}$$

description). The second rule, $(\iota_{KM}I)$ ensures that all improper descriptions are equivalent.

$$\frac{\neg(\exists y \forall x.\varphi(x) \leftrightarrow x = y)}{\iota x.\varphi(x) = \iota y.\neg y = y} \ (\iota_{KM}I) \tag{12.4.83}$$

In addition to inheriting the limitation of (ιI_{HB}), the approach of Kalish and Montague may lead to undesired results. For example, one can derive (the regimentation of) the natural number between 1 and 2 is 0.

12.4.2.3 Tennant

Tennant, in [218] (elaborating upon in [214]), proposes I/E-rules for a contextual proof-theoretic definition of ι. He does this within a uniform proof-theoretic definition of a class of subnectors he calls *abstraction operators*. He certainly shares (at least a version of) the view of PTS as a theory of meaning. However, the contexts he allows are only *identity statements* of the form $t = \iota x.\varphi(x)$ (where t is any term in the underlying object language). His rules are formulated in a *free logic* framework, where existential commitments are made explicit. He uses $\exists!t$, defined by $\exists!t =^{df.} \exists x.x = t$. The rules are presented below, in my notation.

$$\frac{\begin{array}{ccc} [\varphi(a)]_i, [\exists!a]_i & & [a = t]_j \\ \vdots & & \vdots \\ a = t & \exists!t & \varphi(a) \end{array}}{t = \iota x.\varphi(x)} \ (\iota_T)I^{i,j} \tag{12.4.84}$$

The E-rules derive each of the premises of $(\iota_T I)$ from the major premise $t = \iota x.\varphi(x)$.

This I-rule does not allow to infer directly a predication of the form $\psi(\iota x.\varphi(x))$. Instead, one has to infer $\psi(\iota x.\varphi(x)) \wedge \psi(t)$; for example,

$$\text{Louis is the (current) king of France and he is bald} \tag{12.4.85}$$

Representing nested definiteness is even more cumbersome.

12.4.3 Harmony and stability of the absolute definiteness rules

Below are the three reductions establishing intrinsic harmony (contexts omitted).

$$\frac{\begin{array}{ccc} \mathcal{D}_1 & \mathcal{D}_2 & \mathcal{D}_3 \\ \exists x.\varphi(x) & \forall u.\forall v.\varphi(u) \wedge \varphi(v) \to u = v & \forall w.\varphi(w) \to \psi(w) \end{array}}{\dfrac{\psi(\iota x.\varphi(x))}{\exists x.\varphi(x)} \ (\iota E_1)} \ (\iota I) \qquad\qquad \rightsquigarrow_r \quad \begin{array}{c} \mathcal{D}_1 \\ \exists x.\varphi(x) \end{array}$$

$$\tag{12.4.86}$$

$$
\dfrac{\dfrac{\mathcal{D}_1 \qquad \mathcal{D}_2 \qquad\qquad \mathcal{D}_3}{\dfrac{\exists x.\varphi(x) \quad \forall u.\forall v.\varphi(u)\wedge\varphi(v)\to u=v \quad \forall w.\varphi(w)\to\psi(w)}{\psi(\iota x.\varphi(x))}\,(\iota I) \qquad \dfrac{\mathcal{D}_4 \quad \mathcal{D}_5}{\varphi(u)\quad\varphi(v)}}{u=v}\,(\iota E_2)
$$

$$
\rightsquigarrow_r
$$

$$
\dfrac{\dfrac{\mathcal{D}_4 \quad \mathcal{D}_5}{\dfrac{\varphi(u)\quad\varphi(v)}{\varphi(u)\wedge\varphi(v)}\,(\wedge I) \qquad \dfrac{\mathcal{D}_2}{\dfrac{\forall u.\forall v.\varphi(u)\wedge\varphi(v)\to u=v}{\varphi(u)\wedge\varphi(v)\to u=v}\,(\forall E)\times 2}}{u=v}\,(\to I)
$$

$$\tag{12.4.87}$$

$$
\dfrac{\mathcal{D}_1 \qquad \mathcal{D}_2 \qquad\qquad \mathcal{D}_3}{\dfrac{\dfrac{\exists x.\varphi(x) \quad \forall u.\forall v.\varphi(u)\wedge\varphi(v)\to u=v \quad \forall w.\varphi(w)\to\psi(w)}{\psi(\iota x.\varphi(x))}\,(\iota I)}{\forall w.\varphi(w)\to\psi(w)}\,(\iota E_3)} \qquad\qquad \rightsquigarrow_r \qquad \forall w.\varphi(w)\to\psi(w)\;\;\dfrac{\mathcal{D}_3}{}
$$

$$\tag{12.4.88}$$

The harmoniously-induced GE-rule for ι is presented in (12.4.89).

$$
\dfrac{\Gamma:\psi(\iota x.\varphi(x)) \qquad \Gamma,\exists x.\varphi(x),\forall u.\forall v.\varphi(u)\wedge\varphi(v)\to u=v,\forall w.\varphi(w)\to\psi(w):\xi}{\Gamma:\xi}\,(\iota GE)
$$

$$\tag{12.4.89}$$

The three regular (ιE)-rules are easily derivable from (ιGE).

Below is the expansion establishing local completeness of the ι-rules.

$$
\dfrac{\mathcal{D}}{\psi(\iota x.\varphi(x,\overline{b}),\overline{a})} \;\rightsquigarrow_e
$$

$$
\dfrac{\dfrac{\mathcal{D}}{\psi(\iota x.\varphi(x,\overline{b}),\overline{a})}\,(\iota E_1)}{\exists x.\varphi(x,\overline{b})} \qquad \dfrac{\dfrac{\mathcal{D}}{\psi(\iota x.\varphi(x,\overline{b}),\overline{a})}}{\forall u.\forall v.\varphi(u,\overline{b})\wedge\varphi(v,\overline{b})\to u=v}\,(\iota E_2) \qquad \dfrac{\dfrac{\mathcal{D}}{\psi(\iota x.\varphi(x,\overline{b}),\overline{a})}}{\forall w.\varphi(w,\overline{b})\to\psi(w,\overline{a})}\,(\iota E_3)
$$
$$
\dfrac{}{\psi(\iota x.\varphi(x,\overline{b}),\overline{a})}\,(\iota I)
$$

$$\tag{12.4.90}$$

12.4.4   Reified sentential meaning for definiteness

I now turn to the definition of the (reified) sentential meanings for sentences constructed with ι-terms, applying the ideas in Chapter 9 to such sentences.

For sentential operators, the I-rule determining canonical derivations is that of the operator dominating φ. For subnectors like ι, it is the I-rule of the subnector.

Definition 12.4.82 (reified meanings). *The* (reified) meaning *of (a compound)* φ *is given by*

$$
[\![\varphi]\!] =^{df.} \lambda\Gamma.[\![\varphi]\!]^c_\Gamma
$$

$$\tag{12.4.91}$$

Recall that meanings of *atomic* sentences are assumed given externally to the ND-system used for conferring meaning on compound sentences and the convention about them. Here, even if $\psi(x)$ is atomic, $\psi(\iota x.\varphi(x))$ is compound.

$$\frac{\Gamma :_c \exists x.\varphi(x) \wedge \Gamma_{c,x}(x) \quad \Gamma :_c \forall u.\forall v.\varphi(u) \wedge \Gamma_{c,x}(u) \wedge \varphi(v) \wedge \Gamma_{c,x}(v) \rightarrow u = v \quad \Gamma :_c \forall w.\varphi(w) \wedge \Gamma_{c,x}(w) \rightarrow \psi(w)}{\Gamma :_c \psi(\iota x.\varphi(x))} \ (\iota I_C)$$

$$\frac{\Gamma :_c \psi(\iota x.\varphi(x))}{\Gamma :_c \exists x.\varphi(x) \wedge \Gamma_{c,x}(x)} \ (\iota E_{C,1})$$

$$\frac{\Gamma :_c \psi(\iota x.\varphi(x)) \quad \Gamma :_c \varphi(u) \wedge \Gamma_{c,x}(u) \quad \Gamma :_c \varphi(v) \wedge \Gamma_{c,x}(v)}{\Gamma :_c u = v} \ (\iota E_{C,2})$$

$$\frac{\Gamma :_c \psi(\iota x.\varphi(x))}{\Gamma :_c \forall z.\varphi(z) \wedge \Gamma_{c,x}(z) \rightarrow \psi(z)} \ (\iota E_{C,3})$$

Figure 12.3: I/E-Rules for contextual definiteness

12.4.5 First-order logic with contextual definiteness

As mentioned in the introduction, the natural use of definiteness in natural language (when interpreted as Russell's existence and uniqueness) is by relativizing uniqueness to some context. I would like the incorporation of definiteness into FOL to allow also modelling this contextualization. This is the purpose of this subsection. The R-contexts used are the same as for the QDR case.

Once again, the main idea, to be captured by the rules for contextual definiteness below, is that a R-context provides a contextual assumption, here restricting the free variable in the premise of the I-rule of the ι operator.

12.4.5.1 An ND-system for contextual definiteness with ι-terms

The contextual definiteness I/E-rules are presented in Figure 12.3. For simplicity, parameterization is left implicit.

Remarks:

- The (ιI_C) rule ensures the existence and uniqueness of the x satisfying $\varphi(x)$ in the context c defined by Γ_c; thus there may exist other xs, satisfying $\varphi(x)$, but *not* satisfying $\Gamma_{c,x}(x)$, namely, not being in the context c.

- Similarly, the ιE-rules are strengthened so as to infer from $\Gamma :_c \psi(\iota x.\varphi(x))$ the existence *in the context* Γ_c of an x satisfying $\varphi(x)$ as well as its uniqueness in that context. Furthermore, note that this x also satisfies $\psi(x)$.

To understand the way the rules in Figure 12.3 work, inspect the way a famous example regarding uniqueness is handled. Consider

$$\text{The Russian voted for the Russian} \qquad (12.4.92)$$

which may be regimented using parallel definiteness as

$$V(\iota x.R(x), \iota y.R(y)) \qquad (12.4.93)$$

Recall that for absolute definiteness, $\iota x.\varphi(x)$ and $\iota y.\varphi(y)$ have the same meaning, attributing uniqueness to some Russian. The two applications of (ιI) introduce the two terms into $V(x, y)$ under the *same conditions*. This renders (12.4.92) as having the same meaning as the reflexive sentence

$$\text{The Russian voted for himself} \qquad (12.4.94)$$

This holds for general parallel definiteness, where in $\psi(\iota x.\varphi(x), \iota y.\varphi(y))$, the two definite descriptions are introduced into $\psi(x, y)$ under the same conditions, and mean the same.

Now, consider a context c of a boxing competition, where there are judges and competitors, and judges vote for competitors. Suppose that c specifies that there is a unique Russian judge and a unique Russian competitor. In that context, (12.4.92) might mean the same as

$$\text{The Russian judge voted for the Russian competitor} \qquad (12.4.95)$$

This *non-reflexive* contextual reading of (12.4.93) may be derived using the contextual definiteness rules by letting Γ_c have two conditions: $\Gamma_{c,x} \equiv J(x)$ and $\Gamma_{c,y} \equiv C(y)$ (for *judge* and *competitor*, respectively). When the rule ιI_C is used twice in order to introduce $\iota x.R(x)$ and $\iota y.R(y)$ into $V(x, y)$, it attributes uniqueness in two different ways. To introduce $\iota x.R(x)$, uniqueness is attributed to $R(x) \wedge J(x)$, while for the introduction of $\iota y.\varphi(y)$, uniqueness is attributed to $R(y) \wedge C(y)$.

In general, within contextual parallel definiteness such as $\psi(\iota x.\varphi(x), \iota y.\varphi(y))$ the two definite descriptions, differing in the name of the bound variable, may have different meanings!

12.4.6 Harmony of the contextual definiteness rules

Intrinsic harmony is established by reductions similar to those for absolute definiteness, under the important proviso that *reductions are performed in the same context*. We exemplify the first reduction.

$$
\cfrac{
\cfrac{
\begin{matrix}\mathcal{D}_1 \\ \exists x.\varphi(x)\wedge\Gamma_{c,x}(x)\end{matrix} \quad
\begin{matrix}\mathcal{D}_2 \\ \forall u.\forall v.\varphi(u)\wedge\Gamma_{c,x}(u)\wedge\varphi(v)\wedge\Gamma_{c,x}(v)\to u = v\end{matrix} \quad
\begin{matrix}\mathcal{D}_3 \\ \forall w.\varphi(w)\wedge\Gamma_{c,x}(w)\to\psi(w)\end{matrix}
}{\psi(\iota x.\varphi(x))} \ (\iota I_C)
}{\exists x.\varphi(x)\wedge\Gamma_{c,x}(x)} \ (\iota_{C,1} E)
$$
$$\rightsquigarrow_r$$
$$\begin{matrix}\mathcal{D}_1 \\ \exists x.\varphi(x)\wedge\Gamma_{c,x}(x)\end{matrix} \qquad (12.4.96)$$

The other reductions are modified accordingly.

The expansion for the contextual definiteness rules is adapted similarly from the one for the absolute definiteness ones. Below is the expansion establishing local-completeness of the ι-rules.

$$
\frac{\begin{array}{c}\mathcal{D}\\ \psi(\iota x.\varphi(x))\end{array}}{\exists x.\varphi(x)\wedge\Gamma_{c,x}(x)}\ (\iota E_{C,1}) \qquad \frac{\begin{array}{c}\mathcal{D}\\ \psi(\iota x.\varphi(x))\\ \rightsquigarrow e\\ \hline \psi(\iota x.\varphi(x))\end{array}}{\forall u\forall v.\varphi(u)\wedge\Gamma_{c,x}(u)\wedge\varphi(v)\wedge\Gamma_{c,x}(v)\rightarrow u=v}\ (\iota E_{C,2}) \qquad \frac{\begin{array}{c}\mathcal{D}\\ \psi(\iota x.\varphi(x))\end{array}}{\forall w.\varphi(w)\wedge\Gamma_{c,x}(w)\rightarrow\psi(w)}\ (\iota E_{C,3})
$$

$$
\psi(\iota x.\varphi(x)) \qquad (\iota I_C)
$$

$$(12.4.97)$$

12.5   Absolute definiteness in the NL-fragment

I now augment E_1^+ with another determiner, the definite article 'the' (keeping the fragment under the same name). Typical sentences now include

> the girl smiles/is beautiful, every boy loves the girl, the girl who smiles sings
> $$(12.5.98)$$

Note that nested definite descriptions are present too, as in

> the girl whom the boy loves smiles $(12.5.99)$

Again, we are choosing a Russellian interpretation of 'the' because it lends itself conveniently to contextual definiteness (as in (12.4.1)). Note again that in the fragment there is no unrestricted quantification, only restricted one, as in some X. Thus, the I-rule below has as its underlying Russellian contextual definition (12.4.51), the latter avoiding unrestricted existential quantification.

The I/E-rules for 'the' are presented below.

$$
\frac{\Gamma:S[(\text{some } X)_{r(S[\mathbf{j}])+1}] \quad \Gamma,\mathbf{j}\text{ isa }X,\mathbf{k}\text{ isa }X:\mathbf{j}\text{ is }\mathbf{k}}{\Gamma:S[(\text{the } X)_{r(S[\mathbf{j}])+1}]}\ (\text{the}I)
$$

$$(12.5.100)$$

$$
\frac{\Gamma:S[(\text{the } X)_{r(S[\mathbf{j}])+1}]}{\Gamma:S[(\text{some } X)_{r(S[\mathbf{j}])+1}]}\ (\text{the}E_1) \qquad \frac{\Gamma:S[(\text{the } X)_{r(S[\mathbf{j}])+1}] \quad \Gamma:\mathbf{j}\text{ isa }X \quad \Gamma:\mathbf{k}\text{ isa }X}{\Gamma:\mathbf{j}\text{ is }\mathbf{k}}\ (\text{the}E_2)
$$

$$(12.5.101)$$

Remarks:

- The first premise of (the$I$) assures the existence of an X satisfying $S[-]$. The second premise assures the uniqueness of X: if both $\mathbf{j}$ and $\mathbf{k}$ are X, they are the same. Here the use of restricted quantification manifests itself. Note that the assumptions are discharged by the rule.

- There are two E-rules for the. The first infers the existence of an X satisfying $S[-]$, and the second the uniqueness of X.

Example 12.5.70. *Below (12.5.70) is a derivation establishing*

⊢the girl is happy, every girl who is happy smiles : the girl smiles

By applying the technique presented in Chapter 9, one can extract from the sentential meaning of $[\![\,S[(\text{the }X)]\,]\!]$ a *lexical* meaning $[\![\text{the}]\!]$. I skip this issue here, as it involves a lot of machinery orthogonal to contextualizing meaning.

12.6 Contextual definiteness in the NL-fragment

In this section, I develop the proof-theoretic semantics for contextual definiteness in the fragment E_1^+.

An *NLR-context* (NL restricting context) c is again a finite collection Γ_c of pseudo-sentences with one parameter only, where $\Gamma_{c,\mathbf{j}}$ is the sub-collection with the parameter $\mathbf{j}$. Let CNL be the collection of all NLR-contexts.

Recall that the *NLR*-contexts $\Gamma_{c,\mathbf{j}}(\mathbf{j})$ can be of one of the forms $\mathbf{j}$ isa X (X a noun), $\mathbf{j}$ is A (A an adjective) and $\mathbf{j}$ V (V a verb-phrase). Note that compound contextual restrictions can also be imposed. For example, $\Gamma_{c,\mathbf{j}} = \mathbf{j}$ isa man whom every girl loves.

I start with the (the$I$)-rule for introducing $S[(\text{the }X)]$. The idea is to have the premises justifying the existence and uniqueness of an X in the context c. This is expressed by a relative clause X who $\hat{\Gamma}_{c,\mathbf{j}}(-)$, where $\hat{\Gamma}_{c,\mathbf{j}}(-)$ is defined as follows.

$$\hat{\Gamma}_{c,\mathbf{j}}(-) =^{\text{df.}} \begin{cases} \text{is } A & \Gamma_{c,\mathbf{j}}(\mathbf{j}) = \mathbf{j} \text{ is } A \\ \text{isa } X & \Gamma_{c,\mathbf{j}}(\mathbf{j}) = \mathbf{j} \text{ isa } X \\ V & \Gamma_{c,\mathbf{j}}(\mathbf{j}) = \mathbf{j} \, V \end{cases} \qquad (12.6.102)$$

Note that for adjectives, we could avoid relativization by using adjectival modification. Thus, if X is girl and the restricting condition is $\mathbf{j}$ is small, we could use the small girl. To keep things simple, we use here relativization too, namely, the girl who is small. Note that the I/E rules are structured similarly to their absolute definiteness counterpart, with relativization expressing contextual restriction incorporated. In the conclusions of the E-rules, the existence and uniqueness are assured for an X *in the context* c. The derivations are in (12.6.103)–(12.6.105).

12.7 Summary

In this chapter, I have considered a PTS for contextual meaning variation, for two variants of such a variation.

$$\frac{\Gamma : S[(\text{some } X \text{ who } \Gamma_{c,j}\hat{}(-))_{r(S[\mathbf{j}])+1}] \quad \Gamma, \mathbf{j} \text{ isa } X \text{ who } \Gamma_{c,j}\hat{}(-), \mathbf{k} \text{ isa } X \text{ who } \Gamma_{c,j}\hat{}(-) : \mathbf{j} \text{ is } \mathbf{k}}{\Gamma : S[(\text{the } X)_{r(S[\mathbf{j}])+1}]} \ (\text{the}I_c)$$

(12.6.103)

$$\frac{\Gamma : S[(\text{the } X)_{r(S[\mathbf{j}])+1}]}{\Gamma : S[(\text{some } X \text{ who } \Gamma_{c,j}\hat{}(-))_{r(S[\mathbf{j}])+1}]} \ (\text{the}E_{c,1})$$

(12.6.104)

$$\frac{\Gamma : S[(\text{the } X)_{r(S[\mathbf{j}])+1}] \quad \Gamma : \mathbf{j} \text{ isa } X \text{ who } \Gamma_{c,j}\hat{}(-) \quad \Gamma : \mathbf{k} \text{ isa } X \text{ who } \Gamma_{c,j}\hat{}(-)}{\Gamma : \mathbf{j} \text{ is } \mathbf{k}} \ (\text{the}E_{c,2})$$

(12.6.105)

- Contextual quantifier domain restriction

- Contextual definiteness (existence and uniqueness)

The PTS employs an explicit representation of a context as an open formula, guaranteeing that such contexts are always expressible in the object language.

In both cases, an intermediate step is used, an augmentation of first-order logic with a contextual meaning variation, for a clearer exposition of the ideas involved. The meaning-conferring ND-system is shown to be harmonious, thereby qualifying for its task.

Chapter 13

Afterword

In this book, I have tried to present Proof-Theoretic Semantics, both for logic and for (fragments of) NL, in a coherent way, but *the way I see it*. In several respects, my personal view differs from the more common views in the literature. Below is a partial list of the main deviation from the more traditional presentations.

Point of Departure: Most discussions of PTS originate from some philosophical disputes, disputes leading to some differences in the presumed meanings attached to various expressions, mainly to those known as 'logical constants'. Examples of such disputes are those over metaphysics – realism vs. anti-realism –, or the epistemological role meanings need to play.

While I do relate occasionally to such philosophical issues, mainly when presenting main stream motivations for PTS, I tried to avoid having them as *driving* the presentation. Rather, my point of departure was more mathematically oriented, even more computational, attempting to understand PTS as a *universal definitional tool* for a theory of meaning. This led to a stronger emphasis on the technicalities involved in the development of PTS, that to a large extent can be seen as independent of any underlying philosophy.

Mode of definition: While it is a fundamental characteristic of PTS that 'rules determine meaning', there is a prevailing view (with some exceptions mentioned in the text) that the meaning determined this way is *implicit*. On the other hand, I opted for this determination to result in an *explicit* definition, in the form of a *proof-theoretic semantic value*. One major motivation for this (others having been mentioned in the text) is to be able to "place" meanings in a lexicon of a grammar.

This view could be seen as "denotation" in a sense, but the "denotations" are all proof-theoretic, collections of canonical derivations in meaning-conferring proof systems, not to be conflated with traditional conceptions of denotation, the latter model-theoretic in nature.

Flow of meaning: This is an aspect of PTS hardly ever considered explicitly in the literature. It is reflected in my emphasis of the distinction between *sentential meanings* and *sub-sentential meanings*. Traditionally (for PTS of logic), rules are conceived to determine *directly* the meaning of the logical constants, the only interesting subsentential phrases. On the other hand, I presented a two-layered conception of the determined meanings:

- Sentences (formulas) are the *primary* carriers of meaning, and it is *their* explicit meanings that are determined by the rules of a meaning-conferring proof-system.
- Subsentential phrases are *secondary* carriers of meaning, their meanings being *derived* (in a way specified in the text) from the sentential meanings. For logic, those are the meanings of the logical constant; for NL, on the other hand, there is a wealth of "interesting" subsentential phrases, not only single words.

Proof-Theoretic consequence: Traditionally, also for adherers of PTS, proof-theoretic logical consequence is identified with *derivability* (some exceptions are listed in the text).

I want to claim, generalizing from the model-theoretic conception of logical consequence as propagation of truth in models, that for any theory of meaning, logical consequence consists of propagating the *central concept* of that theory. the truth-propagation underlies consequence because truth (in a model) is the central concept of model-theoretic semantics. So, for PTS, logical consequence needs to be seen as propagation of *its* central concept, namely *canonical derivability*, grounds for assertion.

Clearly, not all the goals of PTS have been achieved yet, and much remains to be done. A formost task is to convince practitioners, if not to adhere to it, at least consider it as a viable alternative to other theories of meaning.

However, even when viewed "from the inside", there are still problems awaiting better resolution, if not ultimately solved. In a recent paper, Schroeder-Heister [197] identified three such major problems, a view I certainly adopt myself. The problems are not open in the mathematical sense, having no solutions yet, but in the sense that the community has not settled yet on a definite view of the issue involved.

The nature of assumptions: As presented in the test, there are various views of the role of open assumptions in PTS; starting from the more traditional vie of 'assumptions as placeholders', and ending with my own view, inspired by data-bases hypothetical reasoning.

The nature of harmony: Again, there are several notions of harmony, as discussed in the text. Each of them still suffers from some problems. In particular, the "stability part" of harmony needs a more definite definition. There is also some new work, unpublished yet, by Schroeder-Heister and Tranchini, on an intensional notion of harmony.

Extending the scope of PTS: As described in the text, formal system like Definitional Reasoning, not exactly logic, has been suggested as an appropriate means to express PTS. Of cowrse, my own work, extending PTS to NL, is in this direction too, and certainly is in need of expansion.

To these, I would like to add another problem, the **proof-theoretic meaning of atomic sentences**, a good solution of which I view as paving the way to **Proof-Theoretic Lexical Semantics**, which was hardly touched upon in the literature. Again, its main importance is in the PTS for NL.

Appendix

Type-Logical Grammar

The (associative, product-free) Lambek-calculus **L**, on which the TLG considered here is based[1], is presented (in its natural-deduction presentation) in Figure 13.1 (cf. [123], p.120). Categories $\mathcal{C}$ are formed from a given finite set $\mathcal{B}$ of *basic categories* by[2] *directed implications*. Note that contexts Γ here are *sequences* (not sets or multi sets) of *signs*, pairs $\mathbf{c} : x$ ($\mathbf{c}$ – a category, x – a variable), where $subjects(\Gamma)$, the variables in Γ, are pairwise distinct.

A *type-logical grammar (TLG)* G over a set Σ of *terminal symbols* (*words* in case of natural language), is given by a *lexicon* α_G, assigning to each $\sigma \in \Sigma$ a finite set of pairs of categories from $\mathcal{C}$, and meanings (proof-theoretic meanings here), and a designated category $\mathbf{c}_0$. The lexical assignment is naturally lifted by concatenation to $\alpha[[w]]$, $w \in \Sigma^+$ by setting $\alpha_G[[\sigma w]] = \alpha_G[[\sigma]]\alpha_G[[w]]$. The *interpreted* language

---

[1]Strictly speaking, both grammars G_{prop} and G_{FOL} use only the applicative fragment **AB** of **L**.
[2]In Lambek's original notation, \ and / are used instead of directed arrows.

$$\overline{\mathbf{c} : x \rhd \mathbf{c} : x} \ (Ax)$$

$$\frac{\Gamma_1 \rhd \mathbf{c}_1 : N \quad \Gamma_2 \rhd (\mathbf{c}_1 \to \mathbf{c}_2) : M}{\Gamma_1 \Gamma_2 \rhd \mathbf{c}_2 : (MN)} \ (\to E) \quad \frac{\Gamma_2 \rhd (\mathbf{c}_2 \leftarrow \mathbf{c}_1) : M \quad \Gamma_1 \rhd \mathbf{c}_1 : N}{\Gamma_2 \Gamma_1 \rhd \mathbf{c}_2 : (MN)} \ (\leftarrow E)$$

$$\frac{[\mathbf{c}_1]_i : x, \ \Gamma \rhd \mathbf{c}_2 : M}{\Gamma \rhd (\mathbf{c}_1 \to \mathbf{c}_2) : \lambda x.M} \ (\to I_i) \quad \frac{\Gamma, \ [\mathbf{c}_1]_i : x \rhd \mathbf{c}_2 : M}{\Gamma \rhd (\mathbf{c}_2 \leftarrow \mathbf{c}_1) : \lambda x.M} \ (\leftarrow I_i) \ , \ \Gamma \text{ not empty}, x \text{ fresh}$$

Figure 13.1: The **L**-calculus

$L[[G]]$ defined by G is:

$$L[[G]] =^{df.} \{(w, M) \epsilon \Sigma^+ \times Term \mid \exists \Gamma \epsilon \alpha_G[[w]]. \Gamma \vdash_\mathbf{L} (\mathbf{c}_0, M)\}$$

where M is a linear λ-term over $subjects(\Gamma)$. The actual meaning of w is obtained by substituting M_i, the meaning term of $\alpha_G[[w_i]]$ for x_i in M, where $w = w_1, \cdots, w_n$. It is common to define a mapping from syntactic categories to semantic types, $CtoT$, where for an atomic category $\mathbf{c}$, $CtoT(\mathbf{c})$ is an arbitrary type, while

$$CtoT(\mathbf{c}_1 \to \mathbf{c}_2) = CtoT(\mathbf{c}_2 \leftarrow \mathbf{c}_1) = (CtoT(\mathbf{c}_1), CtoT(\mathbf{c}_2))$$

that can be shown to maintain the invariant $\alpha(w) \vdash \mathbf{c} : M^\tau$ implies $CtoT(\mathbf{c}) = \tau$. See [123] for a detailed exposition of TLG.

Bibliography

[1] Alan R. Anderson and Nuel D. Belnap. *Entailment: The Logic of Relevance and Necessity*, volume I. Princeton University Press, Princeton, 1975.

[2] Arnon Avron. Tonk - a full mathematical solution. In Anat Biletzki, editor, *Hues of Philosophy*. College Publications, 2010. In memory of Ruth Manor.

[3] Arnon Avron and Iddo Lev. Canonical propositional gentzen-type systems. In Rageev Gorè, Alexander Leitch, and Tobias Nipkov, editors, *In Proceedings of the 1st joint conference on automated reasoning (IJCAR 2001, Sienna)*, pages 529–544. Springer, LNAI 2083, 2001.

[4] Henk P. Barendregt. *The Lambda Calculus, its syntax and semantics*. North Holland, 1984.

[5] Chris Barker. *Possessive descriptions*. CSLI publications, Stanford, CA, 1995.

[6] Aaron Barth. A refutation of frege's Context Principle. *Thought*, 2012.

[7] Jon Barwise and Robin Cooper. Generalized quantifiers and natural language. *Linguistics and Philosophy*, 4(2):159–219, 1981.

[8] Jon Barwise and John Perry. *Situations and Attitudes*. MIT Press, 1983.

[9] Nuel Belnap. Tonk, plonk and plink. *Analysis*, 22(6):30–35, 1962.

[10] Hanoch Ben-Yami. Bare quantifiers? *Pacific Philosophical Quarterly*, 2013.

[11] Kent Bendall. Natural deduction, separation, and the meaning of logical operators. *Journal of Philosophical Logic*, 7:245–276, 1978.

[12] Kent Bendall. Negation as a sign of negative judgement. *Notre Dame Journal of Formal Logic*, XX(1):68–76, 1979.

[13] Gavin M. Bierman and Valeria de Paiva. On an intuitionistic modal logic. *Studia Logica*, 65(3):383–416, 2000.

[14] Denis Bonnay. Tonk strikes back. *Australian journal of Logic*, 3:33–44, 2005.

[15] Denis Bonnay and Maikaël Cozie. Which logic for the radical antirealist? In Shahid Rahman, Giuseppe Primiero, and Mathieu Marion, editors, *The Realism-Antirealism Debate in the Age of Alternative Logics*. Springer, 2012. Logic, Epistemology and the Unity of Science vol. 23.

[16] Branislav Boricic. On sequence-conclusion natural-deduction systems. *Journal of Philosophical Logic*, 14:359–377, 1985.

[17] Laurent Kieff Cedric Dégremont and Helge Rückert, editors. *Dialogues, Logics and Other Strange Things: Essays in Honour of Shahid Rahman*. College Publications, 2008.

[18] Carlo Cellucci. Existential instantiation and normalization in sequent natural deduction. *Annals of pure and applied logic*, 58:111–148, 1992.

[19] Patrizio Contu. The justification of the logical laws revisited. *Synthese*, 148(3):573–588, February 2006. Special issue on *Proof-Theoretic semantics*, Reinhard Kahle and Peter Schroeder-Heister, eds.

[20] Roy T. Cook. WhatÕs wrong with tonk(?). *Journal of Philosophical Logic*, 34:217Ð226, 2005.

[21] Roy T. Cook and Jon Cogburn. What negation is not: intuitionism and '0 = 1'. *Analysis*, 60(1):5–12, 2000.

[22] Haskel B. Curry. *A theory of formal deducibility*. Notre Dame, Indiana, 1950. Notre Dame Mathematical Lectures, Number 6.

[23] Donald Davidson. Truth and meaning. *Synthese*, 17(3):304–323, 1967.

[24] Rowan Davies and Frank Pfenning. A modal analysis of staged computation. *Journal of the ACM*, 48(3):555–604, 2001.

[25] Sjaak de Mey. 'Only' as a determiner and as a GQ. *Journal of Semantics*, 8:91–106, 1991.

[26] Ruy J. G. B. de Queiroz. Meaning as grammar plus consequence. *Dialectica*, 45(1):83–85, 1991.

[27] Ruy J. G. B. de Queiroz. On reduction rules, meaning-as-use, and proof-theoretic semantics. *Studia Logica*, 90(2):211–247, 2008.

[28] Kosta Došen. Logical constants s punctuation marks. *Notre Dame Journal of Formal Logic*, 30(3):362–381, 1989.

[29] Kosta Došen. Logical consequence: A turn in style. In ria Luisa Dalla Chiara, Kees Doets, Daniele Mundici, and Johan van Benthem, editors, *Logic and Scientific Methods*, volume 259, pages 289–311. Springer, 1997. Synthese Library.

[30] Kosta Došen. A prologue to the theory of deduction. In M. Pelis and V. Puncochar, editors, *The Logica yearbook 2010*, pages 65–80. College Publications, London, 2011.

[31] Kosta Došen. Inferential semantics. In Heinrich Wansing, editor, *Dag Prawitz on Proofs and Meaning,*, pages 147–162. Springer, London, 2015.

[32] Michael Dummett. *Frege: Philosophy of language*. Duckworth, London, 1973.

[33] Michael Dummett. *The Interpretation of Frege's Philosophy*. Harvard University Press, Cambridge, MA, 1981.

[34] Michael Dummett. *The Sea of Languages*. Clarendon Press, Oxford, 1993.

[35] Michael Dummett. *The Logical Basis of Metaphysics*. Harvard University Press, Cambridge, MA., 1993 (paperback). Hard copy 1991. All page references are to the paperback version.

[36] Roy Dyckhoff. Implementing a simple proof assistant. In *Proceedings of the Workshop on Programming for Logic Teaching*. Leeds Centre for Theoretical Computer Science Proceedings 23.88, 1988., Leeds, July 1987.

[37] Pascal Engel. Logic, reasoning and the logical constants. *Croatian Journal of Philosophical*, VI(17):219–235, 2006.

[38] Frederic Fitch. *Symbolic Logic, An Introduction*. The Ronald Press Company, 1952.

[39] Andreas Fjellstad. How a semantics for tonk should be. *Review of symbolic logic*, 2015. DOI 10.1017/S1755020314000513.

[40] Graeme Forbes. *Attitude Problems*. Oxford University Press, 2000.

[41] Itamar Francez. Quantified possessives and direct compositionality. In Ed Cormany, Satoshi Ito, and David Lutz, editors, *Proceedings of SALT 19, Ohio State University*, April 2009.

[42] Nissim Francez. Relevant harmony. *Journal of Logic and Computation*, 2013, doi:10.1093/logcom/ext026. Special issue *Logic: Between Semantics and Proof Theory*, in honor of Arnon Avron's 60th birthday.

[43] Nissim Francez. Bilateralism in proof-theoretic semantics. *Journal of Philosophical Logic*, 43(2-3):239–259, 2014. DOI 10.1007/s10992-012-9261-3.

[44] Nissim Francez. Harmony in multiple-conclusions natural-deduction. *Logica Universalis 8 (2)*, pages 215–259, 2014. DOI 10.1007/s11787-014-0103-7.

[45] Nissim Francez. A logic inspired by natural language: quantifiers as subnectors. *Journal of Philosophical Logic*, 3:1153–1172, 2014. DOI 10.1007/s10992-014-9312-z.

[46] Nissim Francez. Views of proof-theoretic semantics: Reified proof-theoretic meanings. *Journal of Computational Logic*, 2014. Special issue in honour of Roy Dyckhoff, doi:10.1093/logcom/exu035.

[47] Nissim Francez. On distinguishing proof-theoretic consequence from derivability. *(Under refereeing)*, 2015.

[48] Nissim Francez. A proof-theoretic semantics for adjectival modification. 2015. (under refereeing).

[49] Nissim Francez. On the notion of canonical derivations from open assumptions and its role in proof-theoretic semantics. *Review of Symbolic Logic*, 8(2):296–305, 2015 (doi:10.1017/S1755020315000027).

[50] Nissim Francez. The granularity of meaning in proof-theoretic semantics. In Nicholas Asher and Sergei Soloview, editors, *Proceedings of the 8th International Conference on Logical Aspects of Computational Linguistics (LACL), Toulouse, France, June 2014*, pages 96–106, Berlin, Heidelberg, June, 2014. Springer Verlag, LNCS 8535.

[51] Nissim Francez and Gilad Ben-Avi. Proof-theoretic semantic values for the logical operators. *Review of symbolic logic*, 4(3):466–478, 2011.

[52] Nissim Francez and Gilad Ben-Avi. A proof-theoretic reconstruction of generalized quantifiers. *Journal of Semantics*, 32(3):313–371, 2015. doi:10.1093/jos/ffu001.

[53] Nissim Francez and Roy Dyckhoff. Proof theoretic semantics for a fragment of natural language. *Linguistics and Philosophical*, 33(6):447–477, 2010.

[54] Nissim Francez and Roy Dyckhoff. A note on harmony. *Journal of Philosophical Logic*, 4(41):613–628, 2012.

[55] Nissim Francez, Roy Dyckhoff, and Gilad Ben-Avi. Proof theoretic semantics for subsentential phrases. *Studia Logica*, 94(3):381–401, 2010.

[56] Gotlob Frege. *Die Grundlagen der Arithmetik*. Georg Olms, 1884.

[57] Dov M. Gabbay. *Labelled Deductive Systems*, volume I. Clarendon Press, Oxford, 1996.

[58] James W. Garson. Natural semantics: Why natural deduction is intuitionistic. *Theoria*, LXVII:114–139, 2001.

[59] James W. Garson. *What Logics Mean: From Proof Theory to Model-Theoretic Semantics*. Cambridge University Press, Cambridge, UK, 2013.

[60] Peter T. Geach. *Logic Matters*. Blackwell, Oxford, 1972.

[61] Gerhard Gentzen. Investigations into logical deduction. In M.E. Szabo, editor, *The collected papers of Gerhard Gentzen*, pages 68–131. North-Holland, Amsterdam, 1934/1935. English translation of the 1935 paper in German.

[62] Gerhard Gentzen. The consistency of elementary number theory. In M.E. Szabo, editor, *The collected papers of Gerhard Gentzen*, pages 493–565. North-Holland, Amsterdam, 1935. English translation of the 1935 paper in *Mathematische Annalen* (in German).

[63] Jean-Yves Girard. Linear logic. *Theoretical Computer Science*, 50:1–10, 1987.

[64] Jean-Yves Girard. Linear logic. *Theoretical Computer Science*, 50:1–102, 1987.

[65] Michael Glanzberg. Context and unrestricted quantification. In Agustin Rayo and Gabriel Uzquiano, editors, *Absolute Generality*. Clarendon Press, 2006.

[66] Valery Ivanovich Glivenko. Sur quelques points de la logique de m. brouwer. *Bull. Acad. Sci. Belgique*, 15:183–188, 1936.

[67] Kurt Gödel. Eine interpretation des intuitionischen aussagenkalk§ls. *Ergebnisse Math. Colloq.*, 4:39–40, 1933.

[68] Mario Gómez-Torrenta. The problem of logical constants. *The Bulletin of Symbolic Logic*, 8(1):1–37, 2002.

[69] Ian Hacking. What is logic? *The journal of Philosophy*, 76:285–319, 1979.

[70] Raul Hakli and Sara Negri. Does the deduction theorem fail for modal logic? *Synthese*, 2011.

[71] Lars Halläs. Partial inductive definitions. *Theoretical Computer Science*, 87(1):115–1142, 1991.

[72] Elena Herburger. Focus and weak noun-phrases. *Natural Language Semantics*, 5(1):53–78, 1997.

[73] David Hilbert and Paul Bernays. *Grundlagen der Mathematik (vol. 1)*. Springer Verlag, Berlin, 1934.

[74] J. Roger Hindley. *Basic simple Type Theory*. Cambridge University Press, 1997.

[75] Jaakko Hintikka. *Logic, Language Games, and Information*. Oxford University Press, IOxford, 1973.

[76] Ole Thomassen Hjortland. *The Structure of Logical Consequence: Proof-Theoretic Consequence*. PhD thesis, University of St Andrews, 2009.

[77] Ole Thomassen Hjortland. Harmony and the context of deducibility. In Catarina Dutilh Novaes and Ole Thomassen Hjortland, editors, *Insolubles and Consequences: Essays in honour of Stephen Read*, pages 135–154. College Publications, London, 2013 (to appear).

[78] Wilfrid Hodges. Formal features of compositionality. *Journal of Logic, Language, and Information*, 10:7–28, 2001.

[79] Lloyd Humberstone. The revival of rejective negation. *Journal of Philosophical Logic*, 29:331–381, 2000.

[80] Lloyd Humberstone. *The Connectives*. MIT Press, Cambridge, MA, 2011.

[81] Luca Incurvati and Peter Smith. Rejection and valuations. *Analysis*, 70(1):3–10, 2010.

[82] Luca Incurvati and Peter Smith. Is 'no a force indicator? Sometimes, possibly. *Analysis*, 2012. To appear.

[83] Stanislaw Jaśkowski. On the rules of suppositions in formal logic. *Studia Logica*, 1:5–32, 1934. Reprinted in S. McCall (1967) Polish Logic 1920-1939, Oxford: Oxford Univ. Press, pp. 232Ð258.

[84] Ingebrigt Johansson. Der minimalkalkul, ein reduzierter intuitionistischer formalismus. *Compositio Mathematica*, 4:119–136, 1936.

[85] Reinhard Kahle and Peter Schroeder-Heister. Introduction: Proof-theoretic semantics. *Synthese*, 148(3):503–506, February 2006. Special issue on Proof-Theoretic semantics, Reinhard Kahle and Peter Schroeder-Heister, ends.

[86] Donald Kalish and Richard Montague. *LOGIC techniques of formal reasoning*. Harcourt, Brace & World, New York Chicago San-Francisco, Atlanta, 1964.

[87] Edward Keenan. Natural language, sortal reducibility and generalized quantifiers. *J. of Symbolic Logic*, 58(1):314–325, 1993.

[88] Edward Keenan. The definiteness effect: semantics or pragmatics? *Natural Language Semantics*, 11(2):187–216, 2003.

[89] Edward Keenan and Jonathan Stavi. A semantic characterization of natural language determiners. *Linguistics and Philosophy*, 9(3):253–326, 1986.

[90] Andrey Nikolaevich Kolmogorov. On the principle of the excluded middle. In Jean van Heijenoort, editor, *From Frege to Gödel*, page 414Ð447. Harvard University Press, 1967. Originally published in Russian, *O principe tertium non datur*, 1925.

[91] Michael Kremer. Logic and meaning: the philosophical significance of the sequent calculus. *Mind*, 97(385):50–72, 1998.

[92] Michael Kremer. Read on identity and harmony – a friendly correction and simplification. *Analysis*, 67(2):157–159, 2007.

[93] Saul Kripke. *Naming and Necessity*. Harvard University Press, Cambridge, MASS, 1980.

[94] Steven T. Kuhn. Embedded definite descriptions: Russellian analysis and semantic puzzles. *Mind*, 109(435):443 – 454, 2000.

[95] Nils Kürbis. Pluralism and logical basis of metaphysics. In Ondřej Tomala and Radek Honzik, editors, *The Logica yearbook 2006*. Filosofia, Prague, 2006. Extract from the author's Ph.D Thesis.

[96] Nils Kürbis. How fundamental is the fundamental assumption? *Teorema*, XXXI/2:5–19, 2012.

[97] Nils Kürbis. What is wrong with classical negation? *Submitted for publication*, 2012.

[98] Nils Kürbis. Proof-theoretic semantics, a problem with negation, and prospects for modality. *Journal of Philosophical Logic*, 2013. DOI 10.1007/s10992-013-9310-6.

[99] Nils Kürbis. Proof-theoretic semantics, a problem with negation and prospects for modality. *Journal of Philosophical Logic*, To appear. DOI 10.1007/s10992-013-9310-6.

[100] Joachim Lambek and Paul J. Scott. *Introduction to Higher-Order Categorical Logic*. Cambridge University Press, Cambridge, UK, 1988. Cambridge Studies in Advanced Mathematics.

[101] Shalom Lappin, editor. *The Handbook of Contemporary Semantic Theory*. Wiley-Blackwell (Paperback edition), 1997.

[102] Richard K. Larson. The grammar of intensionality. In Georg Preyer and Gerhard Peter, editors, *Logical form and natural language*, pages 228–262. Oxford University Press, 2001.

[103] Hughes Leblanc. Two shortcomings of natural deduction. *Journal of Philosophy*, 63:29–37, 1966.

[104] Menno Lievers. Two versions of the manifestation argument. *Synthese*, 115:199–227, 1998.

[105] Per Martin Löf. Intuitionistic type theory. *Notes by Giovanni Sambin of a series of lectures given in Padua, June 1980.*, 1984.

[106] Paul Lorentzen. *Einfürung in die Logic und Mathematik*. Springer, Berlin, Germany, 1955. Second edition, 1969.

[107] Paul Lorentzen and Kuno Lorentz. *Logic, Language Games, an.* Wissennschaftliche Buchgesellschaft, 1978. In Geman.

[108] Zhaohui Luo. Common nouns as types. In Denis Béchet and Alexandre Dikovsky, editors, *Proceedings of Logical Aspects of Computational Linguistics (LACL)*, pages 173–185. Springer Verlag LNCS 7351, 2012.

[109] Edwin D. Mares. Relevance and conjunction. *Journal of Logic and Computation*, 50:38–41, 1985.

[110] Edwin D. Mares. *Natural Semantics: Why Natural Deduction is Intuitionistic*. Cambridge University Press, Cambridge, 2004.

[111] Mathieu Marion. Game semantics and the manifestation thesis. In Shahid Rahman, Giuseppe Primiero, and Mathieu Marion, editors, *The Realism-Antirealism debate in the age of Alternative Logics*, pages 141–168. Springer, 2012.

[112] Lisa Matthewson. Quantification and the nature of cross linguistic variation. *Natural Language Semantics*, 9:145–189, 2001.

[113] Craig Graham McKay. A consistent propositional logic without any finite models. *Journal of Symbolic Logic*, 22:7–21, 2012.

[114] Peter Milne. Classical harmony: Rules of inference and the meaning of the logical constants. *Synthese*, 100(1):49–94, 1994.

[115] Peter Milne. Harmony, purity, simplicity and a "seemingly magical" fact. *The Monist*, 85(4), 2002.

[116] Peter Milne. Existence, freedom, identity, and the logic of abstractionist realism. *Mind*, 116:23–53, 2007.

[117] Peter Milne. Inferring, splicing, and the stoic analysis of argument. In Catarina Dutilh Novaes and Ole Thomassen Hjortland, editors, *Insolubles and Consequences: Essays in honour of Stephen Read*, pages 135–154. College Publications, London, 2013 (to appear).

[118] Peter Milne. Inversion principles and introduction rules. In Heinrich Wansing, editor, *Dag Prawitz on proofs and meaning*. Springer, Dordrecht, 2013 (to appear).

[119] Friederike Moltmann. Intensional verbs and quantifiers. *Natural Language Semantics*, 5(1):1–52, 1997.

[120] Friederike Moltmann. Intensional verbs and their intentional objects. *Natural Language Semantics*, 16(3):239–270, 2008.

[121] Friederike Moltmann. *Abstract Objects and the Semantics of Natural Language*. Oxford University Press, Forthcoming.

[122] Richard Montague. Universal grammar. *Theoria*, 36:373–398, 1970.

[123] Michael Moortgat. Categorial type logics. In Johan van Benthem and Alice ter Meulen, editors, *Handbook of Logic and Language*, pages 93–178. North Holland, 1997.

[124] Enrico Moriconi. Normalization and meaning theory. *Epistemologia*, XXIII:281–304, 2000.

[125] Julien Murzi. *Intuitionism and Logical Revision*. PhD thesis, University of Sheffield, 2010.

[126] Alberto Naibo and Mattia Petrolo. Are uniqueness and deducibility of identicals the same? *Theoria*, 81(2):143–181, 2015.

[127] Sara Negri. A normalizing system of natural deduction for intuitionistic linear logic. *Arch. Math. Logic*, 41:789–810, 2002.

[128] Sara Negri. Varieties of linear calculi. *J. of Phil. Logic*, 32(6):569–590, 2002.

[129] Sara Negri. Proof theory for modal logic. *Philosophy Compass*, 6:523Ð538, 2011.

[130] Sara Negri and Jan von Plato. Sequent calculus in natural deduction style. *Journal of Symbolic Logic*, 66(4):1803–1816, 2001.

[131] Arthur Nieuwendijk. *On Logic*. PhD thesis, ILLC, University of Amsterdam, 1997.

[132] Grigory K. Olkhovikov and Peter Schroeder-Heister. On flattening general elimination rules. *Review of symbolic logic*, 7(1), 2014.

[133] Peter Pagin. Compositionality, understanding, and proofs. *Mind*, 18(471):713–737, 2009.

[134] Francesco Paoli. *Substructural Logics: A Primer*. Springer, 2002. Trends in Logic series.

[135] Francesco Paoli. Quine and slater on paraconsistency and deviance. *Journal of Philosophical Logic*, 32:553–579, 2003.

[136] Francesco Paoli. Implicational paradoxes and the meaning of logical constants. *Australasian Journal of Philosophy*, 25(4):553–579, 2007.

[137] Christopher Peacocke. What is a logical constant? *Journal of Philosophy*, 73(9):221–240, 1976.

[138] Francis J. Pelletier and Allen P. Hazen. Natural deduction. In Dov M. Gabbay and John Woods, editors, *Handbook of the History of Logic*, volume 11 (Central Concepts). Elsevier, 2012. To appear.

[139] Francis Jeffrey Pelletier. Context dependence and compositionality. *Mind & Language*, 18(2):148–161, 2003.

[140] Jaroslav Peregrin. Language and models: Is model theory a theory of semantics? *Nordic Journal of Philosophical Logic*, 2:1–23, 1997.

[141] Jaroslav Peregrin. Inferentialism and the compositionality of meaning. *International Review of Pragmatics*, 1(1):154–181, 2009.

[142] Stanley Peters and Dag Westerståhl. *Quantifiers in Language and Logic*. Oxford University Press, 2006.

[143] Frank Pfenning and Rowan Davies. A judgmental reconstruction of modal logic. *Mathematical Structures in Computer Science*, 11:511–540, 2001.

[144] Jan Von Plato. Natural deduction with general elimination rules. *Archive Mathematical Logic*, 40:541–567, 2001.

[145] Jan Von Plato. Gentzen's proof systems: byproducts of the work of a genius. *The Bulletin of Symbolic Logic*, 18(3):313–367, 2012.

[146] Dag Prawitz. *Natural Deduction: Proof-Theoretical Study*. Almqvist and Wicksell, Stockholm, 1965. Soft cover edition by Dover, 2006.

[147] Dag Prawitz. Ideas and results in proof theory. In J. Fenstad, editor, *Proc. 2nd Scandinavian Symposium*. North-Holland, 1971.

[148] Dag Prawitz. Towards a foundation of a general proof theory. In Patrick Suppes, Leon Henkin, A. Joja, and Grigore C. Moisil, editors, *PLogic, Methodology, and Philosophy of Science IV*, pages 225–250. North-Holland, 1973.

[149] Dag Prawitz. On the idea of a general proof theory. *Synthese*, 27, 1974.

[150] Dag Prawitz. Meaning and proofs: on the conflict between classical and intuitionistic logic. *Theoria*, 43(2), 1977.

[151] Dag Prawitz. Proofs and the meaning and completeness of logical constants. In J. Hintikka, I. Niiniluoto, and E. Saarinen, editors, *Essays in mathematical and philosophical logic*, pages 25–40. Reidel, Dordrecht, 1978.

[152] Dag Prawitz. Meaning approached via proofs. *Synthese*, 148:507–524, 2006.

[153] Dag Prawitz. The epistemic significance of valid inference. *Synthese*, 187:887–898, 2012.

[154] Huw Price. Sense, assertion, dummett and denial. *Mind*, 92(366):161–173, 1983.

[155] Huw Price. Why "not"? *Mind*, 99(394):221–238, 1990.

[156] Arthur N. Prior. The runabout inference-ticket. *Analysis*, 21:38–39, 1960.

[157] Francesco Pupa. Impossible interpretations, impossible demands. *Linguistics and Philosophy*, 2015. DOI 10.1007/s10988-015-9169-9.

[158] Willard Van Orman Quine. *Word and Object*. MIT Press, Cambridge, MA, 1960.

[159] Shahid Rahman, Giuseppe Primiero, and Mathieu Marion, editors. *The Realism-Antirealism debate in the age of Alternative Logics*. Springer, Dordrecht Heidelberg London New York, 2012. Logic, Epistemology, and the Unity of Science 23.

[160] Augustin Rayo and Gabriel Uzquiano, editors. *Absolute Generality*. Oxford University Press, Oxford, 2006.

[161] Stephen Read. Harmony and modality. In *[17]*, pages 285–303.

[162] Stephen Read. *Relevant Logic*. Blackwells, 1988. A revised version available online from the author's homepage.

[163] Stephen Read. Sheffer's stroke: a study in proof-theoretic harmony. *Danish Yearbook of Philosophy*, 34:7–23, 1999.

[164] Stephen Read. Harmony and autonomy in classical logic. *Journal of Philosophical Logic*, 29:123–154, 2000.

[165] Stephen Read. Logical consequence as truth preservation. *Logique et Analyse*, 183-184:479–493, 2003.

[166] Stephen Read. Identity and harmony. *Analysis*, 64(2):113–119, 2004. See correction in [92] and revision in [169].

[167] Stephen Read. Harmony and modality. In Céric Dégremont, Laurent Kielf, and Helge Rückert, editors, *Dialogs, Logics and other strange things: essays in honor of Shahid Rahman*, pages 285–303. College Publications, King's College, London, 2008.

[168] Stephen Read. General-elimination harmony and the meaning of the logical constants. *Journal of philosophic logic*, 39:557–576, 2010.

[169] Stephen Read. Identity and harmony revisited. 2014. Revised version of [166].

[170] Stephen Read. General elimination harmony and higher-level rules. In Heinrich Wansing, editor, *Dag Prawitz on Proofs and Meaning*, pages 293–312. Springer Verlag, 2015. Outstanding contributions to logic series.

[171] Stephen Read. Proof-theoretic validity. In Colin Caret and Ole T. Hjortland, editors, *Foundations of Logical Consequence*, pages 136–158. Oxford University Press, 2015. Outstanding contributions to logic series.

[172] Greg Restall. *Substructural Logics*. Rutledge, London, New York, 2000.

[173] Greg Restall. Multiple conclusions. In Petr Hàjek, Luis Vàldes-Villanueva, and Dag Westerståhl, editors, *Proceedingth of the twelfth International Congress on Logic, Methodology and Philosophy of Science*, pages 189–205, London, 2005. KCL Publications.

[174] Greg Restall. Assertion, denial, accepting, rejecting, symmetry, paradox, and all that. *Studies in Logic*, 1(1):26–37, 2008.

[175] Greg Restall. Assertion, denial and non-classical theories. In *Proceedings of the Fourth World Congress of Paraconsistency*, 2008.

[176] Greg Restall. Truth values and proof theory. *Studia Logica*, 92(2):241–287, 2009.

[177] Greg Restall. proof theory and meaning: on the context of deducibility. In Françoise Delon, Ulrich Kohlenbach, Penelope Maddy, and Frank Srephanl, editors, *Proceedings of the in Logic Colloquium 2007*, pages 204–219. Cambridge University Press, 2010.

[178] Mark Richard. Seeking a centaur, Adoring Adonis: Intensional transitives and empty terms. In P. French and H. Wettstein, editors, *Figurative Language*, pages 103–127. Blackwell, 2001. Midwest studies in philosophy 25.

[179] Ian Rumfitt. 'yes' and 'no'. *Mind*, 169(436):781–823, 2000.

[180] Bertrand Russell. On denoting. *Mind*, 14(56):479–493, 1905.

[181] Nathan Salmon. *Frege's puzzle*. Bradford Books, MIT Press, Cambridge, MA, 1986.

[182] Tor Sandqvist. Acceptance, inference, and the multiple-conclusion sequent. *Synthese*, 2011. to appear, DOI 10.1007/s11229-011-9909-5.

[183] Peter Schroeder-Heister. A natural extension of natural deduction. *Journal of symbolic logic*, 49:1284–1300, 1984.

[184] Peter Schroeder-Heister. Hypothetical reasoning and definitional reflection in logic programming. In Peter Schroeder-Heister, editor, *In Proceedings of the International workshop on Extensions of Logic Programming, Tuebingen, 1989*, pages 327–339. Springer, 1991.

[185] Peter Schroeder-Heister. Uniform proof-theoretic semantics for logical constants (abstract). *Journal of symbolic logic*, 56:1142, 1991.

[186] Peter Schroeder-Heister. Rules of definitional reflection. In *In Proceedings of the 8th annual IEEE Symposium on Logic in Computer Science, Montreal*, pages 222–232, Los Alamitos, 1993. IEEE Computer Society Press.

[187] Peter Schroeder-Heister. On the notion of *assumption* in logical systems. In R. Bluhm and C. Nimtz, editors, *Philosophy-Science-Scientific Philosophy*. Mentis, Paderborn, 2004. Selected papers of the 5th int. congress of the society for Analytic Philosophy, Bielfield, September 2003.

[188] Peter Schroeder-Heister. Generalized definitional reflection and the inversion principle. *Logica Universalis*, 1:355–376, 2007.

[189] Peter Schroeder-Heister. Proof-theoretic versus model-theoretic consequence. In M. Pelis, editor, *The Logica yearbook 2007*, pages 187–200. Filosofia, Prague, 2008.

[190] Peter Schroeder-Heister. Sequent calculi and bidirectional natural deduction: On the proper basis of proof-theoretic semantics. In M. Perlis, editor, *The Logica 2008 yearbook*. London: College Publications, 2009.

[191] Peter Schroeder-Heister. Proof-theoretic semantics. In Eduard N. Zalta, editor, *Stanford Encyclopedia of Philosophy (SEP), http://plato.stanford.edu/*. The Metaphysics Research Lab, Center for the Study of Language and Information, Stanford University, Stanford, CA, 2011.

[192] Peter Schroeder-Heister. The categorical and the hypothetical: a critique of some fundamental assumptions of standard semantics. *Synthese*, 187(3):925–942, 2012.

[193] Peter Schroeder-Heister. The categorical and the hypothetical: a critique of some fundamental assumptions of standard semantics. *Synthese*, 187(3):925–942, 2012.

[194] Peter Schroeder-Heister. Generalized elimination inferences, higher-level rules, and implications-as-rules interpretation of the sequent calculus. In Edward Hermann Haeusler, Luiz Carlos Pereira, and Valeria de Paiva, editors, *Advances in Natural Deduction*. Springer, Heidelberg, 2013.

[195] Peter Schroeder-Heister. The calculus of higher-level rules, propositional quantification, and the foundational approach to proof-theoretic harmony. *Studia Logica*, 102:1185–1216, 2014. Special issue on *Gentzen's and Ja?kowski's Heritage: 80 Years of Natural Deduction and Sequent Calculi*, Andrzej Indrzejczak, ed.. DOI: 10.1007/s11225-014-9562-3.

[196] Peter Schroeder-Heister. Harmony in proof-theoretic semantics: A reductive analysis. In Heinrich Wansing, editor, *Dag Prawitz on Proofs and Meaning*. Springer Verlag, 2014, to appear. Studia Logica Outstanding contributions to logic series.

[197] Peter Schroeder-Heister. Open problems in proof-theoretic semantics. In Thomas Piecha and Peter Schroeder-Heister, editors, *Advances in Proof-Theoretic Semantics*. Springer, 2016. Trends in Logic series.

[198] Peter Schroeder-Heister. Validity concepts in proof-theoretic semantics. *Synthese*, 148(3):525–571, February 2006. Special issue on *Proof-Theoretic semantics*, Reinhard Kahle and Peter Schroeder-Heister, eds.

[199] Wilfrid Sellars. Inference and meaning. *Mind*, 62(247):313–338, 1953.

[200] David J. Shoesmith and Timothy J. Smiley. *Multiple-Conclusion Logic*. Cambridge University Press, Cambridge, UK, 1978.

[201] Alex Simpson. *The Proof theory and semantics of Intuitionistic Modal Logic*. PhD thesis, University of Edinburgh, 1994.

[202] Hartley B. Slater. Harmonising natural deduction. *Synthese*, 163(2):187–198, 2008.

[203] Timothy J. Smiley. Rejection. *Analysis*, 56(1):1–9, 1996.

[204] Scott Soames. Direct reference, propositional attitudes, and semantic content. *Philosophical Topics*, 15:47–87, 1987.

[205] Jason Stanley and Zoltán Gendler Szabó. On quantifier domain restriction. *Mind & Language*, 2-3:219–261, 2000.

[206] Florian Steinberger. Not so stable. *Analysis*, 69:655–661, 2009.

[207] Florian Steinberger. Why conclusions should remain single. *Journal of Philosophical Logic*, 40:333–355, 2011.

[208] Florian Steinberger. On the equivalence conjecture for proof-theoretic harmony. *Notre Dame journal of formal logic*, 54(4):79–86, 2012.

[209] J. T. Stevenson. Roundabout the runabout inference-ticket. *Analysis*, 21(6):124–128, 1961.

[210] Charles Alexander Stewart. *On the formulae-as-types correspondence for classical logic*. PhD thesis, Oxford University, 1999.

[211] Marta Sznajder. *Dynamic Semantics for Intensional Transitive Verbs Đ a Case Study*. PhD thesis, University of Amsterdam, 2012.

[212] Alfred Tarski. Der warheitsbegriff in den formalisierten sprachen. *Studia Philosophica*, 1:261–405, 1935. English translation in A. Tarski, *Logic, Semantics, Meta-mathematics*, Clarendon Press, Oxford, 1956.

[213] Neil Tennant. *Natural Logic*. Edinburgh University Press, 1978.

[214] Neil Tennant. *Anti-Realism and Logic*. Oxford University Press, Oxford, United Kingdom, 1987. Clarendon Library of Logic and Philosophy.

[215] Neil Tennant. *The Taming of the True*. Oxford University Press, Oxford, United Kingdom, 1997.

[216] Neil Tennant. Negation, absurdity and contrariety. In Dov M. Gabbay and Heinrich Wansing, editors, *What is Negation*. Kluwer Academic Press, Dordrecht, 1999.

[217] Neil Tennant. Ultimate normal forms for parallelized natural deductions, with applications to relevance and the deep isomorphism between natural deductions and sequent proofs. *Logic Journal of the IGPL,*, 10(3):299–337, 2002.

[218] Neil Tennant. A general theory of abstraction operators. *The philosophical quarterly*, 54(214):105–133, 2004.

[219] Neil Tennant. Rule-circularity and the justification of deduction. *The Philosophical Quarterly*, 55(221):625–648, 2005.

[220] Neil Tennant. Inferentialism, logicism, harmony, and a counterpoint. In Alex Miller and Annalisa Coliva, editors, *Essays for Crispin Wright: Logic, Language and Mathematics*. Oxford University Press, London, 2007.

[221] Laura Tesconi. *A Strong Normalization Theorem for natural deduction with general elimination rules*. PhD thesis, University of Pisa, 2004.

[222] Mark Textor. Is 'no' a force-indicator?no! *Analysis*, 71:448–456, 2011.

[223] Richmond H. Thomasson, editor. *Formal philosophy : selected papers of Richard Montague*. Yale university press, New Haven, CT, 1974.

[224] Luca Tranchini. Refutations: a proof-theoretic account. In Carlo Marletti, editor, *First Pisa colloquium on logic, language and epistemology*, pages 133–150, Pisa, 2010. ETS.

[225] Johan van Benthem, G. Heinzman, M. Rebuschi, and H.Visser, editors. *The Age of Alternative Logics*. Springer, Dordrecht, 2006. Logic, Epistemology, and the Unity of Science 23.

[226] Dirk van Dalen. Intuitionistic logic. In Lou Gobble, editor, *The Blackwell Guide to Philosophical Logic*, page 224Đ257. Blackwell, Oxford, 2001.

[227] Luca Viganò. *Labelled Non-Classical Logics*. Reidel, Dordrect, 2000.

[228] Kai von Fintel and Lisa Matthewson. Universals in semantics. *The Linguistics Review*, 25(1-2):139–201, 2008.

[229] Philip Wadler. There is no substitute for Linear Logic. In *8th int. workshop on mathematical foundations of programming semantics*. Oxford, UK, 1992.

[230] Steven Wagner. Tonk. *Notre Dame Journal of Formal Logic*, 22(4):289–300, 1981.

[231] Heinrich Wansing. The idea of a proof-theoretic semantics and the meaning of the logical operations. *Studia Logica*, 64:3–20, 2000.

[232] Heinrich Wansing. Connectives stranger than tonk. *Journal of Philosophical Logic*, 35:653–660, 2006.

[233] Alan Weir. Classical harmony. *Notre Dame Journal of Formal Logic*, 27(4):459–482, 1986.

[234] Dag Westerståhl. Determiners and context sets. In Johan van Benthem and Alice ter Meulen, editors, *Generalized Quantifiers in Natural Language*. Foris, Dordrecht, 1985.

[235] Alfred North Whitehead and Bertrand Russell. *Principia Mathematica*. Cambridge University Press, Cambridge, UK, 1910.

[236] Bartosz Wieckowski. Rules for subatomic derivation. *Review of Symbolic Logic*, 4:219–236, 2011.

[237] Ludvig Wittgenstein. *Remarks on the Foundations of Mathematics*. Blackwell, London, 1956.

[238] Thomas Ede Zimmermann. On the proper treatment of opacity in certain verbs. *Natural Language Semantics*, 1:149–179, 1993.

[239] Thomas Ede Zimmermann. Monotonicity in opaque verbs. *Linguistics and Philosophy*, 29:715–761, 2006. DOI 10.1007/s10988-006-9009-z.

[240] Richard Zuber. A class of non-conservative determiners in Polish. *Linguisticae Investigations*, 27(1):147–165, 2004.

Names Index

Subject Index

Languages Index

www.ingramcontent.com/pod-product-compliance
Lightning Source LLC
LaVergne TN
LVHW012326060326
832902LV00011B/1746